FRANCE AUSTRALE

The French search for the Southland
and subsequent explorations
and
plans to found a penal colony
and strategic base in south western Australia
1503-1826
with colour illustrations and explanatory maps.

by
Leslie R. Marchant
Chevalier, Ordre National du Mérite

Published by
SCOTT FOUR COLOUR PRINT
40 Short Street, Perth, Western Australia 6000

National Library of Australia
Cataloguing in Publication data

France Australe
ISBN 0-9588487-1-8

Distributed in Australia and Overseas by the Publisher.

The coast, from the moment we first saw it, exhibited nothing
but a picture of desolation; no rivulet consoled the eye,
no tree attracted it, no mountain gave variety to the landscape,
no dwelling enlivened it; everywhere reigned sterility and death.
If a few birds of prey skimmed, with rapid wing,
the flats washed by waves, where, we asked ourselves,
could they satisfy their hunger?
Where could they quench their thirst?
Do all the beings that dwell in this unhospitable land
drink salt water? Where are their resources?
For they must have wants. Where are their enjoyments?
For they must have desires.
At first view, you take in an immense distance:
but beware of looking for any enjoyment;
the search would be merely wasting your strength,
without finding the least relief.

The French artist Jacques Etienne Victor Arago
on the French corvette *Uranie* in Shark Bay, Western Australia 1818

Explanatory Note

Two different spellings have been used for Western Australia to match political realities. The term "western Australia", which appears most frequently, is used for the region lying west of 135° east which marked the limit of British claimed New South Wales from 1788 until all of Australia was annexed. It was this region, open for international exploration and occupation, that interested France.

The term "Western Australia", is used solely for the region covered by the British colony and later State which was founded in 1829, lying east of 129° east which became the new fixed border.

Acknowledgements for illustrations

The author deeply appreciates the kind assistance given by the Wordsworth family, and the permission they gave to reproduce French artworks and charts in the Wordsworth Collection, and to Steven Marcuson and his staff at the Trowbridge Gallery in Claremont, Western Australia, for their willing help and permission to reproduce the Bünting Map and other works, and to Alison Kershaw, the Librarian at the University of Notre Dame Australia, for permission to reproduce the Ptolemy Map.

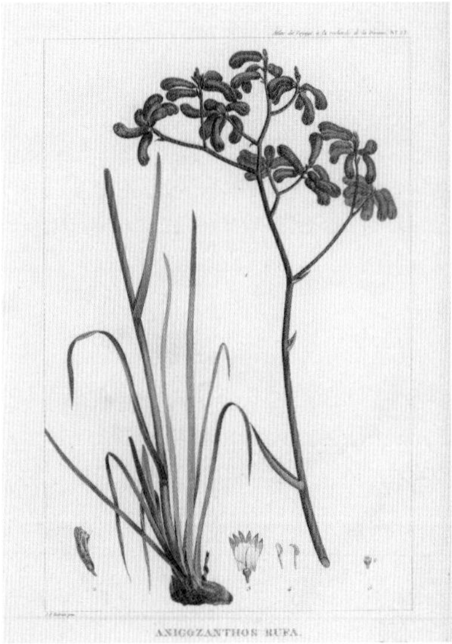

D'Entrecasteaux Expedition, Esperance 1792

ABBREVIATIONS MADE IN THE FOOTNOTES

A.N.	Archives Nationales.
B. Mus. Hist. Nat.	Bibliothèque Centrale du Museum Nationale d'Histoire Naturelle.
B.N.N.A.F.	Bibliothèque Nationale. Nouvelles Acquisitions Françaises.
B. Serv. Hist.	Bibliothèque du Service Historique de la Marine.
B.S. Hyd. Marine	Bibliothèque du Service Hydrographique de la Marine.

Table of Contents

List of maps and illustrations

The botanical scientific drawings in the natural history studies made by the D'Entrecasteaux Expedition

The high quality of the scientific work done on the most impressive European expedition to work in Australia in the Age of Scientific Enlightenment in which France played a leading role, is demonstrated by the botanical drawings in La Billardière's collection.

Three of the botanical discoveries made in Western Australia are reproduced on:

Page iv: The Red Kangaroo Paw (Anigozanthos rufus), a native of Western Australia, is a perennial reaching approximately .8 of a metre in height. The beautiful deep burgundy red flowers appear in summer. This illustration is significant. It records the first Kangaroo Paw discovered in Australia.

Page xvi: The Yate Tree (Eucalyptus cornuta), a native of Western Australia, is an evergreen which reaches a height of approximately 20 metres. The bright yellow flowers appear in summer.

Page 202: The Creeping Banksia (Banksia repens), a native of Western Australia, is an evergreen shrub found on the coastal plain, reaching a span of some 2 metres. It flowers in spring and summer.

Preface

This is a completely refurbished volume rather than a simple new edition of the 1982 book. The text is much the same. It is based on manuscript and other records located, digested and assessed over twenty six years. All of these have been read again, and new information gleaned on the series of expeditions made by the author in the wake of the French explorers has been used, to ensure accuracy in this volume.

There was no reason to make changes. Books based on exhaustive archival researches do not have to be altered. The story recorded in historical records does not change.

The main newness in the book lies with the maps and illustrations. The ones in the 1982 book have been added to, and the original ones enhanced, not only because of the availability of better technology in 1998, but also to impress through the eye, the story of what happened. For the illustrations have been selected not only to parallel the story told in the text, but also to independently show some of the beauty of nature, and activities both in the animal and human worlds as seen by the explorers. At the end of the book, the story told in pictures and the written text come together with a final note about the French shift to New Zealand and then Tahiti, and a painting of the early British settlement at Fremantle and the Swan River, with Rottnest in the background. It was France which was first to scientifically explore there. It considered using the region as a convict settlement. They charted it, but made no painting. An early British painting at the end of the book shows what the region looked like before a city was built. It also marks the end of the French story.

This book fills in two gaps in knowledge, and, in so doing, sets right the story of the past, and revalues the part played by nations and individuals, giving credit where credit is due, and often not previously given. For some explorers who should have enjoyed a deserved fame in their lifetime, were denied this, while others who did less were honoured more, as often happens in life. The historian must not let such falsehood stand, which is easy to do. For it takes skill and patience and long hard work and sound judgment to separate opinion from fact, and to recognize malicious gossip for what it is. It is not the function of historians to parrot gossip written as if it was the true record of the past, to influence posterity to award false honours. For immediate public fame is fickle. When it is given by patronage, it is often short lived, and questionable.

The work of the French astronomer and mathematician Pierre Maupertuis (1698 - 1759), who has a part in this book, for example, needs to be revalued. He was denied the fame he deserved in his lifetime, by malicious enemies, which must have hurt. Their opportunity came when he verified Newton's calculations that the Earth is not a sphere, but an oblate spheroid, which is correct. It is flat at both Poles. For this he was savaged by Voltaire, who knew nothing about the subject, for being a "Flat Earther". The mud stuck. That does little credit for Voltaire who had the ability to be noble.

The aristocrat Yves de Kerguelen (1734 - 1797), in contrast, largely because of his privileged position, was given undeserved public fame which he seemed no more willing to share with others than the girl he had in his bed when he went to sea to explore on his King's ship. For this and other crimes such as deserting his companions, he lost both his liberty and his reputation. But not for long. Time and "fate" were on his side. He was gaoled by the Old Regime. When the French Revolutionaries won power they restored Kerguelen and his reputation, not because of his record, but on the grounds that as he was gaoled by their enemies, he must be good. Fortunately for France, Kerguelen did not live long enough to do further damage to his nation.

La Billardière was somewhat the same. The revolutionary regime that restored Kerguelen, privileged La Billardière above others more competent. Because he was one of them, he was given patronage, favours, grants, and privileged access to the records of the expedition to write its history, which he was appointed to do, although there were others far better qualified and less politically biased. He was too inclined to rely for his information, on politically committed revolutionaries whose records he valued above others. Napoleon, who wanted to be remembered as a patron of science, sought to rectify this. In his reign the more competent de Rossel, a scientist and proficient navigator with a broader outlook and the dedication of a professional, was appointed to write another official history which is more reliable. The fact that his mother had been guillotined while the expedition was exploring in Australia and its waters, did not affect his scientific judgment. However he had been marginalized for too long, and wrote too late in unfortunate circumstances, to win fame. His work came out in wartime and did not make the impact it should have in Europe. La Billardière consequently remains the one more popularly honoured and better known although his expertise is confined to botany.

That is why, to show what really happened and who should be famed eternally, it pays to not write books quickly. Hastily researched books on history are best left unwritten.

The rich collection of historical records which I located and perused over the some twenty six years it took to research and write this book, show that the history of the Great Age of Discovery needs to be revalued and better balanced. Portugal and Spain, both of which won deserved honours, have been given too much of the limelight. That needs to be shared with the other great maritime power, France which explored separately at the same time. For independent France with its independent ecclesiastical Gallicanism, and its early commitment to the principle of the Freedom of the Seas, kept aloof from the partition of the World arranged and approved by Rome, and made its own explorations of the New World and the Old.

It was well placed to do this. France was powerful and populous. It had riches and a blessed position. It was bounded by two seas - the Atlantic and the Mediterranean, both of which combined to create a long tradition of ship building and seamanship. Most important of all, France was better placed than others to benefit from the Renaissance, and early develop a modern scientific outlook, which it did. For it gleaned the classical knowledge which lay at the core of the Renaissance both from the Arabs in neighbouring Spain, and from Byzantium by way of neighbouring Italy. France consequently early flourished intellectually, winning the scientific and mathematical abilities and knowledge needed to make deep sea voyages of exploration.

Like other powers, France at first was misled by Ptolemy's erroneous Map of the World when it searched for the Southland. It believed that if explorers sailed west, they would reach Cathay, and south from there the fabled Indies, or the legendary Southland near where Ptolemy placed the trade mart Cattiga on his map south of the Equator in the Indian Sea. That is why Gonneville, when he reached unknown America, thought he had reached Australia in 1503. After it was realized that America was a New World that barred the way from Europe to the Orient, and that there were two different Indies, an eastern and western one, France played a consistent and important role in American discovery and settlement as well as continuing to explore the Old World and other unknown parts.

After Ptolemy's error was rectified, and an extra quarter added to the Globe, it was generally believed that the Earth consisted of four different parts or worlds or continents with their seas. Dramatists, creative writers, map makers, artists and sculptors all portrayed this in various ways. In the gardens of Versailles, for example, four statues stood for the four parts. The "Old World" consisted of the three separate worlds formed by Europe, Asia and Africa. These were shown by the German writer Bünting as a clover leaf of three parts with Jerusalem at the centre. The New World which made up the fourth part, consisted of America. The manuscript and other sources at the time, refer to a Fifth World where lay the legendary Fifth Continent, the Southland. It was in the search for that world which culminated in the scientific explorations of Australia, that France made its distinctive mark.

The concept of a Fifth World was amorphous in the beginning. It covered all of the southern hemisphere. France consequently searched in the South Atlantic, on the way to India and China, where all it found was a watery waste. It then led the field to scientifically study the Pacific. France had an innovative idea to do this. It created and sent large scale scientific expeditions staffed with civilian scientists. Bougainville's was the first. Britain followed suit, rushing out the indefatigable Cook in

Bougainville's wake with a mini sized team and boat. But with these Cook, more than anyone else, showed that the southern Pacific was also largely a watery waste.

France, as a result, turned to concentrate on the one region left unknown, the southern Indian Ocean. The first part of this book is about France's sustained research effort in that region. The indefatigable Cook of course joined in, making a quick turn around after his first voyage, to go on a new one to explore the southern Indian Ocean. France was already at work there. But Kerguelen let the French side down. Other French explorers nevertheless did much to show that that part also was a watery waste. There was no continent, only a few sparse, course islands all clothed together with the seas, in an unimagined cold where the mountainous seas vie to reach the same heights as the little lumps of land that rise suddenly from the depths. Those are seas only for stout hearted mariners.

The French, like others in Europe, had expected to find a salubrious land, ripe for plantations, mirroring their own nation in the temperate zone, with New Holland, which was known to lie to the north, assuming the role of northern Africa, where often the lands seemed dry and sterile. And that is where France eventually ended up, using its vast scientific resources and talents to explore and make known seemingly barren western New Holland, seeing there was no salubrious Southland where they could grow wheat for their breads, and grapes to make wines to wash them down.

The long sustained explorations France made in western New Holland which Britain had not annexed - the British sector was confined to the old Spanish claimed eastern half of the continent - are described in this book to fill the second gap in knowledge, and correlatively challenge some ingrained beliefs.

It cannot be claimed with certainty, for example, that Australia was first discovered by the Dutch on the *Duyfken* in 1606 as suggested by George Portus in his book *Australia Since 1606*. There is ample evidence to show that this continent was known about, and images of it drawn before the Dutch arrived as is indicated in this book and in works by other scholars such as MacIntyre who wrote freshly on the Portuguese. Torres, incidentally, in the same year as the *Duyfken*, sailed confidently through the gap between Australia and New Guinea which was shown on then existing charts.

Similarly James Cook did not discover Australia as has been popularly claimed. The most correct way to sum up what Cook did in Australia, is to say that he first charted and revealed the shape of the Pacific coast which, incidentally, he did not do very impressively for the time. His longitudes are out, and his compass directions wrong. His Queensland bends too much to the west. Flinders had to re-do that section, for Cook never returned to Australia to check and complete his work. His greatness lies as a Pacific explorer. That is why the French moved in to Australian waters to fill the gap, and in the beginning did that well. Young Beautemps Beaupré, the Chief Hydrographer on D'Entrecasteaux's expedition, was more accurate and skilled than both Cook and Vancouver. He corrected Cook's faulty map of Tasmania, serving in Australian waters the apprenticeship which later won him the title of father of modern cartography. D'Entrecasteaux's survey of Australia made with the new mathematical precision which emerged in France in the Age of Scientific Enlightenment, is more important in the history of science than the British survey of India which has been given the honour of being a starting point.

Another myth which needs to be laid to rest is the claim that Flinders was the first explorer to circumnavigate Australia. Even a superficial view of his log book shows that claim is false. Flinders was ordered to make a circumnavigation as is made clear at the beginning of his journal. But reading past there shows that Flinders never saw or sailed along the western coast, nor much of the northern coast of the continent. He surveyed from near Cape Leeuwin anti-clockwise to the western side of the Gulf of Carpentaria. It was the Australian born hydrographer Phillip Parker King who should be honoured for being the first explorer to circumnavigate and chart the continent.

As is described in the book, France had five important expeditions exploring and surveying in western New Holland in the Age of Scientific Enlightenment and the following Romantic Period, between 1772 and 1826, when Britain warned others off. Others were sent, but did not arrive.

The peak of French scientific attainment was reached with the D'Entrecasteaux Expedition, the last one sent by monarchic France. This was the most extensive and scientifically competent of the missions. The work it did and the results certainly were limited. D'Entrecasteaux was given two conflicting tasks which affected the performance of his expedition. He was sent to search for La Pérouse in the Pacific and also to finish Cook's work by exploring unknown Australia. He did neither well.

The work done by D'Entrecasteaux's scientific team in Australia is exceptionally impressive. It was as a result of that expedition that French scientists acclimatised the Tasmanian gum trees that adorn the Riviera. Banks and his followers never succeeded in that. But D'Entrecasteaux's achievements have never been given the recognition they deserve. The expedition became the victim of politics. It left France when it was being re-organized by revolutionaries. It harboured both republicans and monarchists, and more important, Titans of Science and Satyrs who gave more value to sentiment and preferred to put politics in command. The Satyrs intrigued. There were clashes. The expedition split. The non revolutionaries were reported to the Committee of Public Safety which ran the Reign of Terror at home. In those circumstances, the good research work done was never revealed as it should have been.

The Baudin Expedition which followed, at its beginning looks better on paper than the D'Entrecasteaux Expedition. But that too was rent asunder by the same politics and intrigues which often appear to mar French public and scientific, academic life. The bitter differences which affected the expedition, explain why Baudin's name was not commemorated on the final map. This was drawn, and the names on it were picked by his enemies whose names mainly feature. Not that this was done wholly without reason. Baudin was sent to win scientific honours for France. He failed. A large number of his scientists left the expedition at Mauritius before it arrived in Australia. And those who remained were seldom allowed ashore to do the work they were sent to do. For example, in Geographe Bay, which the French discovered, the scientists were not allowed ashore until the end of the ten day stay. They were then allowed to go ashore briefly just as a gale rose, and all was lost. Baudin incredibly, subsequently reported that Geographe Bay, which is now a major agricultural production region, lacked fresh water and resources. For this and other reasons Baudin was not respected by his scientists who made separate distinctive contributions.

Baudin's was the last expedition to carry civilian scientists. Those which followed were military expeditions under military command. Sending those when the Bourbons were restored to power, is one reason why Britain annexed western New Holland, out of fear.

Britain itself showed little interest in exploring western New Holland during and after the Age of Cook. George Vancouver came of his own initiative, making a running survey along part of the southern coast where he found and explored King George Sound. This survey, which was meant to be completed by Baudin, was done by Flinders who moved in when Baudin failed to carry out his orders, thus leaving a gap.

The main British effort at exploration in western Australia was made during the Peace Conferences which met after the end of the Napoleonic Wars when the restored Bourbons once more took command to win a new prominence for France in Europe. It was because of this, combined with the French plans to recommence exploring western New Holland, that Phillip Parker King was sent on his voyage of circumnavigation, to chart the continent and show the flag. This seemed necessary. For, amongst other things, Heirisson in 1820 drew a new worrying map. On the Baudin expedition he had charted the Swan River, and in Tasmania helped cement relations with the locals after he landed fully uniformed, by willingly going with a Tasmanian Belle and her mother who took him behind some bushes where they removed his colourful garments to see if he looked and functioned like their own men. His new 1820 map of the Fifth Part of the World, drawn at the height of Anglo French tension at a time France planned to colonize south western Australia, showed the footprints left by the French in Australia. Britain reacted by hastening off King on the *Mermaid*, who, amongst other things, recommended the establishment of a British settlement in western New Holland, near Darwin. That is why Britain then moved its border from 135 degrees east to 129 degrees east.

I am deeply grateful for the willing help given by many people and organizations during the years I researched and wrote this book. First I must thank those who laid the early foundations; my teachers who gave me the knowledge of our past and taught me such things as dedication and perseverance, and when the strong use their power to destroy your work, to quietly pick up the pieces and "build again with worn out tools". I also thank the librarians at the State Library who helped when I did my homework; my family, especially my father who fought and lost his health in France, and later introduced me to the places on our coast with French names; and the fine seamen who, when I went to sea, taught me the art of navigation and how to sniff a way through a cyclone to find the safe semi-circle, and how, in the Roaring Forties and the fiercer fifties, to shoulder the mountainous waves or, if they threatened to break themselves and the ship, put about and run with them on the quarter, for there is no shelter in those waters where sailing makes foul Atlantic crossings seem innocuous by comparison.

For the book itself, I must thank for their kindness and help, the many librarians, archivists, curators and their staffs in the many repositories where I worked in France, in particular those at the Archives Centrales de la Marine; the Archives Nationales; the Bibliothèque Centrale du Muséum Nationale d'Histoire Naturelle; the Bibliothèque du Service Historique de la Marine; the Bibliothèque Nationale; the Institut de France; the Musée de l'Homme; and the Société de Géographie in Paris. In the provinces I am indebted to the staffs at the Muséum d'Histoire Naturelle at Le Havre; the Museum at Honfleur and the libraries and archives at Rouen; Evreux; Dieppe; Le Havre and many other places I visited, as well as officials at the churches I visited along the Atlantic seaboard in France.

I must thank in particular my wife Gunhild Marchant (nee Francke) and my sons Nils, Leslie and Francis for their sharing the hardships on expeditions to remote places, and for helping to prepare this manuscript and the maps and illustrations. I also am deeply indebted to the Hon. Mr. David and Mrs Marie Louise Wordsworth for their constant help and support and advice, and for the use of material from the Wordsworth Collection; to Director Dr Lynn Allen and my colleagues at LISWA where I am Visiting Scholar; and to Dr Peter Tannock, the Vice Chancellor of the University of Notre Dame, and to all of my colleagues there for their special help and understanding.

A special note acknowledging the kind help given for the illustrations and maps in this volume appears on page iv.

Leslie R. Marchant
Cottesloe, Western Australia, 1998.

Redouté del.

D'Entrecasteaux Expedition,
Esperance 1792

EUCALYPTUS CORNUTA.

PART I
The Ancien Régime and western Australia

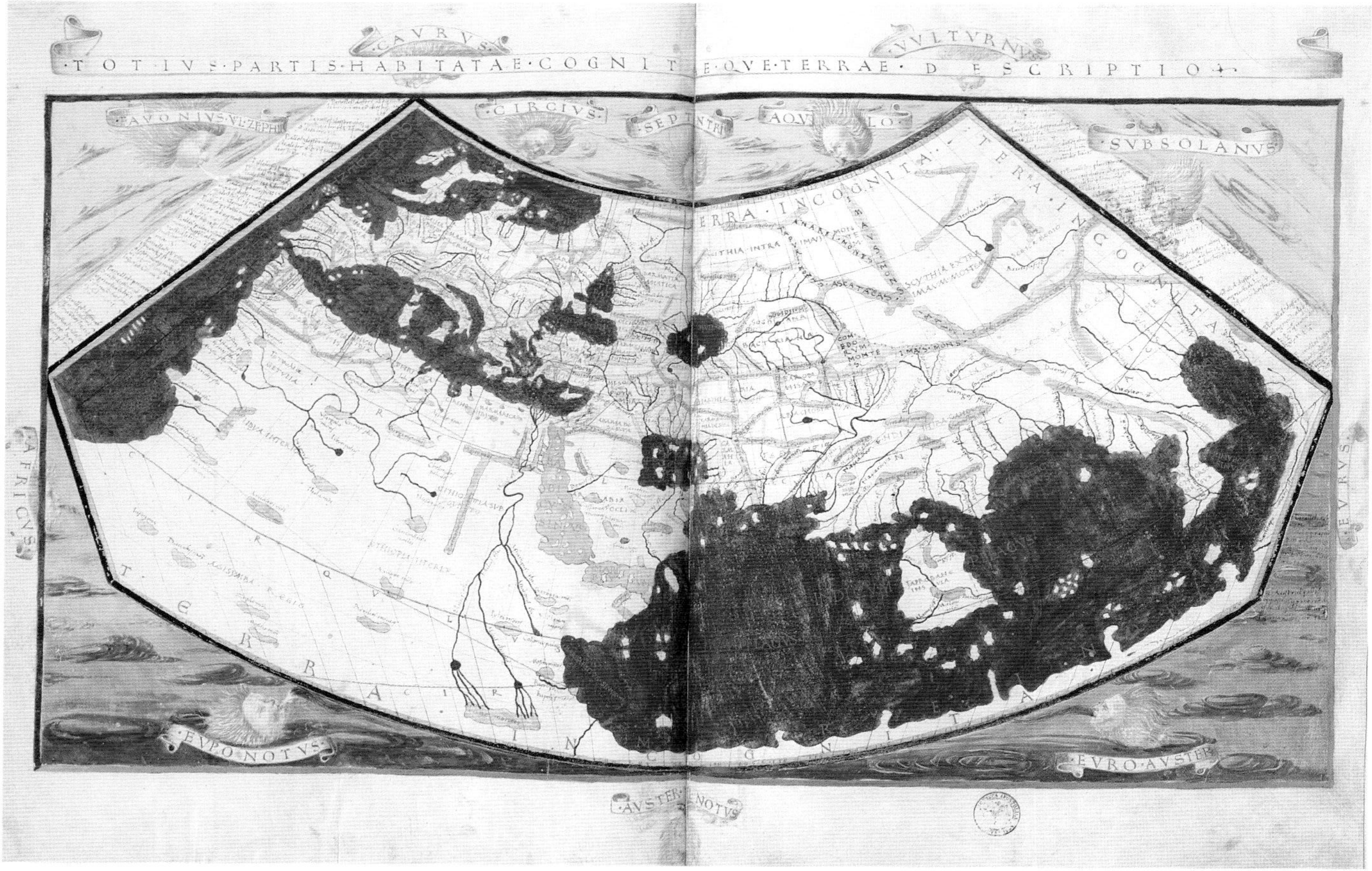

Ptolemy's World Map (2nd century AD), was used for the globes and charts which guided Gonneville and other early maritime explorers. The absence of America led them to believe that after crossing the Atlantic, they were on the east coast of Australia, near the trade mart of Cattiga on the east side of the Indian Ocean. Scholars such as Heinrich Bünting later suggested that region was Biblical Ophir, prompting a series of explorations.

From the Vatican Collection at the University of Notre Dame

CHAPTER I

France and the early search for a route to the Indies down to Paulmier De Gonneville's voyage from 1503 to 1505

France and Early European Maritime Explorations Eastwards

Much of the credit for opening and exploring the sea route east from Europe to the Spice Islands has gone to Portugal and Spain. Yet these two powers did not make the only contribution, although they were given monopoly rights by the Christian Pope to explore and conquer. France was also early and deeply involved supporting independent searches for a viable French route eastwards, seeking trade opportunities of its own; and looking for new lands for the French monarch to acquire and for lost souls for French churchmen to save.

France in fact made not only an early but also a more sustained if not distinctive contribution to knowledge about the Indian Ocean than the Iberian powers. Long after Portugal and Spain had declined, France continued with its efforts at exploration which commenced with the despatch of Gonneville in 1503, and continued throughout periods of monarchy, revolution, restored monarchy and republic to this century.

From the early eighteenth century in particular France played a most distinctive role in the search for the hypothetical *Terra Australis Incognita* believed to exist in the region east of Cape Town. Subsequently, from the latter part of the eighteenth century, France played an equally distinctive and notable part in the exploration of Australia until it was all annexed by Britain. After the mysteries of Australia were revealed, France then made a distinctive contribution to knowledge about Antarctic waters in the region.

France did not only make its mark in the exploration of the Indian Ocean as a whole, and in the further on Pacific. The French also left their impression as colonial expansionists. As a result of discoveries and annexations they soon acquired an extensive empire, in the main insular, which extended from the mainland of Asia to the high latitudes south of the equator. By the time of the French revolutionary wars at the end of the eighteenth century, France was well established as an Indian Ocean power. In the upper part of the Indian Ocean it still had ports and depots in India even though Britain by then dominated the continent. In the lower part, France possessed in the west Madagascar, Reunion, Rodriguez, Mauritius and the Seychelles. Southwards it laid claim to the Austral Islands, now called Marion Island and Prince Edward Island, the Crozets and Kerguelen. To the east it laid claim to south-west Australia from Shark Bay in Endracht Land to Nuyts Land on the south coast. On paper at least, this made France a more extensive Indian Ocean power than Britain was at the time.

It lost some of these holdings as a result of the loss of the revolutionary wars. But in the following peace years France revived and acquired new territories at New Amsterdam, St. Paul, the Comoro Islands and Adélie Land in Antarctica, which have, in the case of the uninhabited islands, remained with France, making it into a power still with Indian Ocean interests today.

Reasons for the French failure to colonize

Yet despite the extensive nature of its holdings in the Indian Ocean, France was never a strong and major colonizing power there. This was primarily for four reasons. Firstly France, although it was one of the first exploring-expansionist powers, never managed to secure a foothold and half-way house near the Cape of Good Hope, to serve as a base for operations. The acquisition of South Africa by the Dutch and later the British subsequently provided these powers with a distinctive strategic advantage over France. Secondly and correlative to this, France preferred and acquired an insular empire. This preference for the most part was for reasons of strategy. Even when France moved after the Napoleonic era, to establish a colony in south-western Australia, the idea of their strategists was to build the settlement on either Rottnest or Garden Island which it was believed would be easy to defend. But these islands acquired by France had no resourceful hinterlands, and apart from Kerguelen island, which is pitted with fiords, no impressive, safe harbours. France's island colonies were, therefore, of no great strategic use and importance. They could not compare with the British settlement at Botany Bay, which the French soon envied. Thirdly, the French were cautious colonizers. Unlike the British they did not rush to establish settlements overseas. Their methods were usually more scientific. Before they sent settlers to live in strange places abroad they invariably made careful methodical surveys of the resources. The trouble with this was, as in the case of western Australia, by the time they acquired the information, others had moved in, even if it seemed to be a desert.

Fourthly, and finally, the reason why France failed to establish a colony in "a vacant territory" such as western Australia, and become a noted colonial power in the Indian Ocean, is because unlike the British they believed in prescriptive rights to overseas territories. According to the French view of international law, once a power discovered and annexed a territory, then that territory belonged to the power which had proclaimed sovereignty. If these claims were not taken away by treaty or other legal means, then the territories remained the property of the original claimant, as happened in the case of western Australia.[1] The explorer St Allouarn annexed this for France on 30 March 1772. This territory was not taken from France by the victors of the various wars in which France was subsequently involved. Therefore, French officials in the post Napoleonic period believed they had the legal right to establish a colony at Rottnest Island or some other suitable place in western Australia.[2] The trouble in that case was Britain did not accept the principle of prescriptive right. It believed in the right to territory by effective occupation. Britain consequently viewed territories which were not occupied and garrisoned as being vacant.[3] Therefore, the French proclamation of annexation of western Australia ended up as a scrap of paper, once the British decided to despatch a garrison to King George Sound in 1826, and to support this act with power.

Many of the colonial differences between France and Britain which emerged in the period in fact stemmed from these differing concepts of the international law of colonial acquisition. But in the end France made some gains without changing its beliefs about the law, or its practice of establishing colonies. France's post-Napoleonic empire, established in the Pacific in particular, was fashioned primarily as a result of "prescriptive right", being made up of islands early discovered and claimed for France by explorers such as Bougainville.

In this context early French explorers and discoverers are of great historic interest. As a result of their efforts there were few places in the Pacific and Indian Oceans which France could not and did not lay claim to in periods of colonial expansion.

Early French penetration of the Indian Ocean

In the case of the Indian Ocean, early penetration was effected by France as a result of three influences in particular. Firstly, France had a long maritime tradition, and the means to participate in long range sea explorations once they commenced. Secondly, France especially under the reign of Francis I, had no desire to let its neighbours, Portugal and Spain, exercise a monopoly in that field. French rulers demanded and saw to it that France acted in its own right. Thirdly, France of all the powers in Europe, had a long and rich tradition of contact and intercourse with the east. In the first instance this was by way of the land routes. But as soon as suggestions were made that there was a way there by sea, France shifted to using this method of contact.

The extent and richness of France's early maritime explorations are not easy to describe and assess. Early records and accounts of explorers and travellers are few. Many of the accounts of early voyages are little more than legends handed down to later generations, or later gleaned and published accounts of earlier voyages which cannot be verified by the historian who likes to rely on first hand accounts and information.

Early voyages southwards from France

There are unconfirmed reports, for instance, that two ships from Dieppe in Normandy sailed down to the Canary Islands in 1364. There are also reports that merchants from Dieppe and Rouen were trading along the West African coast as far south as Sierre Leone at the same time, bartering French goods for oil stone, gold dust and spices.[4]

There is also a report, firmly believed by some in France but as yet unconfirmed, that a Dieppois navigator, Jean Cousin[5] sailed past West Africa and then across the Atlantic to South America in 1488, discovering Brazil before Spain and Portugal.

Unfortunately, first hand accounts of these early voyages are not available. They either were not preserved or have been mislaid. Consequently, without firm supporting evidence, the French claim to be the first Europeans to discover the Americas is tenuous.

This might not always be so. The French Atlantic seaboard and other parts of France are rich in resources for history. Finds of significance can still be made by historians searching for material. It is therefore possible that someone in the future might be fortunate enough to locate missing evidence about the early French voyages of exploration, either in France or even in Portugal or Spain where early librarians and others collected and preserved reports and accounts of explorations so they could be used by their own navigators.

Such early contacts by France with Africa in particular, are not beyond the realms of possibility. The French in the early part of the fourteenth century already had a knowledge of West Africa down to the Niger, gleaned from the Arabs and in particular from the Muslim geographer and explorer Ibn Batutah.[6] His accounts of the lands from India to Timbuctu stimulated French interest and activities.

By the latter part of the fourteenth century, when Ibn Batutah's

Jean De Béthencourt sailed from La Rochelle in May 1402 to trade in the Canary Islands which he subsequently occupied in 1404

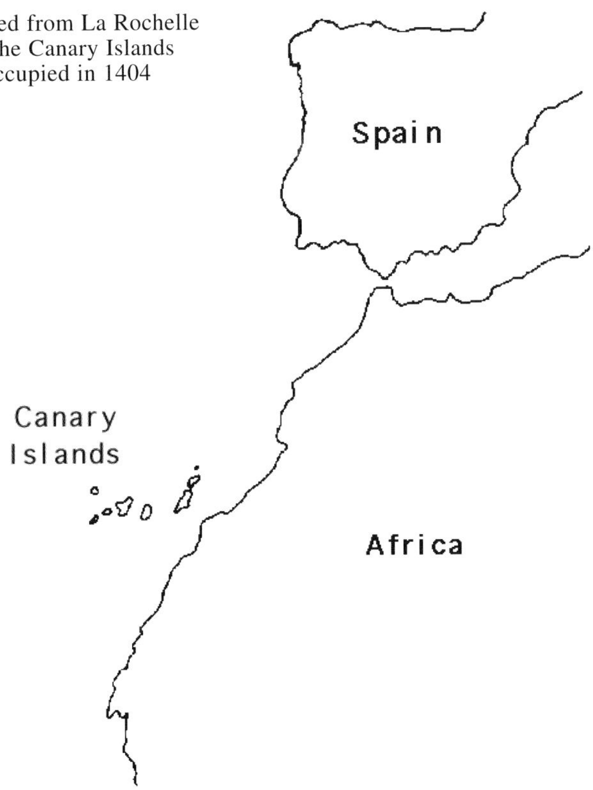

exploration became known, France was already involved in the Canary Islands. Later, in 1402, the Norman, Jean de Béthencourt captured, colonized and ruled the island group.[7] This gave France a base in the Atlantic which stimulated a continued interest to look south even after the Canaries were taken over by Spain, for the French maintained an interest in the islands.

In the same period, in the first years of the fifteenth century, a doctor from Toulouse, Anselme d'Isalguier[8] who was a noted naturalist apparently motivated by an insatiable curiosity, as a result of being stimulated by Ibn Batutah, explored Africa as far as Gao and Timbuctu, helping to fill blank spots in European knowledge of that area.

The French maritime tradition

By the end of the fifteenth century, when the great European maritime explorations began, France therefore already had a long and rich tradition of sea travel for trade and exploration, and was well placed to join in the quest for routes to the "new worlds" and to the east.

A series of factors served to put France in a favourable position among the sea exploring and trading powers of Europe. Firstly, like Spain, France was a two ocean power. It faced the Atlantic Ocean and the Mediterranean, which served to give France a rich combined tradition of seamanship and easy access to the Atlantic sea-routes.

Secondly, while the Mediterranean served well as a training ground for seamen and as a contact point for Arab and other Mediterranean sources of knowledge about the world, the Atlantic and English Channel seas provided an unrivalled training area for sailing skills, more than does the Atlantic coast of Spain. The winds, tides, currents and turbulent seas on the French Atlantic coast serve as challenging classrooms which have helped produce the experienced and talented deep sea fishermen and deep sea long ranging traders who sailed from Dieppe, Harfleur, St Malo and other noted early ports of France. French abilities in this regard were evidenced not only by her sea traders, fishermen and explorers, but also by the emergence of skilled and proficient naval admirals such as Prégent de Bidoux[9] and the later Portzmoguer.[10]

Thirdly, France by the end of the fifteenth century did not have only the trained and able seamen capable of making sustained and difficult voyages. It also had the technical knowledge and capacity to support long ranging sea expeditions.[11]

By then France was not the weak maritime and naval power it had been in the first part of the century in the early period of the hundred years war. Ship construction had improved resulting in the production of large and well constructed ships. By the end of the fifteenth century France had ships such as the *Charente* which were capable of mounting 200 cannon. These ships had the strength to stand up to the buffeting of winds and seas and the capacity to take the stores needed for long, well founded expeditions.

Fortunately for the historian of maritime exploration, a full size ship hull, built by French craftsmen in the great age of discovery, is preserved in the Church of Saint Catherine (Eglise Saint-Catherine) on the market place above the old port of Honfleur, Normandy. This unique all wooden church was built by master shipwrights (Maître de Hache) at the end of the Hundred Years War during a shortage of construction craftsmen. The abilities and skilfulness of the French shipwrights of the time is everywhere evident in the church construction and fittings. But what is of most significance is the domed roof which is in fact a ship hull built and positioned upside down to cover the nave and altar. Standing in the nave and looking upwards is rather like looking into the holds and timbers of a ship. At regular intervals there are hull and deck frames which form the rafters of the church. Between these are neatly placed ribs, and fixed onto these are neatly fitted planks which make the church roof. No doubt the Honfleur shipbuilders found it easier to use their practical knowledge and experience to build the church, rather than to rely on more accepted methods of domed roof construction. But whatever their reason, if what they built as the roof of the church at Honfleur is a typical example of French ship construction, then French sailors at the time had no need to feel apprehensive about the quality of the ships under their feet. The Honfleur hull is well founded and would stand up to all conditions at sea.

Above decks, sail patterns and rigging were also improved, giving French ships a good driving capacity. The old square sails and oblong sails inherited from the north, which are best suited for running before the wind and need to be supplemented by oarsmen to gain distance to windward, were added to by triangular fore and aft sails copied from the Arabs. These allowed ships to sail more directly into the wind, providing them with an independence from oarsmen. In fact the French ships were so well sailed that like the Portuguese they could outsail the Arabs.

French navigators by the end of the fifteenth century, also had the technical support facilities needed to make sustained deep sea voyages far from land. They had the compass to find directions. They had a knowledge of the log which is used to measure distance covered daily, which is essential for navigation. They had the astrolabe to use to make solar and stellar observations. They had a knowledge of astronomy, mathematics and hydrography. And they knew about the movements of winds and waters which are necessary items of information for long distance ocean sailors.

To further support them there were maps and sailing directions that could be used as guides at least as far as the Canary Islands, and there was information contained in the manuscript and newly printed books, gleaned and written about by the emerging authors who were looking at the world beyond Christendom.

Early French contacts with the east

The early use of these capacities by French seamen to find a sea route eastwards to the Indies and Cathay, and to find the location of the mythical Southland on the way is not surprising. France had a long held and deep interest in far eastern lands and in finding ways to reach them, commencing with the reign of Saint Louis [Louis IX, 1226-1270].

This interest was primarily the result of the Mongol invasion of north-east and east Europe. The defeat of a combined and then traditional force of Germans and Poles at Liegnitz on April 9, 1241,

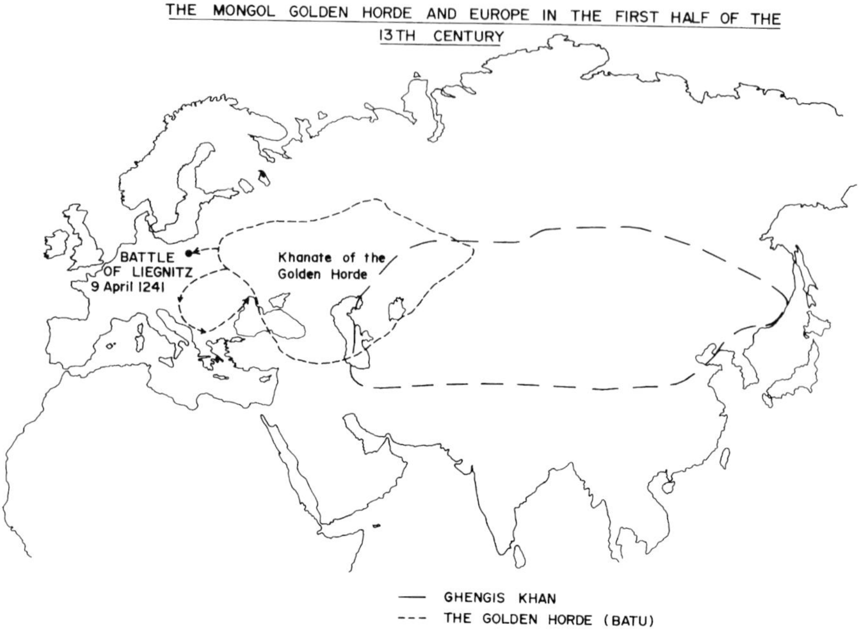

by invading Mongols who were equipped with military rockets, revealed Christendom to be in a perilous position. It was engaged in a war on two fronts. In the south the Christian crusaders were still locked in their early struggle with the forces of Islam. In the north Christendom was confronted by the expanding Mongols under Batu Khan who had been appointed to invade Europe in 1235. After his conquest of Russia by 1240, and his defeat of the Germans and Poles under Henry II of Silesia at Liegnitz, a further force under Batu Khan defeated the Hungarians at Mohi on April 11,1241. Fortunately for Christendom, because of internal problems in the Mongol empire caused by the death of the Great Khan Ogadai in December 1241, Batu Khan did not consolidate his victories by securing the territories he had conquered. He withdrew to Russia to subsequently establish the Khanate of the Golden Horde.

It was at this stage that the Mongol empire of the east came to have a function in European politics, in France and the Papacy in particular, and in concepts of strategy for the war against Islam. Europeans did not view the Mongols as infidels or heretics, as they viewed Moslems. They regarded the Mongols as uncommitted pagans open to Christian influence and conversion. Saint Louis and Pope Innocent IV consequently joined together in an attempt to reach the Mongols in an effort to relieve the pressure on Christendom and to try to turn the tide of battle against the forces of Islam in the Middle East. Part of their design was to have the Mongols join Christendom in a joint attack on Islam, to help save the holy places of Christendom.

In 1247 Saint Louis appointed an experienced traveller and skilled diplomat, Andrew of Lonjumel, to accompany the papal missionary and delegate, Friar Ascelin, to visit the Mongols.[12] These succeeded in their mission. They met the Mongols near Kars and an exchange of ambassadors and emissaries followed. Andrew was consequently appointed by Saint Louis as ambassador to the Great Khan.

He again set out to journey east in 1249 with notes from Saint Louis and the papacy and gifts, reaching the Khan's court near Karakoram a year later. The mission proved a failure. They were regarded as tributaries from France coming to pay homage. The hope for an alliance was consequently aborted.

Despite an insolent letter sent to the French monarch, Saint Louis sent a further mission overland east under William of Rubruquis in 1253. This mission was sent primarily because of a rumour that Batu Khan's eldest son, Sartak, was a baptized Christian. On their journey to Karakoram, the mission came across Nestorian priests who had earlier gone to China from the Syriac church. The significant point of this mission is that it provided Europeans with a knowledge of Cathay, or the Mongol empire in the east, and contributed to the myth about Prester John, the great Christian leader in the east whose forces and wealth could be used to help western Christendom in its time of peril. The land route towards the mythical land of Prester John, and to the Mongols was thus well known to the French before Marco Polo. When the great sea explorations for a route east commenced, the French were not inclined or likely to leave Portugal and Spain to make the quest alone.

Two things, however, stood in the way. When Portugal and Spain commenced their sea searches, France was involved in continental wars and troubles which were capturing its attention and using up its naval maritime and other resources. More significantly, in 1493 Pope Alexander VI established a line of demarcation 100 leagues west of the Cape Verde Islands, giving Spain exclusive rights to the area west of the line, and giving Portugal exclusive rights to lands and seas east of the line. This agreement was confirmed by the Spanish-Portuguese Treaty of Tordesillas signed 7 June 1494, which arranged for the demarcation line to be shifted 370 leagues west of the Cape Verde Islands. This arrangement was given papal sanction in 1506. The significance of this as far as France is concerned is that it deliberately excluded France and the other maritime powers of Christendom. The early French explorers who ventured forth by sea consequently did so without papal sanction or approval, although they were supported by their national monarchy.

French sea routes to the east

Nevertheless, three routes used by French explorers seeking ways to the new world and the east were soon established. One route went westward from France across the north Atlantic to north America. This route was used by Dieppois and Maloines in particular. For instance the Dieppe ship the Dauphine under the command of the explorer Giovanni de Verrazano,[13] was equipped and sent out in 1524 and supported by the noted Dieppois ship owner Jean Ango. Verrazano was sent to explore the east coast of north America. His expedition which discovered present New York, naming it Angoulême after the French capital of the Charente, was followed in 1534 by an expedition from St Malo under Jacques Cartier[14] who sought to find a north-west passage to the Indies. His discovery of the St Lawrence River and his acquisition of the territory about it for France provided the basis for the later French overseas territory founded in Canada by Champlain.

A second route went in a southerly and south-westerly direction from France across the Atlantic to South America. This was the most favoured course for ships making for Cathay, the Spice Islands and other eastern lands. It followed the direction of the trade winds and westward currents. Vessels using it sailed south to pick up the north-east and south-east trade winds, and the Canary and North Equatorial and South Equatorial currents which took them towards Latin America.

From there, once the crewmen were refreshed and the ships refitted, the navigators using this second route, doubled Cape Horn and entered the Pacific. This way to the east was used by Maloine and Dieppois sailors in particular.

A third route used by French navigators was the same as the second route as far as South America. From there vessels following the third route headed south and picked up the westerlies and eastward currents to reach Cape Town, the Indian Ocean and beyond.

It was on these two latter routes, via Cape Horn and via the Cape of Good Hope to the Indies that France suffered directly as a result of the Treaty of Tordesillas. By this treaty and as a result of the sanctioning of it by the Pope, France was denied the right to acquire and establish a refreshment point for its vessels. These were essential if regular sailings, explorations and trade voyages were to be made.

France subsequently for a while established footholds — which could serve the purpose of refreshment points — in South America, which was the most ideal place for a half-way house. In 1555, the French adventurer Nicolas Durand de Villegaignon[15] established the colony of La France Antarctique near present Rio de Janiero. This colony, settled by Huguenots, was short-lived. It was captured by the

FRANCE AT THE END OF THE 15TH CENTURY

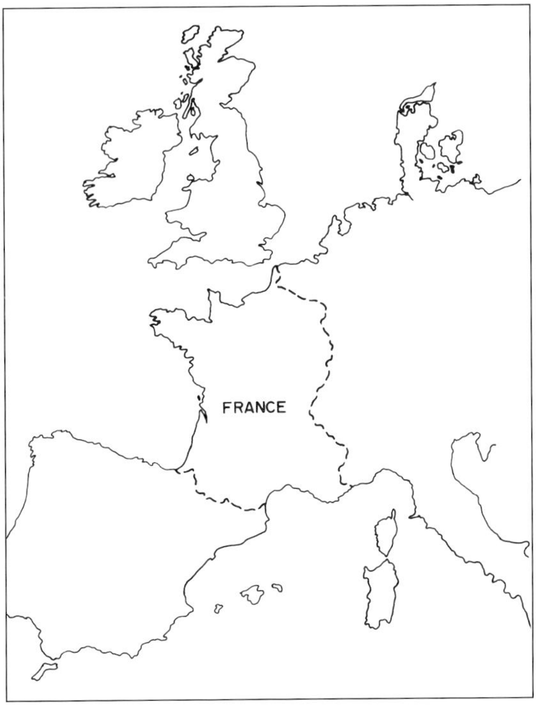

At the end of the fifteenth century, on the eve of the French renaissance when Norman navigators sailed out to compete with Spaniards and Protuguese in the discovery and exploration of 'new worlds and oceans', the later Valois monarchs (1461-1589) had already laid many of the foundations for modern France. The feudal monarchy was replaced by a more absolute variety. Borders were secured. Ambitious provincial leaders were made powerless. Rule was popularly rather than nobility based, and economic conditions were improved.
Previously, under the rule of the early Valois monarchs (from 1328) France was badly situated, and in a seemingly ruinous state. The monarchy was humbled by resounding English victories such as at Agincourt (1415), and by the rebellions of ambitious French provincial princes. The war bled economy was in a poor state. The coinage was debased. Taxation was a burden. Conscription and casualties in the Hundred Years' War cut the work force of the nation.
Under Louis XI (1461-1483) and Charles VIII (1483-1498) the power of the crown was consolidated, the vassal princes were humbled, the merchants and traders were favoured and encouraged instead of the nobles, and economic recovery, made possibly by peace, was effected. Thus by the reign of Louis XII "Le Père du Peuple" (1498-1515) France was well positioned to make a distinctive contribution to the exploration of the world. But although well begun, this drive was not sustained. France became involved in new wars and struggles eastward in Italy and with Spain, which drained resources and taxed manpower. Added to this weaknesses to the state were being made by the rising religious struggles against the Protestants.

Portuguese five years after its founding. Only the French name, Villegaignon Island remains today.

Later, in 1764, Louis Antoine de Bougainville[16], established a French colony further south in the Falkland Islands with the same idea. This effort to provide Frenchmen with a half-way base and a dominating strategic position near Cape Horn was even more short-lived than Villegaignon's. The British established an adjoining settlement and counter claim and Bougainville's colony was ended, the French claim to territory in the Falklands being handed over to Spain in 1766.

France, as a result, had to rely either on conquering or on establishing good diplomatic relations with the Iberian powers, and later with the states of Latin America when they achieved independence, in order to secure refreshment points and reliable bases on the trade wind route to the east. Napoleon's conquest of Spain and his Iberian policy gave France advantages in this regard, at least for a while. By this France acquired bases overseas. But in general France's long term achievements were the establishment of half-way bases in the Indian Ocean close to the Cape of Good Hope especially at Mauritius.

In the meantime, in the eighteenth century, a distinctive French sea route from the South Atlantic to Asia across the Indian Ocean was established. As in the Atlantic this route was planned for ships to use the prevailing winds to advantage. French vessels from the South Atlantic would call at the French base at Mauritius. They would then sail north-west to Asia using the prevailing south-east trade winds and then the monsoon winds to reach India, the East Indies and China and Japan. The constant use of this recommended route

FRANCE LOCATION OF SEAPORTS MENTIONED

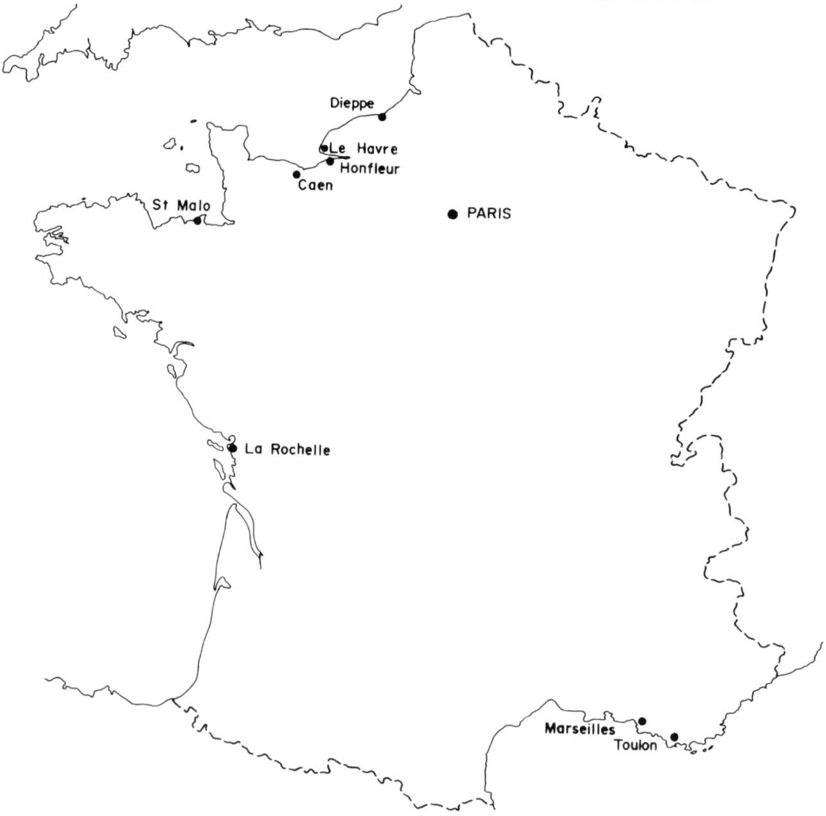

resulted in the French missing the opportunity to engage in the explorations of Australia and the eastern side of the Indian Ocean.

Discoveries in those parts, by the French, were made primarily by explorers who sailed on special voyages south from Mauritius and then eastwards in search of the mythical southern continent. The Frenchman who played a significant part in leading others who followed him to do this was the early navigator Binot Paulmier de Gonneville.

Paulmier de Gonneville

It is not possible to determine from the existing evidence, the real contribution Gonneville made as an explorer and discoverer. The records he left about his voyage do not indicate where he went.

Claims have been made that Gonneville discovered western Australia and the Cape route there before the Dutch. This cannot be definitely proved or disproved. Gonneville lost his ship and his journals on the French Atlantic coast as he was nearing home at the end of his voyage. The only existing account of his expedition is a statement made on 19 June, 1505 by Gonneville to the French Admiralty Court of the Marble Table at Rouen, which had jurisdiction over maritime commerce in that area.[17] This statement contains no sailing directions.

This lack of information does not necessarily mean that Gonneville was a careless log-keeper and a simple minded navigator, or that he was more of a medieval romantic story teller than a modern sea explorer. Gonneville, like other sailors, traders, and explorers at the time had reason to be quiet about the exact location of the new land he claimed for France. France at the time was denied the right to such claims by the Pope and the Treaty of Tordesillas. Moreover, trade rivalry existed and in the circumstances it did not pay those who sought glory and riches to be specific about where sources for these existed. There was a fear that others might benefit.

The statement Gonneville made to the Admiralty Court at Rouen gives a brief summary of the expedition. Gonneville was directly stimulated to explore the route to the Indies by reports of Da Gama's successful voyage, which he heard about in Lisbon. There he engaged two Portuguese pilots, Bastiam Moura and Diegue Cohinto, to guide an expedition. This was equipped and supported by shippers and trade entrepreneurs at Honfleur.

It is not surprising to see men from Honfleur in Normandy in particular play a significant early role in European overseas exploration. Honfleur which is situated on the left bank of the estuary of the Seine, almost opposite Le Havre, was then a sea port of note. It was the sea port for goods coming from Portugal and Spain to Caen and other places in Normandy and the Ile de France. Binot (which could be a local diminutive for Robinet) Paulmier de Gonneville's family had lived nearby for some time. The family is mentioned as living in St Etienne parish at Honfleur in the early part of the fifteenth century. At the time Binot sailed, the family lived at Gonneville near Honfleur, a small village inland from Honfleur.

Binot Paulmier de Gonneville visited Lisbon in 1501, then one of the principle *entrepôts* for Asian and African trade. Impressed by the rare goods found in the markets, and by the talk about new discoveries and new lands, Binot Paulmier took two knowledgeable and experienced Portuguese pilots back with him to Honfleur. There six entrepreneurs fitted out the *Espoir* with a complement of 60 including traders, adventurers and a naturalist, Nicolle Le Febvre as well as sailors and soldiers, thus making it the first known scientific voyage of exploration made by the French.

The ship was loaded with supplies, trade goods including linen cloth, axes, spades, hooks, corn reapers, ploughs, pine cones, mirrors, hardware, glass and a variety of other items as well as a supply of silver to serve as money in the Indies.

The *Espoir* left Honfleur on 24 June 1503. The explorers sailed south to the Canaries, reaching there in eighteen days, and then on to Cape Verde where they traded and replenished supplies, taking on rice and poultry purchased from the Moors. After a ten day stay, Gonneville departed on August 9 and headed for the equator which was crossed on September 12.

After this conditions worsened. Crew members fell sick, and six died. On November 9 seaweed was sighted and birds seen which the Portuguese pilots said was a sign that they were near the Cape of Good Hope. However, they met contrary winds lasting for three weeks during which the chief pilot, Collin Vasseur died. After this they were caught in a tempestuous gale and carried away. The trouble is Gonneville does not say which way the winds blew him.

After being then becalmed for three weeks they sighted land and anchored at the mouth of a large river on January 6, 1504. The expedition stayed there until the following 6 July.

Gonneville's account of the southland

Gonneville has left a brief description of the land and people he discovered and observed while refitting his ship and refreshing his crew. The ship was worm eaten and in need of repairs, and the crew exhausted.

His observations on the people were no doubt coloured by his need for materials and skilled labour. That is what he primarily described. He reported that the inhabitants lived a simple happy life subsisting on hunting, fishing, vegetables and roots which they planted.

Common people wore cloaks of platted rushes or furs or feathers. Men and women were reported to wear identical clothing, but with the men wearing shorter cloaks. Women's cloaks went down to the knees. The women, however, were adorned with necklaces and bracelets made of bone and shell. The men wore long hair with a band of feathers.

The men were reported to have been equipped with bows and arrows and wooden spears with hardened tips.

The land they lived in was regarded as fertile, providing fish, birds, beasts and trees. The naturalist Nicolle le Febvre made a description of the products of the land but unfortunately for historians, this journal was apparently lost with the ship.

Gonneville conducted several expeditions inland. He described the country as being peopled at intervals. The inhabitants lived in villages or camps consisting of 30 to 80 cabins. These were made of stakes covered with rushes with a hole in the roof for a chimney, and had doors which were opened with wooden keys. The beds were made of woven reeds and soft leaves. The household utensils were of wood, but were hardened in fires like pottery.

The country was seen to be politically disunited. It consisted of a series of small territories or cantons under chiefs who owed allegiance to the local king. The local chiefs and king were described as despotic, and the people obedient by necessity.

The tribes or groups lived mostly in peace while Gonneville was at anchor. Two wars broke out while he was there, but were not large scale. The main point made in this regard is that the local king tried to enlist the aid of the French with their sophisticated weapons. This was refused by Gonneville.

Gonneville has left a description of the local king Arosca. His territory consisted of about one dozen villages. He was about 60 years old, a widower with six children. He had a grave appearance, was of medium height and portly.

Arosca's son Essomericq and a guardian, Namoa, sailed with the expedition back to France. First, before he left, an inscribed cross was erected by Gonneville and the land claimed in the name of France and the Pope, Alexander VI.

Gonneville's expedition left the "south Indies" on 3 July, 1504. Unfortunately, for historians, he again does not say which way he sailed. All he states is that the trip was difficult and unpleasant. The first months were marred by bad weather and sickness. Fever broke out not long after leaving land, and a number of the complement died, including the ship's doctor and the "Indian" Namoa.

The ship must have been heading in a northerly direction, at least at the end of the first months at sea. Before the next landfall was made the ship crossed the Tropic of Capricorn. After this, in October 1504, another country was discovered. The expedition put in there on October 10. The crew was badly in need of fresh supplies and recuperation and "a breath of land air". But the landfall made was unfortunate. Gonneville states in a brief description of the place that it was inhabited by "churlish" Indians who were anarchic and cannibalistic. Several men of the expedition lost their lives there, including the curious and adventurous naturalist Le Febvre.

They therefore moved on a further "hundred leagues" (approximately 500 kilometres). There the inhabitants were observed to be the same "savages". After refreshing, the ship left at the end of December 1504 and arrived at the Canary Islands in March 1505. Gonneville then sailed for France.

On 17 May, off the Channel Islands, they were intercepted by two pirates Edward Blunt or Blunth of Plymouth and Mouris Fortun of Brittany. Rather than fall into their hands, Gonneville put his ship onto the reefs off the French coast. He arrived back in Honfleur on May 20 with less than half the number who had set out with him.

Among the survivors reaching Honfleur was the "Indian" Prince Essomericq. Gonneville had captured two other "savages" from his later refreshment point to take back to France. But these jumped overboard as the ship was leaving. Essomericq alone came to France providing evidence that a French navigator had been to the Indies. Essomericq subsequently became a curiosity and was taken into the Gonneville family. His descendants, in particular the Abbe Jean Paulmier de Courtonne, were in part responsible for the later French quest for Australia. In 1663 he urged France and the church to send out missions to christianize his forbears.

The trouble is, neither he nor anyone else knew where to go. Various suggestions were offered. The Abbe Jean Paulmier and the prominent eighteenth century geographer de Brosses believed Gonneville had discovered the *Terre Australe.* Subsequent claims were made that he had discovered Madagascar, and Brazil.[18]

The location of Gonneville land

The most general current belief is that Gonneville in fact landed on the east coast of South America, in the vicinity of San Francisco do Sul situated at about the 10° parallel of latitude south, and that King Arosca was one of the Latin American Indians. The land of "savages" he discovered is held to be the Bahia region of Brazil.[19]

This interpretation would fit in with the description Gonneville gives of his rough trip home. Sailing from there he would have pushed into contrary currents and would have had to reach across the trade winds, and experience the discomforts he mentions in his report on the voyage.

But the exact location of Gonneville's land remains a mystery. However, this might not always be the case. Three ways lay open for the historian of sea explorations to acquire more definite knowledge about Gonneville's discoveries. Firstly, further historical records might be located to help solve the mystery. Some historian in the future might locate an account of the voyage written by the Portuguese pilots or others who went on or were associated with the expedition; or they might find some of the Gonneville family papers. These do not exist in the repositories along the French Atlantic coast, nor in Rouen or Paris. But this does not mean they have disappeared for all time. Gonneville's account and log book have not been found in the Channel Islands or in Plymouth where the "pirates" who chased him operated from. If they took papers from Gonneville's ship, which is doubtful seeing it was reported to have been lost, then they do not seem to have reached either England or the Channel Islands, or at least to have been kept there. However, they might have been preserved somewhere.

Secondly, anthropologists might obtain some evidence from "tribal areas" in South America or elsewhere, indicating that a long lasting visit was made by Gonneville, who left a cross and took away a prince.

There is certainly no evidence of this happening to the tribes of south-west Australia; nor of their being contacted by an expedition which brought shovels, ploughs and other items from Europe, which helps discount the belief that Gonneville landed there. In any case Gonneville's description does not match what exists in western Australia. The aborigines in Australia, for example, did not have houses with doors that had latch keys made of wood, as used in Brittany, which Gonneville describes as being used in the land he discovered.

Thirdly, it might be possible to determine from the description Gonneville gave of the Indians, who in fact they were. It is this evidence in part which has produced the current belief that Gonneville contacted the Curijos or Caryo Indians in South America.

Nevertheless, for the next centuries the French became convinced that Gonneville had in fact gone to the south Indian Ocean, and they consequently identified Gonneville Land with *Terra Australis Incognita* which thus to them became synonymous with France Australe.[20]

Jean Ango's renaissance manor house at Varangeville, Normandy. *Photograph: Gunhild U. Marchant*

CHAPTER 2

French explorations, intellectual curiosity and utopian literature about Australia from Gonneville to Bouvet de Lozier (1505-1740)

Decline of French involvement in Indian Ocean exploration

French curiosity about the unknown *Terra Australis* discovered by Gonneville increased after his return to Honfleur. But until Bouvet de Lozier set out in 1738 to explore the south Atlantic and south Indian Ocean to look for France Australe for the French East Indian Company, the continued curiosity was primarily maintained by French literary and scholarly efforts and the reports of discoveries by other European seamen rather than by the activities of French explorers. Until Bouvet went, France played little part in the continued search for the southland, although it continued with maritime explorations.

Consequently it was Portuguese and Spanish explorers joined later by British and Dutch navigators who pushed forward the frontiers of knowledge about *Terra Australis* in the early period down to the eighteenth century. The Spanish explorer Magellan, for example, discovered the passage into the Pacific by way of the strait that bears his name, revealing that the Atlantic and Pacific Oceans were linked, and that it was thus possible to circumnavigate the world. He and those who followed also revealed to Europeans that the Pacific and the size of the world was larger than believed. European geographers at the time of Magellan had for some time past accepted Ptolemy's incorrect calculations about the length of the circumference of the earth instead of accepting the more accurate earlier calculations of Eratosthenes. Eratosthenes' assessment of the world's circumference is close to the present assessed figure. Ptolemy's figure is approximately 25 percent too short of the total. His globe left no room for a larger Pacific Ocean. The rectification of this error, because of practical work and measurements made by earlier circumnavigators such as Magellan, thus provided map makers with the room to add the Americas to the globe, which previously was thought to be not theoretically possible. According to calculations based on Ptolemy's figures, early geographers concluded the lands Columbus discovered had to be the Indies in Asia.

More significant as far as the search for *Terra Australis* is concerned, the discovery of a viable sea route from Europe to the Pacific by way of South America helped further the search for land in southern latitudes as the Pacific was being explored.

Several early sightings of land in the south were reported by ships going that way. Juan Fernandez, the Spanish explorer, reported sighting land at 40° south beyond Cape Horn in 1576. The English voyagers John Davis and Richard Hawkins found the Falkland Islands in 1592 and 1594 respectively. Theodor Gerards earlier reported seeing a snow capped land south of Cape Horn, when making into the Pacific.

It would be incorrect to claim that Frenchmen saw nothing of these new discoveries. European ship captains at that time often carried cosmopolitan crews of experienced seamen. Gonneville, for example, took Portuguese along with him. Magellan is also reported to have had some nineteen Frenchmen in his crew on his voyage of circumnavigation. There could have been other Frenchmen sailing on other ships under other flags.

Besides this there was a growing number of French "privateers" sailing into oceans claimed as monopoly areas by Spain and Portugal, including the South Seas which became economically attractive to the French as to others. But for the most part these privateers, unlike Dampier, kept silent about their discoveries and exploits.

The shift of French interest to America

More significantly others continued to explore and make contributions to European knowledge, and established French bases abroad which served to expand the rule of the Bourbons. But in this period these efforts were concentrated in the main on the Americas, especially North America; to a lesser extent Southeast and East Asia, and not on the southern part of the Indian Ocean, nor on the search for *Terra Australis*.

The role of Jean Ango

The efforts at exploration in selected areas away from Australia were not made by individual explorers acting independently as Gonneville and his partners had. The man most responsible for keeping France to the forefront among the maritime exploring powers in the great age of discovery was the Norman renaissance ship owner Jean Ango[1] (c1480-1551).

Ango's father was early involved in the Atlantic fish trade and in the import and export business at the growing commercial port of

Dieppe. By the time Gonneville sailed to achieve fame for himself and France, the Ango family not only had extensive shipping interests, but also played a prominent role in seeking to establish the principle of "the freedom of the seas" at the time Portugal and Spain were promoting monopolies which the Pope was willing to sanction.

When Jean Ango took charge of the Ango family shipping business, with the aid of Francis I he carried out his policy of seeing that French vessels freely sailed, traded and explored and annexed lands as widely as possible about the newly discovered sea ways.

Ango had the resources to do this. By the second decade of the sixteenth century his company controlled in the vicinity of seventy vessels and extensive resources. Equally significantly, as a result of his correlative efforts to make Dieppe into a French Florence, he attracted the close interest and support of the French monarch, Francis I, which helped further his own and France's interests abroad. Most prominent in this regard were the Florentine renaissance buildings he constructed as show places in and near Dieppe. Right on the quay in Dieppe harbour he built an impressive Ango company headquarters in the Florentine renaissance style. This ornate building with Italian tiles and bas reliefs, which was aptly named Pensée (Thought) after his father's ship, reflecting the spirit of the times and Ango's intellectual interests, attracted widespread attention including that of Francis I. Unfortunately for posterity the building was destroyed by the British in 1694 in the course of their war with France. But the equally ornate and impressive renaissance style manor house Ango built for the family at Varangeville outside Dieppe, which Francis I visited, remains to show the wide interests and impressive qualities of this renaissance French ship-owner.

Voyages Portrayed in Church Art

While Ango's manor house stands as a permanent memorial to the part played by Dieppois navigators in the renaissance in Europe, artworks in churches in the ancient province of Normandy bear testament to the Dieppois quest for discoveries abroad and to the rising Dieppois interest in exotic places. The impression one gets after making a tour of the whole Atlantic coast of France in fact is not unlike that gleaned from a study of the engravings and lithographs made as a result of the tales and accounts of the later French scientific expeditions which stirred the imaginations of Parisiens and others. For example, there is a large bas relief prominently placed near the chancel in the Church of St. Jacques situated close by the old port at Dieppe. This is supposed to have been influenced by Ango's voyages, and portrays in a vivid way the strange peoples and scenes encountered by explorers. Further south along the Atlantic coast, in the small church above the sea at Varangeville, even more interesting sculptures of exotic peoples have been made on the church pillars, showing how the local imagination must have been stimulated by the tales of the travellers.

It is exceedingly difficult to say if the sculptures are accurate and reliable records of what the voyagers saw, or if they are the results of the artistic imagination. If they are anything like the later French produced lithographs and engravings of the age of scientific discovery, then they are more artistic than realistic. But even so, it would take more than a considerable stretch of the imagination to conclude from the sculptures that the French visited and had seen Australian Aborigines at the time of the renaissance. The figures are all well clothed and bedecked and in no way resemble the Aborigines.

Explorations by Ango's Ships

Ango equipped and despatched significant exploratory voyagers to the Americas and the east. In 1524 he sent Giovanni da Verrazano to North America to explore the east coast there. Verrazano subsequently discovered the area where New York now stands, which he named the Land of Angouleme, which is an historic country near the Charente, but which also obviously refers to his patron, Ango.

In 1529 Ango equipped two ships, the *Pensée* and the *Sacré,* and sent them to Sumatra and elsewhere in the Spice Islands, under the command of the sea-faring brothers Jean and Raoul Parmentier,[2] who were later immortalized in local and national French poetry for their achievements, which were compared with those of the Argonauts.

In the case of the Americas the early French efforts at independent exploration made by Ango were carried on by others after Ango's death and the demise of his company. Jacques Cartier, who was among the most prominent of these, left in 1534 to look for a northwest passage past Canada to Asia, and explored the Atlantic coast of America there, commencing the French interest in Canada. His and later efforts by other French explorers and adventurers played no little part in establishing French rule in the Americas which by the end of the sixteenth century included both Canada and Florida.

Although the voyages made by Ango's captains and other French seamen provided France with an assured placed in the histories of early European maritime exploration, her overall efforts and achievements were limited to select areas, and even there her efforts diminished as time wore on.

This was because France was disadvantaged. She suffered from three weaknesses in particular which adversely affected her explorers and colonial expansionists. Firstly and most significantly, France was still inhibited by the provisions of the Treaty of Tordesillas and the papal pronouncement which approved the Spanish and Portuguese monopoly of the New World and its resources. Frenchmen had neither the legal nor religious moral right to explore seas and lands claimed by the Iberian powers. Nor did the French crown have the right to claim lands there; nor to save souls in the name of the catholic church, nor to exploit the resources.

Reasons for the limited achievements of France in Exploration and Expansion

France, in effect, never ever enjoyed legally the freedom of the seas desired by its renaissance ship-owners. This had become an early issue for the Ango family, as has been indicated, and for other French seamen and fishermen on the Atlantic coast after the rise of the cod-fish trade.[3] Rich fishing grounds for cod were found near Canada in the New World. By treaty, supported by papal religious sanction, these cod-fish resources legally belonged to Portugal. But this situation was not accepted by French merchants, adventurers and fishers, nor by Francis I. They moved in to use these resources and Portugal could do little about this. However, French authorities never succeeded in having the treaty provisions and papal sanctions altered to give Frenchmen the legal right to participate. Those French who explored, exploited and traded did so as privateers. In the reign of Francis I there was little trouble about this. He gave French traders his full support. But after his death in 1547 French seamen and merchants trading overseas could not rely on being supported by their government when they were challenged or opposed by Portugal and Spain.

Secondly and correlatively, although France had emerged as a powerful continental nation by the time the great maritime explorations were made, she lacked the opportunity to use this power to advantage against Spain and Portugal. Her policy towards the self-interested Iberian powers was confused and unrewarding while they remained great powers. And when they declined, France found itself confronted by an expanding Britain which proved to be as difficult an enemy.

Thirdly, for the first half of the sixteenth century France continued to be embroiled in the problems of wars about Italy. Later on in the second half of the century, after the reformation became prominent, France was also adversely affected by civil war and religious troubles.

Added to this, after the reign of Francis I ended, France tended to become politically unstable. As a result, even when Spain suffered defeat by Britain at the time of the Armada, France made none of the gains it had long sought. Nor, later on when the French Duc D'Orleans assumed the throne of Spain in 1700, did France benefit then.

Thus by the end of the sixteenth century France was either expelled from or withdrew from its New World bases in America. The French establishments in Florida and Brazil were taken out by Spain, and Canada was abandoned, so Portugal and Spain consolidated their positions while France receded.

The apparent abandonment of expansionism by France was not left uncommented on by French observers. The calvinist historian La Popelinière[4] writing at the time this happened, while praising his countrymen for their initiatives in exploration, reproached them for not carrying out their national designs.

Cartographic Achievements by the Dieppe School

While this was going on France did make a distinctive and significant contribution to the subject of *Terra Australis,* even though it was not directly involved in explorations searching for it. This contribution was also made by Normandy. In the decades after 1540 a series of unique charts and portulans were produced there by scholars of the so called "Dieppe School of Cartography"[5]. These portulans are of particular interest to those interested in the discovery of Australia by Europe. The Dieppe maps or portulans, unlike other maps of the time, show an island with coasts positioned where Australia is situated. Maps produced by other European cartographers at the time contained in contrast, only an imaginatively drawn large continent of *Terra Australis* covering a great part of the southern hemisphere.

Prior to the appearance of the "Dieppe School", European geographers generally accepted the Ptolemaic theory about oceans and continents in the southern hemisphere. Ptolemy held that the Indian Ocean, like other oceans, was bounded in the south by a great land mass. This theory which was based on the earlier Greek idea that the continents of the world were bounded by a confluent ocean and which was widely accepted in pre-modern Europe, fitted in with the then emerging belief in a balanced Earth, that is that the land masses or continents in the northern hemisphere were balanced with equal land masses in the south.

The main fault with Ptolemy's description of the Indian Ocean region is that he did not extend it beyond 16 degrees south latitude. As a result later map makers made *Terra Australis,* the mythical southern continent, cover a large area of the Indian and other southern hemisphere oceans.

Yet after Marco Polo returned to Europe from his visit to East Asia, his information gleaned from Southeast Asian and other sources indicated that there was a large island south of Java called "Beach" or "locach". This later became known as Greater Java or Java la Grande. It is this island of Greater Java which is interestingly featured in the portulans of the Dieppe School of Cartographers.

Scholars investigating the unique school and their maps have concluded that these were the result of a combination of literary researches in Europe and practical observations abroad by seamen. The outstanding question in this regard is whether or not French seamen were directly involved in the discovery of Greater Java, which, when it appeared on the French maps, seemed to give France a realistic knowledge of the Australian region before it was charted by the Dutch.

The uniqueness of French knowledge is best revealed in the Roze map dated 1542, although Pierre Desceliers is acknowledged as the greatest of the Dieppe mapmakers and the father of French hydrography.

Jean Roze, sometimes written Rotz or Ross, was a widely travelled mariner, a competent hydrographer and cartographer.[6] He served for a period as hydrographer to Henry VIII of England, contributing to the British knowledge of hydrography and the geography of the world.

Unlike the Greater Java which is featured on other Dieppe portulans, which is made to extend from just below Java proper down to the Antarctic, Jean Roze's map of the island is abruptly shortened. The west coast is drawn down to approximately 35^0 south latitude where it ends. The east coast is drawn to approximately 60^0 south where it ends. This implies that there could be a sea passage between Greater Java and the mythical *Terra Australis* which was then still believed to exist. That is Roze indicated that Greater Java was an island and drew the then known part as something like Australia. Moreover, Roze's Greater Java is in somewhat the same position as Australia. Cape Leeuwin, where the west coast of Australia ends in the south is not far from 35^0 south latitude, as shown by Roze as being the southern limit of Greater Java. In the case of the Pacific coast which he draws down to 60^0 south, there are islands in the Pacific which could be used to suggest the island of Australia extended that far.

It is not clear if Roze himself saw Greater Java. But there were others in and about Dieppe who could have had extensive information about the area for him to use. Ango's captains, the Parmentier brothers when they visited Sumatra on the trip they commenced in

JEAN ROTZ' MAP OF JAVA AND THE SOUTHERN CONTINENT, 1542

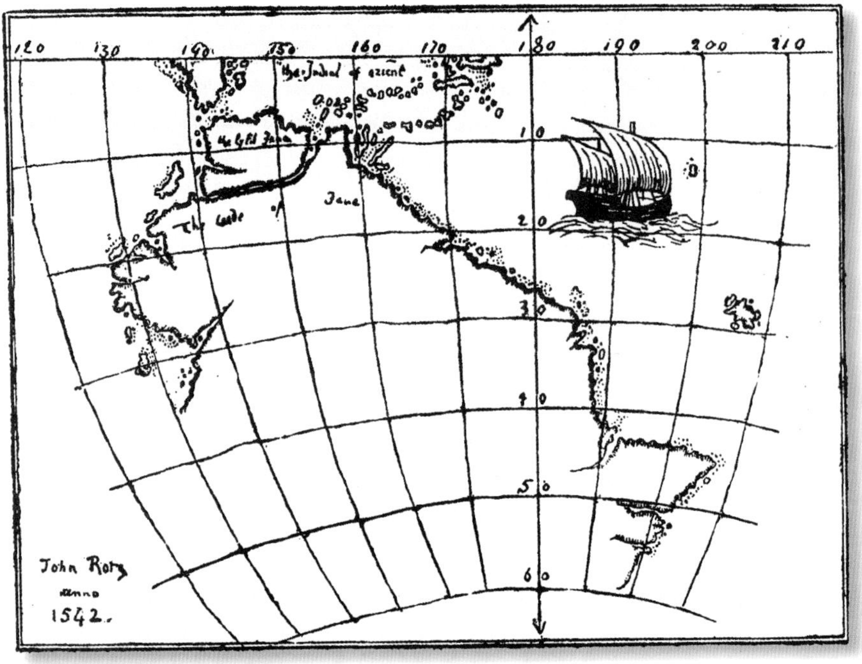

A main point of interest in Rotz' map is the shape he gives to the west coast of the large Southern Continent below the East Indies. This coast is shown as going south-east from Java to approximately 20° south latitude, and then south to the approximate latitude of Cape Leeuwin where the continent ends. This information raises the question of whether or not French cartographers of the Dieppe School knew, eighty years before the Dutch discovered Cape Leeuwin in 1622, that the west coast of Australia ended at approximately 35° south. The actual latitude of Cape Leeuwin is 35° 22' south.

1529, spent time in the region. There they could have gleaned local information about lands in the proximity of the Spice Islands. Such information would have served to clarify the situation resulting from Marco Polo's earlier report.

More significant as far as the historian is concerned is Jean Fonteneau or Alfonse[7] as he is often called. Alfonse was a French pilot and cosmographer who spent the 48 years after 1496, travelling the newly discovered oceans. He is supposed to be the model Rabelais used for his renowned traveller Xenomanes in his *Gargantua and Pantagruel* which was written at the time. Alfonse married a Portuguese, and often used the Portuguese name of Alfonse and passed himself off as that nationality, no doubt to help him avoid the troubles caused to the French by the Treaty of Tordesillas. In his writings Alfonse claims to have seen Greater Java, although he claimed it went as far south as the ice cap.

PIERRE DESCELIER'S MAP OF THE SOUTHERN CONTINENT, 1550

Whatever the source of knowledge, the Dieppe maps are unique and of interest as is Dieppe itself in the history of French explorations and geographic knowledge.

Although the thirst for knowledge about the world continued in France, the Dieppe school declined in influence after the emergence and spread of Mercator's cartographic ideas. This influence once more gave rise to the belief that a large continent existed in the southern hemisphere. As a result, Beach or Greater Java was removed from the maps and portulans being produced after Mercator, to be replaced once more with the mythical continent of *Terra Australis.* This situation persisted until the British explorer Cook and the French explorer Du Fresne revealed in the 1770s that in the mid latitudes of the southern hemisphere, and south of there, was little but water.

Further proposals to explore

Frenchmen made several attempts to follow up Gonneville's work and rediscover his lost land from the end of the sixteenth century. When the expansionist minded French Huguenot leader and admiral, the Comte de Coligny[8] gained a position of influence under Charles IX, he encouraged exploration. In 1571, in part as a consequence of this, the Albacque brothers submitted a proposal to explore "the Third World" as *Terra Australis* then came to be popularly named, so that France could exploit its riches. At the time there were already stories circulating in Europe, about the riches of *Terra Australis,* which had become associated in the minds of the people with the lost mines of King Solomon and their treasures. Coligny's period of influence and French designs in the "Third World" were short lived. Coligny was assassinated in 1572 and civil strife continued in France.

Other attempts were made to mount expeditions. The historian and traveller Voisin de la Popelinière was himself also unsuccessfully involved, planning an expedition. But no French seamen managed to locate Gonneville's lost land, or sail in his wake until the late 1730s. During this interim period in particular the contribution made by the French to the discovery of *Terra Australis* and knowledge about it, was primarily in the field of literature.

Literary interest in *Terra Australis*

In the seventeenth century and the first decade of the eighteenth century, before Bouvet set out on his voyage of exploration, three types of literary works appeared in France dealing with the subject of *Terra Australis.* There were serious pleas presented for France to locate, settle and christianize *Terra Australis.* There were informative books written about the discoveries made by travellers in the southern hemisphere. And there were imaginative French novels set in "Australia" produced as part of the utopian literature then being written and published in Europe.

Gonneville the younger's interest

Foremost of interest among the pleas made to locate the southern continent is the book written by a direct descendant of the "Indian" Prince Essomericq who was brought back to France by Gonneville.

This was published in Paris in 1663.[9] The author, Jean Paulmier de Courtonne, who bore Gonneville's family name Paulmier as a result of Essomericq having been adopted by him, was Canon of St. Peter's Church at Lisieux, a town just south of Gonneville's home town of Honfleur in Normandy, on the road to Caen. Paulmier de Courtonne, in his book, urged France to locate the land of his ancestors, and to establish France there in order to christianize his people.

Although this was a period when French christian missions were being extended abroad to Vietnam, China and other places, Paulmier's plea brought no response. Nevertheless, the book was significant. It recreated an interest in Gonneville, and led to the belief that in actual fact Gonneville had visited *Terra Australis* which Paulmier de Courtonne believed was the home of his ancestors. The book also served to highlight the early contribution France had made to European maritime exploration. And as it was written by a descendant of an "Indian" it raised curiosity about him and his ancestral homeland. The author, as a result, became a popular figure in Europe, receiving the noteworthy appointment of representative of the Danish monarch in France.

As a result of the revived interest in Gonneville, other proposals were made at the end of the century, to go to find the missing southland. De Beaujean or Beaujeu proposed a voyage of circumnavigation in 1698. This was followed by a similar proposal by Voutron. Neither of these proposals was supported, and the expeditions did not take place. France in fact made no further active moves until Bouvet de Lozier's proposal was accepted and supported. In the interim period French seamen, adventurers and explorers tended to visit the South Seas which were then being opened up, joining in the search for the lost goldmines of the "Solomon Islands" and other mythical places believed to have treasures and riches.[10]

Notice of the Solomon and other islands discovered by explorers such as Quiros were being brought to the attention of the French by travel books and geographical works. Many of the French travel books produced at this time were about China and other eastern lands, and the Americas. However, Thevenot's collection of curious voyages referred to *Terra Australis,*[11] which the author urged France to find and settle.

These collections were added to in the following century with works such as de Brosses, which were specifically written on the Australian region.

In the meantime, before specialist studies of *Terra Australis* and the south Indian Ocean appeared, the general travel books which referred to the region were added to by imaginative novels about the southern continent which became equally popular, attracting attention in France to the area.

A series of utopian novels, in the form of More's *Utopia* appeared in France during the last quarter of the seventeenth century and the beginning of the eighteenth century. These all had for their settings, strange and distant places. But like More's book and to a lesser extent the more ill constructed works of Rabelais, this type of literature was more concerned with the rectification of society at home in Europe than with the actual style of life and culture in lands abroad. The new lands discovered and brought to notice in Europe, for these authors, had a didactic function. They were imagined to be and presented as perfect societies with perfect peoples. These created models in consequence were used to criticise life and society in Europe; and became the basis of programmes for political activism aimed at "progress", "salvation", "reform" and redemption in Europe, and at providing "decadent" Europeans with an idyllic "alternative life style". Already French contact with the "indians" in America had produced the idea of the "noble savage", who lived in a simple state of splendour in a natural unpolluted environment, and also had produced the correlative idea that civilization, trade and industry created an evil state for man, and bound him up in chains.

Utopian novels

Among the variety of utopian novels written by authors such as Fénelon, Cyrano de Bergerac and others, two were set in Australia. The others were usually set in North America and other exotic places in the New World,[12] except in the case of Cyrano de Bergerac's works which were more science fiction than terrestial utopian works of fiction. They had their setting on either the Moon or the Sun.[13]

De Foigny

Gabriel de Foigny, a reform-minded religious Frenchman who moved to Geneva which was then a noted centre for evangelical Christian beliefs, wrote two fictional accounts set in Australia featuring a fictional European travel hero, Sadeur. Of these his *La Terre Australe Connue,* printed in 1676 is of most interest.[14]

In this novel de Foigny portrays Australia as an idyllic, well-governed communal state inhabited by perfect immortals.

The novel opens with the traditional shipwreck. But instead of being cast straight on the shore of Australia, like Ferdinand in Shakespeare's *Tempest,* Sadeur, after his ship is wrecked off the coast, has to fight enormous birds which guard Australia, to get to the land. After reaching the shore he discovers that there is a penalty of death for people landing. However, the Australians pardon Sadeur because they find out that like the Australians, he is an hermaphrodite.

After describing the difficulties of landing in Australia, de Foigny then goes on to give a detailed description of the land and its people. The countryside is described as flat, but surrounded by lofty mountains and shallow seas which make it virtually inaccessible.

The Australians are described as being eight feet tall, red in colour, hermaphrodite and immortal. But there is death. However, this is voluntary. When the Australians become weary of life they eat a poison fruit.

They are described as vegetarian; their food consists of fruit. They live in geometrically-shaped houses. They work for the common good and are bound together by mutual affection or brotherhood.

In the novel Sadeur is at first bitterly critical of this alternative life style he finds. However, after an old Australian reasons with him, Sadeur changes his mind completely and becomes an enthusiastic supporter and an enthusiastic propagandist for the Australian way of life which is the purpose of the novel.

De Foigny then goes on to describe the Australian religion and life style. Each inhabitant has to present the state with one child. This means that all are "born free" of family and other ties. The working day of these "free men" or family free men is divided into three equal parts. But every fifth day is spent in meditation at the temple. On the work days the first part is spent in the school of science. There, whenever a discovery is made, it is recorded and used for social development. The second part of the day is spent by all in cultivating flowers. The final part of the day is spent in exercise, military training and in demonstrating achievements made in school.

Naturally de Foigny urges Europeans to forsake their individualist family life styles in Europe and adopt the Australian alternative life style.

De Foigny's work in actual fact excited little attention. It is not a well-constructed novel. It appeals more to the fantastic than to the imaginative and it contains views which were then unpopular if not heretical. De Foigny supports a type of silent, meditational deism which was not an acceptable form of religious devotion in the then existing Christian church. He also opposed the type of free science and the scientific method then emerging in Europe, and added to this a support for bi-sexualism as a way of life which did little to make him and his novels popular.

Vairasse

More skilfully and tightly presented is Vairasse's *Histoire des Sevarambes* printed in 1766.[15] This book made a more significant and lasting impact in Europe. It attracted widespread attention.

In this book the author gives a detailed succinct description of Australia, comparing it favourably with Europe as de Foigny had done.

The Australian nation in the southland is described as having been founded by Sevarais, a Persian traveller who was accompanied by a learned Venetian.

These together created the highly sophisticated Australian civilization with a way of life based on justice, love, charity and good government. The result is a state where crime does not exist, and where there is no disorder, primarily because the three main vices of man — pride, avarice and laziness — have been eliminated.

As in de Foigny's Australia, there are no property rights and all are equal and "free". Each citizen is employed by the state, making things useful for the people as a whole. Each receives all the necessities of life including education and culture.

Religion is in the form of sun-worship, perhaps an allusion to Louis XIV. But freedom of conscience is respected.

Every effort is made to achieve equality. Money is absent. Work is delegated by authority, and goods are distributed equally. But although Vairasse attempts to make Australia seem a democratic republic, the book creates the impression of the country being an authoritarian slave state where the lives of the individuals are ordered and planned by an authority which offers them what it considers to be the necessities of life needed to maintain a perfect society of equals.

French seamen adventurers appeared to have been no more influenced by the imaginative books which created fanciful pictures of Australia, than were English seamen adventurers by Swift's *Gulliver's Travels*[16] and Henry Neville's *The Isle of Pines*[17] which were set in Australia. Other more realistic reports available at the time indicated that the western part of Australia at least, did not appear to be a land of opportunity. The Dutch navigators gave a bad impression of it. This was later reinforced in clearly presented terms by Dampier who was an experienced naturalist and a careful observer. Asia, the Pacific or the South Seas continued to be the main attractive areas for travellers seeking the exotic, and for fortune hunters. French efforts at maritime discovery therefore continued to be spent in those directions.

Growing need for a half way house to India

These continued efforts at the time were added to by new initiatives in empire building, part of which served to further and stimulate French efforts to explore Australia. During the seventeenth and eighteenth centuries, France founded and extended colonial areas in North America and India. New France, founded in Canada and developed by Samuel de Champlain in the early part of the seventeenth century, became most closely attached and of greater interest as it raised few problems of communication. The sea route from France to Canada was short. Supply depots were not required on the way. Canada was thus very easy to get to and from, from France. This was not the case in regard to more distant India. Ships making that voyage needed to take in fresh supplies en route, and to do this they required a base where equipment could be replaced and the men refreshed. It was still widely believed in the years before Cook, that sea air was bad for the human constitution if taken in large quantities and that sailors needed rest periods on land to prevent sickness such as scurvy, which in actual fact is the result of diet deficiencies. Consequently, a quest began for half-way houses to the Indies which stimulated a great amount of rivalry among the maritime powers of Europe.

The ideal place for a half-way house, in the French view, was in southern latitudes equivalent to Europe, that is the temperate, continental zone of the southern hemisphere. This was believed to be the most salubrious climate for Europeans, and the area most suited to meeting the requirements of the commissariat officers. Temperate lands offered the easy production of meat, wheat and wine and cheese which were used as staple diets by the seamen, and offered tall trees with flexible timbers suited for masts and spars, which are not to be found in tropic or tundra regions.

Unfortunately for France, Portugal and Spain controlled the temperate areas of Latin America, and the Portuguese and Dutch by then controlled temperate South Africa. France consequently annexed and established bases on a series of island outposts in the south Indian Ocean to serve as half-way houses and strategic points designed to protect the French trade routes east. Ile de France, or Mauritius as it came to be called after Britain took it from France, became the main post. This was ideally situated not far from the Cape of Good Hope. It offered a salubrious climate suitable for European settlers, and was easy to defend. The main French base there was established to face the wind. This made the passage into the harbour time consuming, but meant that the town could not be easily taken by surprise by wind-driven vessels.

The French base at Mauritius was well-situated for sailing ships making for all parts of the Indian Ocean and beyond. Ships setting out northwards from there can ride the trade winds to reach East Africa, the Middle East, India, Southeast and East Asia and the Pacific beyond. Just south of the island, the prevailing westerly winds can be easily reached and used to sail southwards and westwards. For this reason this island was used continually by the

MAURITIUS IN THE
LATE EIGHTEENTH CENTURY

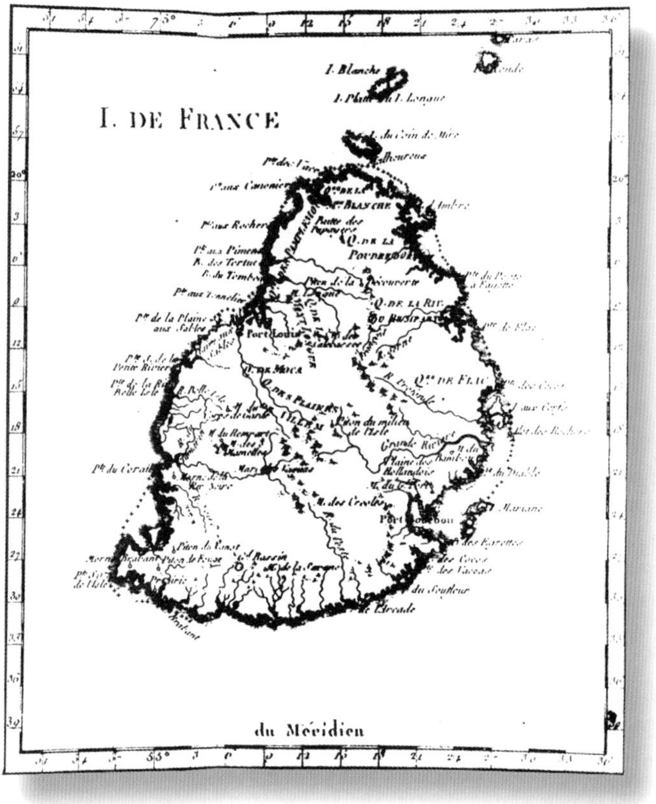

Maritius was taken by France in 1715 after a Dutch occupation from 1598 to 1710. In 1715 the French changed the Dutch name for the island, Mauritius, to the Ile de France. Britain captured the island from France during the Napoleonic Wars (1810) and acquired it in 1814.

French as the base for their Antarctic and Australian explorations. There was only one main problem. It had little in the way of a hinterland like the Dutch enjoyed at Cape Town, and could not be relied upon to support a large garrison or fleet.

The island, like the rest of France's Indian Ocean empire, was acquired as a result of the efforts of the French East India Company. This was one of the significant chartered companies created by different European governments to stimulate national trade and explorations. Founded during the reign of Louis XIV, in 1664 as part of Colbert's programme for financial reorganization, it secured a series of bases around the Indian coast, and a series of island outposts on the way to there.

The French East India Company experienced varying fortunes. It

PORT LOUIS, MAURITIUS

Port Louis in the eighteenth century was far from being an ideal harbour. Large vessels came to anchor in the open roadstead which could get quite choppy, as the view indicates. The lack of good facilities for war and other ships as one of the reasons which stimulated France to search for Terra Australis in the Indian Ocean, and to later explore western Australia.

suffered in the financial failures of 1720, brought about by the policies of the imported Scots adviser John Law.[18]. It later revived and was absorbed into the General Company of the Indies which controlled French territories in India until much of these were lost to Britain, together with Canada, as a result of the Treaty of Paris which ended the Seven Years War in 1763.

Bouvet de Lozier's search for Gonneville Land

The first major French mission which was sent under Bouvet de Lozier to previously unexplored southern waters to search for Gonneville's lost land, indicates both the many misconceptions the French had about *Terra Australis,* the wasteful methods used to locate it, and the role played by the French East India Company in French explorations and expansionism.

At the time the mission left, in 1738, it was believed in France that Gonneville had in fact discovered *Terra Australis,* and that the place where he landed was situated in the mid forties south latitudes in the

South Atlantic. This was equal to the latitude of the Bordeaux, Nice, Genoa area in Europe, and it was expected that Gonneville's land in the southern ocean therefore enjoyed a similar salubrious climate to the Mediterranean. This misconception was held by France to the end of the eighteenth century. It was not until then that they realized that the weather pattern in the southern hemisphere differed from the north; that snow, icebergs and fog were to be found in lower latitudes closer to the equator than in the northern hemisphere, and that the men they sent had to be equipped with winter clothing even in summer. It was this misconception which resulted in the first modern French explorer, St. Allouarn, coming to western Australia. He was forced to leave the snow and cold of the latitudes of the forties for the warmer latitudes of Cape Leeuwin.

Bouvet de Lozier had no experience of sailing in the mid latitudes of the South Atlantic before he went on his mission. Born at St. Malo, Brittany, on 14 January, 1706, the son of a barrister, Jean Baptiste Charles Bouvet de Lozier, sometimes referred to as Lozier Bouvet, joined the French East India Company navy in 1731.

It was then not unusual for talented and ambitious men keen on seafaring and travel, to join the French East India Company navy. Opportunities existed there for ambitious and talented men to serve as officers and to command ships irrespective of their social status. The French navy at the time did not offer such positions to the sons of commoners. To serve as naval officers in France was one of the privileges of the nobility. There were officers not of the nobility, but these wore special coloured uniforms, and except in war-time enjoyed few of the opportunities offered to men of their social status in the French East India Company navy.

Bouvet de Lozier was early attracted to exploring the southern oceans for Gonneville's land. Not long after he joined the Company navy he commenced to make a series of proposals to the directors to send him on a special mission of exploration to locate the lost continent.[19] Eventually, after receiving the support of Joseph François Dupleix, the active and expansive-minded governor general of the French establishment in India, the directors agreed to provide ships for the mission. If located, and if the climate was as salubrious as suggested and the products of the land as bountiful as indicated, then the land discovered and early claimed by Gonneville for France would make a valuable half-way house for company ships going to India and other parts of the east.

Bouvet set out from Lorient on 19 July 1738, with two ships, the *Aigle* and the *Marie,* equipped with three small vessels designed for close to shore survey work; 160 men, and a large supply of provisions and trade goods.

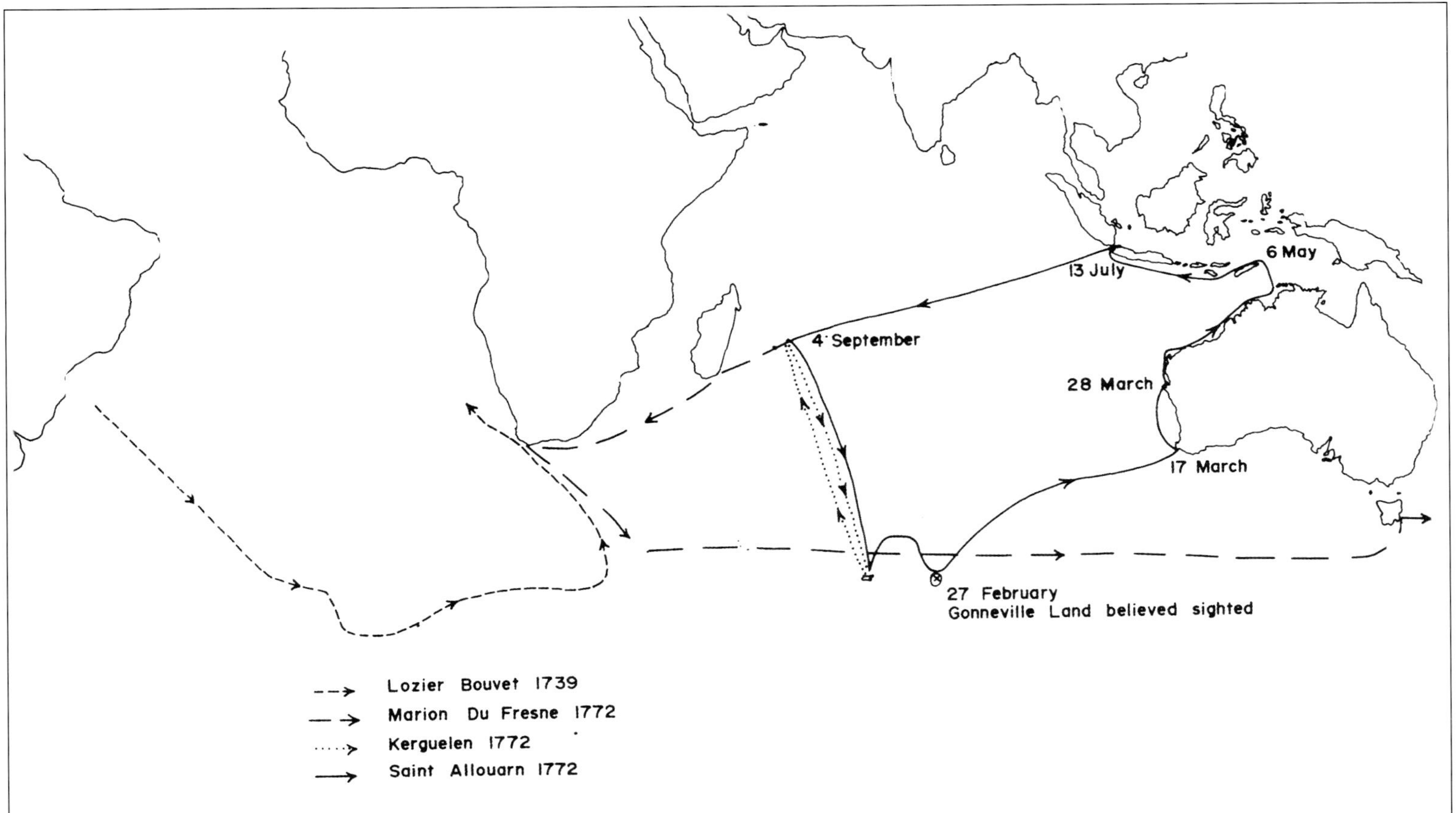

THE SEARCH FOR TERRA AUSTRALIS BY FRENCH EXPLORERS IN THE 18TH CENTURY

Bouvet was given specific, detailed orders. He was to survey in forty degrees south latitudes in the South Atlantic. Any place he discovered suitable for a Company establishment, was to be taken possession of. When the survey was completed the *Aigle* was to go to Mauritius, while Bouvet was to continue to make further searches eastward from the Atlantic and then make for Cape Town on the *Marie*.

The expedition reached southern waters in December 1738, at the beginning of the southern warm season.[20]. Having sailed south-west in the Atlantic, he reached the mid 40 degree south latitudes at approximately 15 degrees west and searched the seas between 40^0 and 55^0 south there, moving gradually eastward to approximately 30^0 east.

Throughout this survey trip Bouvet de Lozier met the bad weather usually found in those waters. He encountered fog, winds, rough seas, snow and icebergs.

His sightings of the latter and comments upon these and on various amphibious animals, birds and seaweeds which he made in his journal, indicates further misconceptions the French had about the physical world at the time. He believed, as was then the fashion, that ice was formed only on the land from land water, and that the sightings of icebergs meant land was nearby. Fog was believed to be somewhat the same in origin. It was then believed to be land formed and a sign that land was close. He also regarded the sighting of seals and penguins, birds and seaweed as further signs of land close by. In fact all that he saw in the South Atlantic can be seen far away from land. There are penguins, seals, birds, seaweed, fog and icebergs to be seen in the vast expanse of water between the ice-cap and the continents of Africa, Australia and South America. And the icebergs found in Antarctic seas can be encountered often as far up as the roaring forties, and occasionally even further north, well away from land.

Discovery of Bouvetoya

Bouvet de Lozier found the only piece of land which exists in the waste of sea he sailed. On 1 January 1739, the day of the Feast of the Circumcision he sighted a snow-covered land surrounded by ice. He reckoned this to be at 54^0 south and 540 to 560 leagues or 26^0 to 28^0 east of the Azores, and was convinced it was a part of Gonneville's land. He named the place Cap de la Circumcision. In actual fact what he had discovered was what is now called Bouvet Island, or Bouvetoya which belongs to Norway, and is actually situated at 54^0 26'S and 3‘ 24' E or 31^0 24' east of the Azores which are at 28^0 W indicating the degree of error of Bouvet de Lozier's calculations of longitude. He was more than 3^0 to 5^0 out to the west.

Bad weather prevented the expedition from approaching close to the land discovered. Bouvet consequently did not discover that it merely was a small island, and not a cape. He then continued to run before the wind sailing east and venturing no further south. Because of the continued bad weather, the cold and sickness, he headed north when the ships were just past the longitude of Cape Town, therefore just missing the opportunity of discovering the western most islands in the south Indian Ocean, the Prince Edward Islands which were left for the next French explorer, Marion Du Fresne, to discover.

The *Aigle* sailed north reaching Mauritius on 9 March, 1739. Bouvet de Lozier reached Cape Town on the *Marie* on 28 February. He left on March 31 to make a further search of the South Atlantic on the way home. He eventually reached Lorient on 24 July 1739, without making any more discoveries.

Bouvet de Lozier's report indicates he returned to France as a more experienced man with changed impressions about the southern ocean. His ship had been continually fog bound and iced-up, even though it was the warm season. The continent he believed he had discovered there, he considered because of the severe weather, to be of no use to Europeans. He had had no opportunities to use his trade goods in the area.

Bouvet de Lozier spent the remainder of his Company naval career primarily in Indian waters, protecting French commerce in the wars with England, and aiding Dupleix carry out his expansionist political schemes. When France lost its Indian empire he retired to France, where he was ennobled and subsequently acted as adviser to the government on the subject of the exploration of the Senegal area in Africa.

The results of his mission to the South Atlantic and South Indian Ocean were negative. Bouvet de Lozier proved that there was no land in the part of the South Atlantic he traversed at least down to 54^0 south. This reinforced the discoveries made on a previous voyage. Another French East India Company ship, the *Saint Louis,* had earlier sailed along the mid thirty to forty degree south parallels of latitude from South America to South Africa in 1708, proving there was no land to be found between these two capes on the south continents. But these findings did not appear to have attracted notice.

Following Bouvet de Lozier's abortive efforts, the French took no further initiatives in the search for Gonneville's land or *Terra Australis* for just over thirty years.

This was not only because of Bouvet de Lozier's failure to achieve results for the Company. Just after he returned with his dismal news,

France became embroiled in another struggle with England, in what was in effect part of a new hundred years war between the two contesting powers. Confrontations between France and Britain lasted from early in the eighteenth century until after the Napoleonic wars in the early nineteenth century.

There was still a lot of exploring work to be carried out in southern waters. Bouvet de Lozier had searched only approximately 30 of the 360 degrees of longitude down to 54° south. The remaining area still remained a mystery to be solved.

France's defeat by England and its loss of India and Canada stimulated the next move to find answers to the riddle of the southern continent.

The next direct search was a by-product of the efforts of the eminent French explorer and colonist, Bougainville, whose work helped open the Pacific to France, and whose discoveries and annexations for the French monarch served to shape the form of France's later empire in the Pacific.

Bougainville and the way via the Pacific

Louis Antoine de Bougainville, a knowledgeable mathematician who alternately served France as army commander, naval commander, colonial administrator and explorer had distinguished himself as aide to the Marquis of Montcalm in the campaigns against the British in Canada. After the loss of Canada to Britain, he determined to establish a French colony in the Falkland Islands, which would serve France as a half-way house base for Pacific and other trade. Although he felt secure in doing this, in that the Falklands were both isolated and distantly removed from British spheres of interest, he and his colonists were evicted when the British under Commodore John Byron hastened down to establish an English colony in the same islands. France in the interim acted swiftly against Britain by suddenly ceding the islands to Spain, as part of its policy of seeking co-operation with the Iberian powers, thus creating a controversy about sovereignty there, which lasts to this day.

Partly as a consolation, Bougainville was given command of an expedition charged with making a world voyage. The bulk of the exploring work on this voyage was done in the Pacific, where Tahiti, Samoa, the New Hebrides and other islands were visited and surveyed and annexed for France.

Bougainville brought back with him from the Pacific, a Polynesian named Aoutouron or Ahu-toru who gained considerable attention in France and Europe. Unlike the case of Gonneville's exotic companion Essomericq, the French knew exactly where Aoutouron came from and where he should be returned to after his visit to France. It was Marion Du Fresne's offer to do this which gave him the opportunity to make a further search for Gonneville's lost land, and which led him to Australia in 1771.

In the meantime, further developments had taken place among intellectuals in France, in regard to *Terra Australis*. The quest for systemized and extensive knowledge by French savants and encyclopedists who emerged in the enlightenment in France, stimulated a French interest in *Terra Australis* and the mysteries of the southern oceans which they believed had to be solved for the benefit of science.

Maupertuis and the quest for the southland

Two of the emergent philosophers in particular turned their attention specifically to Australia. The French astronomer and mathematician, Pierre Louise de Maupertuis, who introduced Newton's ideas on gravitation to the French, in 1752 produced a paper on the progress of science.[21]. This was written for the enlightened Prussian monarch, and patron of scholars, Frederick the Great. In this paper Maupertuis drew the Emperor's attention to the lack of knowledge about the southern seas and the need to explore there.

Maupertuis' paper has never been given the place it deserves in the history of the scientific exploration of Australia. His enemies, in particular Voltaire, launched a bitter personal attack on him, and succeeded in wrongfully discrediting him. This was primarily done because of his work in Lapland. He led an expedition there in 1736, primarily to measure the length of one degree along a meridian, seeking to verify Newton's conclusion that the world is an oblate spheroid, flattened at the poles, which was then not acceptable.

Although he was subsequently made a member of the Academy of Science at Berlin and served as president of it there, and contributed to the Academy of Science in Paris, his views on this were specifically misrepresented by his enemies in Paris, where he was incorrectly branded "a believer in the flat earth". Like other scholars who had been abused and misrepresented by enemies for personal venom, Maupertuis found no champion to support him, and found others professing the same ideas he produced, being favoured instead.

De Brosses and the new drive to explore the Indian Ocean

De Brosses, a savant who was prominent in Dijon intellectual circles, in 1756 produced a work similar to that of Maupertuis, but with additions. De Brosses' two volume book *Navigation Aux Terres Australes,*[22] draws attention to the lack of knowledge about the southern oceans, makes a plea to rectify this and urges France to annex Gonneville's southland to serve as a French base.

The book is divided into three sections. One part deals with voyages made, and the then existing state of knowledge about "Magellanie", that is the region around Cape Horn. Another part deals with "Polynesie". The remaining section is about "Australasie" which, together with "Australia", had become a popularly accepted term in France for the Australian region south of Java and Asia.

The last section is of significant interest for the historians of Australia and the southern Indian Ocean. It commences with a study of Gonneville, indicating that De Brosses was firmly convinced that Gonneville in fact had discovered Australia. He then surveys the voyages made by different Spanish and Dutch explorers and by Dampier.

De Brosses' work directly affected efforts subsequently made by Europeans to make known the south seas, and to explore Australia. But he did not stand alone in this regard. The savants around him who were thirsty for knowledge in encyclopedic proportions, supported the move to get to know *Terra Australis* and its people and products. Naturalists such as Buffon, social philosophers, scientists and systematizers of knowledge lent their support. The entries on Australia and different objects found there, contained in the encyclopedias produced by Diderot and others are primarily the results of this early stimulus and quest for accurate information designed to replace the then current myths. It was these works and the demand for knowledge which brought private and political support, which led to a new drive by the French to discover *Terra Australis.*

This unique map of western Australia with Ceylon or Sumatra on the World Map in Heinrich Bünting's 1581 book *Itenerium sacrae scriptorae,* was identified by him with Biblical Ophir. This follows Bernhard von Breydenbach's 1486 book on classical sources for geography. It was drawn 40 years after the Dieppe maps of *Terra Australis* and a year after Portugal was absorbed by the Habsburg Empire which changed the control of records. Locating Bünting's sources to see how imaginative or informed he was, is difficult due to destructions wrought in Germany in the Thirty Years War.

CHAPTER 3

Marion Dufresne's and Kerguelen-Tremarec's search for Gonneville Land, and St. Allouarn's survey of south western Australia in 1772

In the latter part of 1771 two maritime expeditions, one a local effort under Marion Dufresne,[1] the other sent from France under Kerguelen, were at Mauritius preparing to sail south in the Indian Ocean to search in the mid forty degree latitudes for Gonneville's lost land.

Marion Dufresne's search for Gonneville Land

There has been some confusion about which Marion Dufresne led the local expedition. As in the case of William Dampier, there was more than one person of that name living at the same time. Manuscript records of the voyage kept in the Bibliothèque Nationale and other repositories refer to him only as Marion Dufresne. Unfortunately he did not write an account of his voyage and his name therefore does not appear on the title page of a book. He was killed and eaten by Maoris during his voyage before he could do this.

In encyclopedias he is usually listed as Nicholas Thomas, born at St. Malo in 1729. This was suggested by a local British consul who conducted researches in later years. Subsequent researches by French scholars conducted in France and Mauritius, indicate the explorer was in fact Marc Joseph sometimes known as Marc Marion Dufresne.

Born in 1724 Marc Joseph Marion Dufresne had a long and eventful career at sea, commencing in 1733. He served in naval ships in the wars against England. Among other exploits, he was involved in the French effort to get Prince Charles out of Scotland. He later voyaged to China and other parts of East Asia for the French East India Company. When he was commander of the Company vessel the *Comte d'Argenson,* he was selected to take the famous French astronomer Alexandre Pingré[2] to the island of Rodriguez in the Indian Ocean, to observe the transit of Venus in 1761. He later settled in Mauritius where he became a trader and land owner and a prominent member of local society.

His early links with scientists thirsting for knowledge continued to stimulate his own desires in that direction. He developed a passion to explore the seas for undiscovered lands. The seas to the south of Mauritius were then still unknown. Bouvet had sailed in the forty and fifty degree latitudes only as far as approximately 32° east of Paris, that is to the meridian running east of the east coast of South Africa. All to the east of that was very much unknown in those latitudes eastwards as far as Cape Horn.

The arrival of the Polynesian Aoutouron (Ahu-toru) at Mauritius, on his way back home from Paris where he had been taken by Bougainville,[3] provided Marion Dufresne with the opportunity to set out on a voyage of exploration. Aoutouron had been sent from France to Mauritius, with orders for the authorities there to send him to Tahiti. Bougainville offered to pay the cost of chartering a vessel to do this. Other prominent French citizens also contributed money to help pay these costs and to send supplies of useful goods to Tahiti. This meant funds were therefore readily available for an expedition such as Dufresne hoped to make.

Aoutouron arrived at Mauritius in October 1770. Marion Dufresne took the opportunity of his temporary presence at Mauritius, waiting for transport east, to submit a proposal to the local governor, Pierre Poivre,[4] to send an expedition to Tahiti which could explore the southern oceans on the way, to try to locate Gonneville's lost lands.[5]

Governor Poivre approved the plan. He was similar minded to Marion Dufresne in regard to exploration. He was himself a widely-travelled navigator. He was responsible, among other things, for establishing direct trade between France and Cochin China. He also was a competent naturalist. In his period as governor he introduced the cultivation of spices to Mauritius to improve the economy. As a man of science he enthusiastically supported the quest for knowledge and desired to help achieve a name for France and for Mauritius in this regard.

Marion Dufresne was consequently given command of two ships, the *Mascarin* and the *Marquis de Castries*. At first the expedition was planned to be broadly scientific. The French astronomer Alexis Marie de Rochon[6] was at Mauritius at the time, waiting to join the other expedition under Kerguelen. However, differences between these two men after Kerguelen arrived led to the suggestion that Rochon join Marion Dufresne. In the end Rochon did not go. Governor Poivre kept him back, hoping to patch up the difference between him and Kerguelen. Instead Marion Dufresne was accompanied by the navigator Crozet and a young captain Du Clesmeur who commanded the second vessel.

Departure from Mauritius

Marion Dufresne, after hasty preparations, sailed from Mauritius on 18 October 1771, to forestall Kerguelen who was still getting ready to go. Dufresne's plan was to sail to Tahiti, and on the way search for Gonneville's land which was believed to be south-east of Cape Town, then sail along Tasman's route south of New Holland to see if this was a practical route for French merchantmen to use to reach the South Pacific. Marion planned to call at Van Diemen's land and New Zealand. After this he intended to explore in the Pacific and then return home by way of the East Indies.

His first port of call was nearby French controlled Reunion.[7] There it was discovered that Aoutouron had contracted smallpox at Mauritius where there was an epidemic of the disease. Desiring to quarantine the vessel, he departed from Reunion which was free of the disease and anchored off St. Louis in Madagascar. There Aoutouron died. This removed the basic reason for the trip. Marion Dufresne, however, decided to continue. But he altered his plans. He set sail for Cape Town, staying there until after Christmas 1771. The delay drew heavily on their provisions and time, and in the opinion of some of his officers adversely affected the outcome of the voyage. It caused Dufresne to hurry and to miss sighting Kerguelen Island and to fail to solve the mystery of Gonneville's land supposed to be in the vicinity.

The expedition set sail from Cape Town on 28 December 1771, and headed southeast, with the purpose of searching eastward in the mid forty latitudes from the 33° meridian of longitude east of Paris.

Dufresne's first search was for two supposed islands, Denia and Marseveen which were shown on old Dutch charts to be at latitude 40° or 41° south, below Africa. But no land was sighted there. All that the expedition encountered was fog, cold, and rough weather. In fact these islands do not exist.

Discovery of Prince Edwards Islands

On July 13 land was sighted further east. But the fog and mist were so bad that the islands could not be easily seen. However, it was named Terre d'Espérance. This was Marion Island which is part of the Prince Edward Isles so named by Cook. Another island in the group was later seen. This appeared to contain a large cavern and was named Ile de la Caverne.

As the weather appeared to improve the vessels then turned back towards the first island sighted to see if they could examine the resources offered here. Dufresne was searching in particular for timber resources. There was little of this commodity available in

Mauritius, which was needed for ship spars and repairs. But unfortunately for the expedition, when the ships were just off the north-east of Marion Island, they collided causing extensive damage. Repairs took several days to complete, but thereafter the sailing performance of the vessels was impaired, especially that of the *Marquis de Castries*. This further impeded the expedition.

Discovery of the Crozet Islands

Because of the violent winds which arose, coming from the west, the ships departed without making the planned survey of the islands. They proceeded along the 46° parallel of latitude. On January 21 this course brought them to the Crozet Islands estimated by Marion Dufresne's calculations at 46° south and 42° east of Paris. This latitude is approximately correct, which says something for Dufresne's abilities as a navigator. The islands are on the 46° parallel. But they are on the 52° east meridian from Greenwich, or 47° 40' east of Paris which means he was a little over 5 ° out in longitude, which then was not an unusual sort of error.

This island group was more closely explored on January 24th. Three islands were located and named — the Ile Froid which is now called Hoy Island and the nearby Apostles; Ile Aride which is now called East Island; and Ile Prise de la Possession which is now called Possession Island. The expedition was more impressed with this group than the previous islands seen. Large white ice areas on Possession Island indicated to them that there was a glacier and this to them meant the existence of a large river.[8] It was then believed that glaciers were frozen river mouths. In fact there is no such river on the island. The glacier there is the result of snow deposits.

A small cove was found on the eastern side of the island which is sheltered from the prevailing westerlies. There a young officer Roux was sent ashore to take possession of the group for France and to leave a bottle with a manuscript claiming possession. The expedition then departed, noting that "the Austral Islands" as they were collectively named, offered only coarse plants and were not habitable. They contained no trees which could be used as timbers and spars. The expedition in fact had to wait until it reached New Zealand before adequate supplies of timber and other commodities needed to keep a sailing vessel in good repair, were found, which indicates the later importance of that country for the French and others at the time.

Passage to New Zealand

The earlier intention of sailing further south into the 50° of latitude was abandoned because of the bad sailing performance of the vessels after the collision, and the bad weather encountered. Marion Dufresne feared getting surrounded by ice, as had happened to Bouvet de Lozier in those latitudes. He therefore continued eastward, sailing in the vicinity of the mid forties, thus bypassing Kerguelen Island which was left to be discovered by the Kerguelen expedition less than two weeks later. Marion Dufresne was just north of the island on February 1. Kerguelen discovered it on February 13.

Marion Dufresne then sailed on to Tasmania and then to New Zealand, which to the French appeared to be more salubrious than the lands discovered in the same latitudes in the Indian Ocean. The reports of the remnants of his expedition which arrived back at Mauritius on 9 March 1773, did nothing to stimulate excitement about the south Indian Ocean. Instead it turned French eyes much further east to New Zealand and the Pacific which seemed to offer better weather, more prospects and better attractions.

Kerguelen-Tremarec

Of all the French explorers who were sent to solve the mystery of the southern Indian Ocean and of the location of *Terra Australis,* Kerguelen-Tremarec was given the best opportunity to make a most distinctive contribution and bring honour to France for being the first to make the region known. Kerguelen-Tremarec enjoyed both the support of the French court and the local authorities in Mauritius who were charged with helping make the mission he led a success. He consequently lacked little. The only item he was short of in his supplies was winter clothing. But this was an oversight made by many at the time. The expedition left for the 40° latitudes south in the southern summer and did not expect to find the cold, bad weather which exists there.[9]

But despite all the assistance he was given, Kerguelen-Tremarec's mission was a failure. He spent only three days looking at a small part of Kerguelen Island from a distance before hastening back to Mauritius, leaving his companion vessel behind. The survey he made was consequently superficial and his maps and reports were inconclusive and misleading. He wrongly jumped to the conclusion that he had discovered the rich and salubrious Gonneville Land, and hastened home to report the find.

Cook's more impressive explorations

It was therefore left for Cook and the British to unravel the mysteries of the southern Indian Ocean and reap the honours as first explorers, as had happened previously in the Pacific at the expense of the French. Britain was already prepared to do this just after the French court took the initiative to have the southern Indian Ocean explored by French seamen. Challenged by this and pressured by interested groups, the British government mounted a second expedition under Cook, after he returned from his successful Pacific voyage. This second expedition on the *Resolution* was planned to make known the whole of the waters in the southern hemisphere. It was despatched from Britain on 13 July 1772, with Cook on the *Resolution* and Furneaux on the accompanying vessel, the *Adventure*. Kerguelen had set out from France the year before, on 1 May 1771, but due to delays did not leave Mauritius for his southern Indian Ocean voyage until 16 January 1772, and was back there reporting his find in March. Cook consequently had very much an open field. He reached the southern Indian Ocean in the summer at the end of 1772 and spent the period from December to March 1773 making a systematic survey which revealed that the Indian Ocean was a waste of water down to the Antarctic polar ice cap and that the few islands which existed there were barren and inhospitable, while Kerguelen was loudly proclaiming the virtues of the salubrious land he saw briefly through the bad weather.

On paper, Yves Joseph de Kerguelen-Tremarec seemed a good choice to lead the expedition which could have done much to restore French prestige.[10] He was a competent, well-thought-of seaman. Born at Finisterre, Britanny, to a noble family on 3 February 1734, he joined the navy in 1750 wearing the special uniform reserved for privileged officers. He soon engaged in scientific research and surveys which were then becoming popular among naval officers as the Enlightenment progressed. At this time he was in particular engaged in a survey of the French coast. In 1755, because of his distinctive efforts, he was made a member of the Academy of the French Navy. He then served as a commander of a vessel in the Seven Years War against Britain. After this he went to Iceland for fisheries protection work. While there he made scientific observations about the sea which were published as a book. Back in France he engaged in further survey work and research which was designed to help with the erection of lighthouses on the French coast. His strong field of research, however, was naval construction. It was Kerguelen-Tremarec who was primarily responsible for the design and production of the "corvette-cannonière" class of vessel after the older style sloops of war had been destroyed in the fighting at Algiers.

Kerguelen-Tremarec's proposals

After Bougainville returned to France in triumph from the Pacific in 1764, Kerguelen-Tremarec, like many other Frenchmen, developed an interest in that region. In 1770, moved by the reports of Bougainville, Kerguelen-Tremarec used his family connections at Court to get approval to take an expedition there. He subsequently secured the attention and support of intellectuals, in particular that of the eminent French hydrographer Après de Mannevillette.[11] Manevillette, who was closely associated with the French East India Company, and who had written sailing guides and drawn maps of the route from Europe to India and the East Indies, succeeded in turning Kerguelen-Tremarec's eyes and interest to the southern Indian Ocean where Mannevillette believed Gonneville's missing fifth continent was located, somewhere south beyond St. Paul's and New Amsterdam.

After delays caused by the threat of renewed conflict with Britain, Kerguelen was given a ship, the *Berryer,* and a company to survey to the south-east of Africa. The aim of the voyage was not only to discover the missing continent of *Terra Australis.* There was a strategic-economic motive as well. France still lacked a useful base on the way to India, and was particularly short of wood supplies for Mauritius. Kerguelen who himself wrote a memorandum on the need for France to acquire a temperate colony in the southern hemisphere, consequently had this quest added to his objectives.[12]

Besides this, he was given the more immediate, practical task of reporting on the utility of the new route from Mauritius to the Coromandel coast as suggested by Grenier.[13] Jacques Raimond Vicomte du Giron Grenier, an hydrographer and ship commander had explored the Indian Ocean in the 1760s and, as a result, recommended that a sea route using trade winds and monsoons be used to reach India directly and quickly from Mauritius. Some controversy developed about the practicality of this. Kerguelen was consequently requested to report on it and make recommendations.

Departure from France

Kerguelen sailed from France on the *Berryer* on 1 May 1771, and arrived at Mauritius on 20 August. Finding the *Berryer* unsuited for survey work he succeeded, with the help of Pierre Poivre, the governor of the island, in exchanging this for two other vessels — the *Fortune,* a 24 gun ship with a crew of 200 and the *Gros Ventre,* a 16 gun ship with a crew of 120 which was placed under the command of Louis François Marie Alleno de Saint Allouarn.

These two vessels were ready to sail in September 1771. As it was still spring in the southern hemisphere, Kerguelen decided to first sail north and report on the practicability of the Grenier route. The survey work he did on this in the following months is impressive, indicating his qualities as a seaman and surveyor. He completed this work by 8 December, when he returned to Mauritius having discovered the Fortune Bank and having determined that Grenier's route was practicable.[14]

Departure from Mauritius

After hastily provisioning and preparing his ships, Kerguelen left Mauritius on 16 January 1772 to seek Gonneville Land, where he was ordered to establish good relations with the natives. It is difficult to tell from his log book the exact course he followed.[15] In the part of the Indian Ocean he traversed there was then and usually is a large variation of the compass which alters rapidly as a ship heads south, and at that time there was no variation chart. This lack of information which meant the ships' compasses were not being properly corrected, together with the effect of the west wind and drift could explain why Kerguelen-Tremarec made some easting as he pushed south which led him to come directly on Kerguelen Island.

Discovery of Kerguelen Island

After an uneventful voyage, apart from the fact that the weather became increasingly worse and colder as the vessels made south, Kerguelen-Tremarec's men came across seals, penguins and seaweed after three weeks' sailing. Just afterwards, at 5 p.m. on 11 February, land was sighted to the west. This was the Ile de la Fortune which is to the west of Kerguelen Island.

As the winds were strong and the seas high, the vessels moved cautiously south and raised further land. By 13 February they were off the south-west corner of Kerguelen Island itself, and thus free of an immediate lee shore. With the wind to the west the ships made into the south coast of the island between Cape Bourbon and Cape Dauphin which offered some refuge. When they were close in, a small bay was identified, lying between the two capes, but close to Cape Dauphin.[16] This bay was named Sea Lion Bay (Baie du Lion Marin) by Kerguelen, and separately named Gros Ventre Bay (Baie du Gros Ventre) by St. Allouarn. The present name of the bay is the Anse du Gros Ventre.

In order to get a closer look at the bay and the island, Kerguelen lowered his dinghy which was placed under the command of Ensign Rosily who was sent off to make soundings in the bay opening which is clearly visible from a ship lying south of the Pic St. Allouarn, and just to the east of that mountain.

St. Allouarn on the *Gros Ventre* also launched a dinghy which was put under the charge of Du Boisquenneux. Like Rosily, he found dif-

KERGUELEN'S EXPEDITION IN THE SOUTHERN INDIAN OCEAN

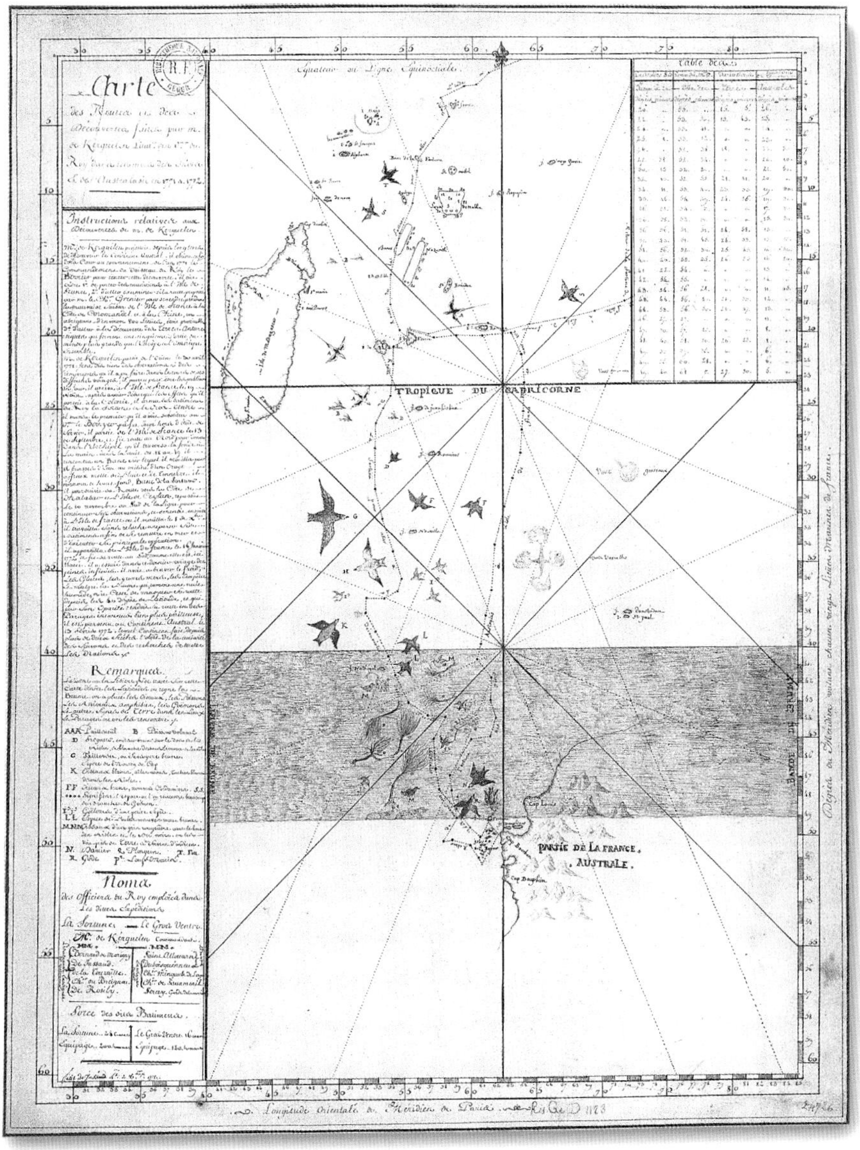

This map shows the course taken by Kerguelen's expedition from Mauritius (Ile de France) to Kerguelen Island, called on the map partie de la *France Australe*.
The map is unique and interesting. It contains sketches of the birds, sea animals and sea weeds seen on the voyage and has a belt of fog extending from 40°S to 50° south, which it was then believed had to be sailed through to get to the clear atmosphere where Terra Australis Incognita was situated.

ficulty in making the coast across the open sea in the high wind and waves. It took two and a half hours for him to get to the bay. In the meantime the wind continued to rise and the *Fortune,* rolling in the big swell and waves, had its foremast damaged. Kerguelen by this time had gathered the impression that he was at the entrance to a big gulf faced by the land in sight to his north, and unsighted land to the east and to the south. In fact, there is no land in the last two directions. The land to his north, the only land in sight, forms the south coast of Kerguelen Island, and this coast runs almost due east-west. Nevertheless, Kerguelen considered he was on a lee shore and made off westward, beating into the wind and punching into the waves, with a damaged foremast. If he had run the other way, eastwards, he would have come across the fine harbours and safe conditions which exist on the east side of Kerguelen Island.

Annexation of Kerguelen Island

The dinghy under the command of Duboisquenneux entered the bay later in the day, at 4 p.m. The bay was sounded and found deep and safe although it had a rock bar extending a little way into the sea on the east side. The men in the dinghy landed near there, and as ordered, took possession of the land in the name of the King of France. They hoisted a French flag, fired a volley from the muskets, gave three cheers and then buried a parchment in a bottle to record the act of proclamation.

The landing party was not impressed by what they saw. There was no sign of human habitation, it was cold and the land was seemingly barren with stunted growth.

While this was being done, and Rosily was active with his survey, fog came in. Fortunately the *Gros Ventre* under St. Allouarn was still close by in sight and the small boats made for there. Safe aboard, but with Rosily's small boat abandoned, the *Gros Ventre* set sail in the rising wind with all of the rescued men aboard, making for deep water westward in the direction that Kerguelen had taken. The next morning, with weather conditions somewhat improved, the *Gros Ventre* put back into the south coast of the island with Rosily still on board,[17] and moving further east than before, discovered the Baie d'Audierne and the Baie de la Mouche. However, as the weather again closed in and there was no sign of Kerguelen, St. Allouarn did not explore further, but made off westward again to search for his companion.

EXPLORATION OF KERGUELEN ISLAND

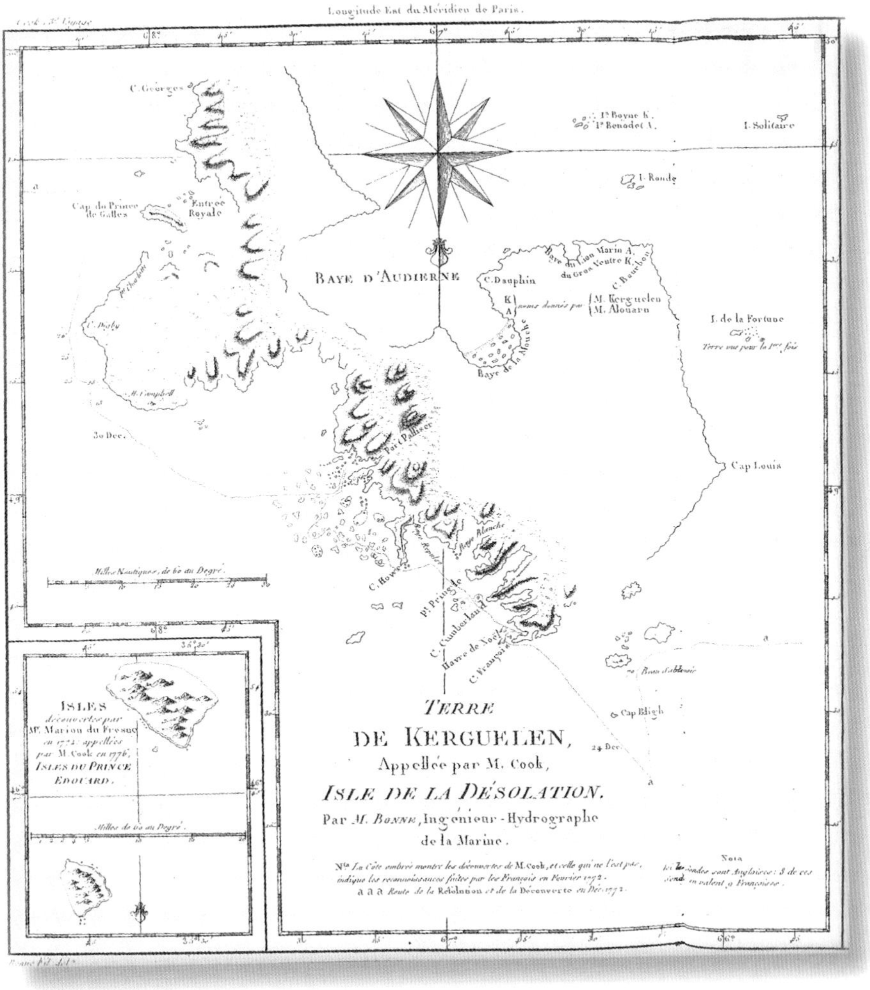

South is at the top of this map. The discoveries made by Kerguelen and St. Allouarn are shown by the continuous line starting on the right hand side (the west side of the island) and going to the top of the map (the south side of the island). Kerguelen made his landfall near the Ile de la Fortune, which is noted on the map as being the land first sighted, and then moved into the Baye d'Audierne. The coast in fact does not go south from there, to make a large continent as Kerguelen supposed. Kerguelen is a small island.

The north coast, partly mapped by Kerguelen in a later voyage in February 1772, was more thoroughly mapped by Cook in December 1772. His discoveries are shown at the bottom of the chart (the north side of the island).

Separation from St. Allouarn in the *Gros Ventre*

Kerguelen in fact was in that direction, but had moved far out to sea. There, seeing no sight of the *Gros Ventre,* Kerguelen headed for home on 16 February without going back to look.

It is difficult to comprehend why Kerguelen took this action which cost him and France dearly. Only two explanations are possible. Kerguelen could have acted as he did because he thought that St. Allouarn, once he was separated, would ignore his orders and go straight back to the base at Mauritius. Alternatively, Kerguelen could have been so excited about his discovery, which he believed to be the long-missing Gonneville land, that he rushed back to make this known and reap the honours. He certainly tried to do this as soon as he landed, creating the impression that France had made a great find and gain.

Inexplicably, although he only saw Kerguelen Island from a distance and had no first hand reports on it — his men had to make for the *Gros Ventre* after he sailed off — Kerguelen gave a most enthusiastic and misleading report about the newly-discovered land.[18] He claimed it was the cape of an extensive temperate land, that it was rich in resources, that it seemed to be cultivated and would be a valuable acquisition which would compensate France for the recent loss of Canada.

Kerguelen-Tremarec's subsequent voyages of exploration in search of Gonneville Land

This resulted in France losing further time and allocating scarce resources to follow up this find with further expeditions. Kerguelen himself led two more, both of which were similar failures. His second expedition made at the end of 1773, after Cook's visit to Kerguelen, reached the north part of the island but again he did not stay after again taking possession of the land for France on 6 January 1774. This voyage was of little importance. By then it was known that Kerguelen Island, which Cook named the Island of Desolation, was barren, inhospitable, uninhabited and more famous for coarse cabbages than trees.

Back in France in September 1774, he scurrilously attacked one of his officers and came off worst. Arrested for cowardice and other serious charges including trading and carrying an illegal passenger on a King's ship which were banned by the naval reforms after the disastrous Seven Years War. The passenger, Louison, was a young girl Kerguelen slept with on the voyage. He and the clerk who rigged the crew list were found guilty. Kerguelen was jailed for 20 years, reduced to 6. Released in 1778, he fell into his old ways, leading a third failed expedition. Rehabilitated and employed by the Revolutionaries as a victim of the *ancien régime,* he died in 1797 before he could do much damage.

Kerguelen-Tremarec's failings

It is difficult for the historian to revalue Kerguelen-Tremarec and give him more praise than he received by his contemporaries, many of whom were his enemies and envious of him. No matter how much the historical evidence is sifted, the fact remains that Kerguelen was an unimpressive, incompetent and impetuous explorer. He was more like a 'Foul Weather Jack Byron' than a Cook. He did not appear to have the tenacity required to be a good explorer.

However, he cannot be wholly condemned. The survey he made of Grenier's route was impressive. Moreover he was not disliked nor given little respect by those who went with him. The competent seaman Rosily who later rose to prominence and fame, although he had been apparently deserted by Kerguelen-Tremarec at Kerguelen Island, had sufficient confidence in his former commander to apply to accompany him on his second expedition.

Certainly he does not deserve to be condemned for his survey work in the southern Indian Ocean, which admittedly does not appear to be of the same quality as his work in the north part of that Ocean. He certainly made mistakes, for which he has been taken to task.[19]. He calculated the western extremity of Kerguelen as being 61° 10' east of Paris, that is 63 ° 30' east of Greenwich. This is 5 ° 10' out to the west. The actual longitude of the extremity of the west coast of Kerguelen at Cape Louis is 68° 40' east. Cook, on his second voyage, had to zig-zag to locate the recently discovered island, which in fact lies between 48° 40' south and 49° 45' south and 68° 40' east and 70° 45' east.

The condemnation of Kerguelen for making such erroneous calculations needs some qualification. Kerguelen's instruments might have been to blame, rather than he and his officers' navigational abilities. He was sailing in unknown seas where the compass variation is frequently considerable, and was then uncharted. More significantly, it is frequently difficult in the waters in which he sailed in the southern Indian Ocean to get an accurate compass correction by the normal method of amplitude then used. The horizon at sunrise and sunset there, which needs to be seen for eastern or western amplitudes, is often unclear. Besides this, the compasses then used were not wholly accurate on shipboard. They had a simple ferrous needle, and considerable moments of inertia, added to by the rolling ship. These imperfections were not corrected until the compass was set in a bowl of liquid, and multiple steel needles were used in the nineteenth century to reduce inertia. An error of variation in a westward direction in particular would result in the ship's easting being wrongly estimated. It would appear to go more to the south than to the east, and result in a wrongful longitude assessment of a newly discovered coast. Kerguelen's second-in-command on the *Gros Ventre* was 10° out in his longitude when he arrived at Flinders Bay. The only accurate way to calculate positions in unknown places then, was to set up observatories on shore. Kerguelen did not do this because of the bad weather.

As far as his impetuous judgement about the quality of Kerguelen Island is concerned, no excuse can be offered for Kerguelen-Tremarec except to say that he was very much the victim of the preconceived view and prejudices of his age. He misread the signs he saw on the island. He saw large glaciers and believed this indicated large rivers and, therefore, fertility, very much as Dufresne had done further west. Yet, like there, the glaciers on Kerguelen are very short, consisting of deposits of snow caught on the mountain slopes. He used observations of these and a distant view of the vegetation which is sparse and stunted and could seem to be cultivated, to conclude that the land had qualities it in fact did not possess.

St. Allouarn's departure for Western Australia on the *Gros Ventre*

After Kerguelen disappeared from sight at Kerguelen Islands, St. Allouarn continued his search for him north along the inhospitable west coast. Failing to find a trace of the *Fortune* there, he decided to follow his original orders and sail eastward to look for Gonneville Land in that direction, believing he would meet up with his commander in western Australia, where the expedition was to call.

St. Allouarn made his easting in the mid 40° south latitudes with little that was eventful happening, except the continued bad weather, until 27 February. On that day the crew sighted seals, sea lions and birds and concluded that they were in the vicinity of "Gonneville Land". However, the weather in those parts, at approximately 90° east and 48° south, was too bad for him to stay and investigate in what in fact is empty ocean.[20] With his crew cold and discomforted, he therefore turned north towards Cape Leeuwin.

Arrival at Flinders Bay

The *Gros Ventre* approached the western Australian coast a little to the east of Cape Leeuwin on 16 March 1772, in fine clear autumn weather. The ship ran before a fresh south south-west to south wind with all sails set, steering just to the east of north-east.

ST. ALLOUARN EXPEDITION, 1772

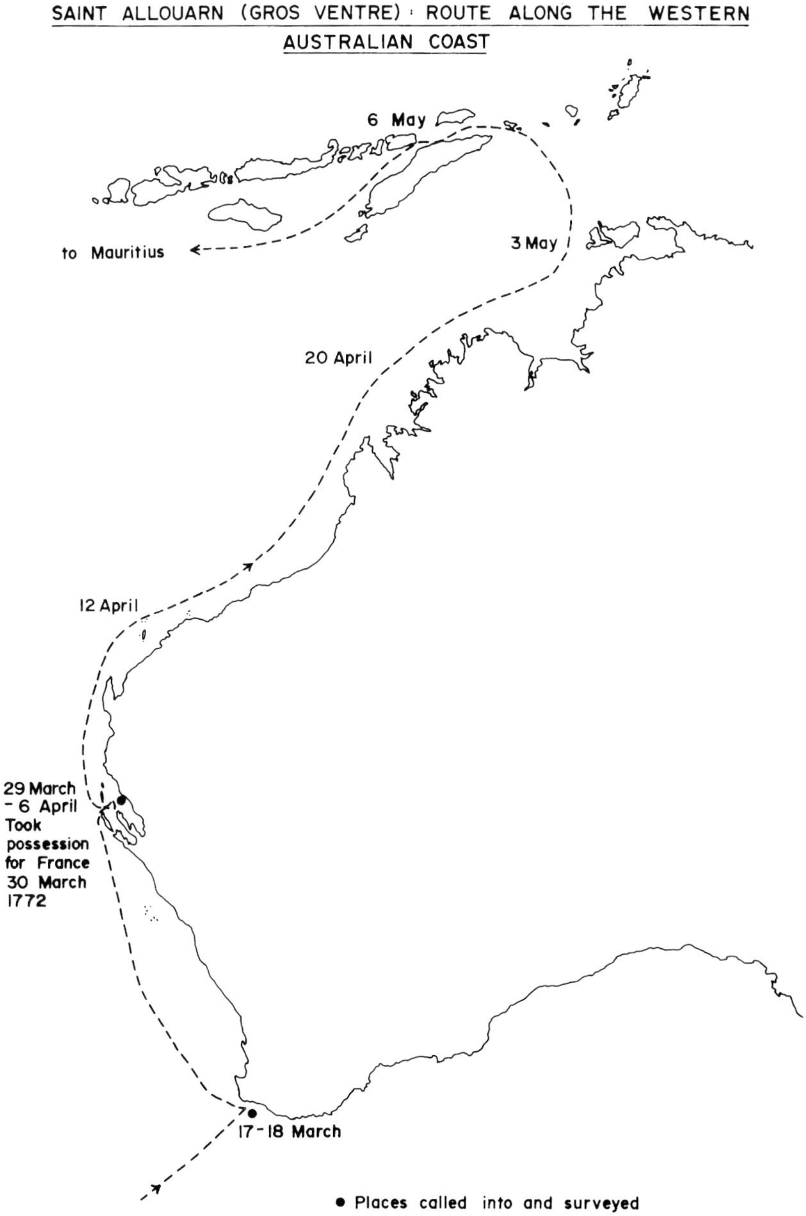

Location of anchorages and areas surveyed:
Flinders Bay, 17-18 March 1772.
Shark Bay, 29 March-6 April 1772.

Sail was shortened at sunset as land was believed to be close by in the vicinity. At 11 p.m. the breeze dropped, as it often does during those conditions in that area.

During the calm conditions at 2 a.m. on 17 March in the moonlight, the second gunner sighted land a short distance away to the north north-west and to the north. It was estimated to be one and a half leagues or less than five miles away which is uncomfortably close for a sailing ship approaching a new landfall on the lee.

The vessel was then put on a starboard track and sail was further shortened, as it headed slightly into the wind on a south-east course. Later the ship was turned to follow a south-west route as the wind shifted eastward.

For the rest of the night the ship was held just off the coast, awaiting the dawn and an opportunity to make a survey in good light conditions. Soundings were made regularly; the leadman finding first grey sand, then coral and shell "containing" many objects, in depths increasing from 25 or 30 (French) fathoms. Converted to the shorter English fathom this would be from just over 28 to just less than 34 fathoms.

The sun rose at 6 a.m. and the vessel was edged gradually towards the coast with soundings being taken each ten minutes. By fortune St. Allouarn had struck the coast at the only safe anchorage in the region. Flinders Bay is tucked behind Cape Leeuwin. It is open to the south, but if a boat goes in close enough, the water is calm and reasonably sheltered and has a good holding ground, suited for almost all conditions. The main good feature is, Flinders Bay is sheltered from the large swells which come from the west to hit the western coast of Australia in those latitudes.

St. Allouarn did not take his ship too close to shore, nor too close to the Cape Leeuwin side of Flinders Bay. He anchored 3 miles out from the shore in 16 French fathoms, estimating his position to be about 7 leagues or 21 miles from the furthermost point of Cape Leeuwin. In fact he was lying out to sea almost south of the head of the bay, east of Ledge Point.

Anchor was dropped at 11.30 a.m. The small boat was lowered. Soundings were taken in all directions around the ship. The boat was then sent to make a close inshore survey. It headed directly for the shore line towards the north and not towards the distant lee side of Cape Leeuwin where there are sheltered sandy bays, good landing places and a river. These were not seen during the survey. The party in the boat did not land at all. They encountered a heavy backwash near the beach which nearly swamped them. This indicates they were in the eastern part of the bay, where there is a surge and a surf. But they made a close observation of the land along the coastline. They

ST. ALLOUARN'S ANCHORAGE ON THE GROS VENTRE
IN FLINDERS BAY 17th MARCH 1772

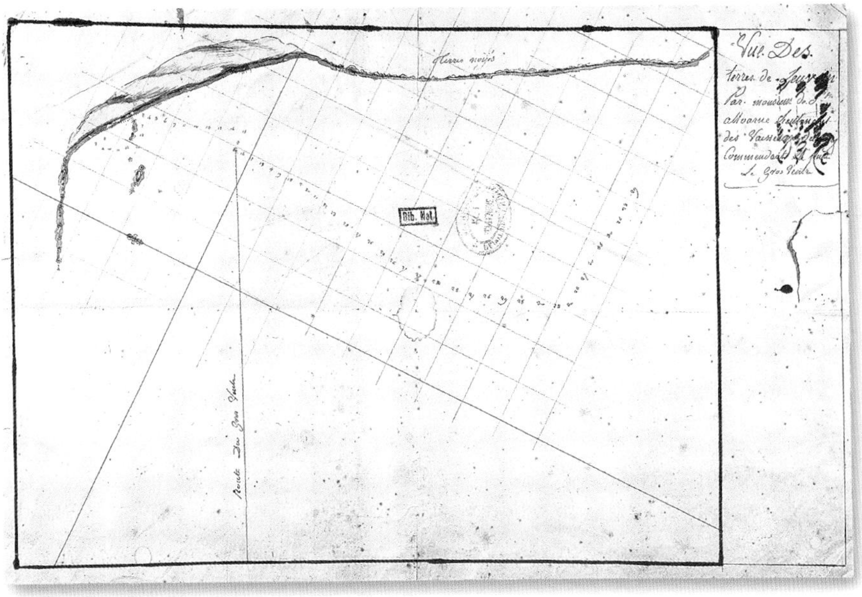

wondered if it was man-made and decided it was a natural formation. Despite a close search, they found no evidence of human habitation. This became a feature of French explorations. They found western Australia to be very sparsely inhabited. Despite attempts to meet Aborigines, the French did not closely encounter these until 1817 when Freycinet's party was at Shark Bay.[21]

From the ship a false impression of the land was gained. They believed that Cape Leeuwin was an island off the mainland, and that other islands were also evident. This is an impression easily gleaned from off the south-west coast of Australia. The hills are mostly round and slope gently. From the distance these seem like isolated pieces of land, although in fact they are all part of the mainland. Rosily, however, made a most accurate first map of Flinders Bay.

The next day, the 18th was uncomfortable as far as the weather was concerned. The wind which continued from the south became squally. Nevertheless, the crew fished to find out about and sample the local produce of the sea. They reported good catches of *"sarde"* which could have been pilchards or the local variety of herring. No description is given of them, but they were reported as being very plentiful right along the coast. Most impressive were the sharks. These were a variety they had not encountered before, with two rows of teeth at the top and the bottom, and with a head and tail like a *"ton"* (a tuna). One they caught was 500lbs which they described as monstrous and no less dangerous than the sharks elsewhere.

At 6 p.m. on the 18th March 1772, the anchor was raised and the ship left Flinders Bay to go to Shark Bay. The south-east wind which was blowing as the ship departed later shifted to the north-east and pushed the ship seawards south away from Cape Leeuwin, but not before further land was sighted towards the east in the direction of present day Point D'Entrecasteaux.

The next day the weather worsened. Squalls blew from the south-west with the suddenness often experienced on the south-west coast of Australia in Cape Leeuwin waters. The seas rose and the winds increased to an uncomfortable strength. As he made south on a port tack St. Allouarn recorded "I think it very lucky we left our anchorage which offered protection only from the north-east".

In these circumstances it took St. Allouarn several days to round Cape Leeuwin, driving south in order to make his westing by taking a safe starboard tack.

On the northward passage, after leaving Cape Leeuwin well to starboard, the *Gros Ventre* kept the coast approximately a hundred and fifty miles to the east, and made for Shark Bay. No more land was seen until they made a landfall there after travelling north for nearly 700 sea miles.

Arrival at Shark Bay

It is generally claimed by Western Australian and other English language writers that St. Allouarn sighted the coast at Shark Bay on 29 March 1772. In actual fact he made the landfall on the afternoon of March 28.

The error previous writers have made is an easy one to make. The log books kept on French ships at the time of St. Allouarn, seem as if they are day to day records, but in fact they are not. French seamen at the time worked and made their entries in the log book according to a twelve hour and not a twenty-four hour clock. The twenty-four hour periods into which their log books are divided do not go from midnight to the next midnight. They go from midday to midday. Therefore a double date entry is madc at the top of the official log entry. For instance, the entry for the twenty-four hour period when they made the landfall near Shark Bay is headed March 28/29. When they recorded sighting land at 2.30 in the afternoon, this was in the first watch after midday on the 28th.

St. Allouarn expected to make his landfall before this. He sounded all of the way along the coast from Cape Leeuwin, making certain that he kept far enough out from shore to dodge the infamous Houtman Abrolhos. Once past there, on 26th March, he made gradually towards the coast, seeking to land at the bay discovered by Dirk Hartog and explored later by William Dampier.

He was held off shore by east winds on the 26th, but managed to make some easting after that. St. Allouarn, still navigating along the coast by Après de Mannivillette's chart, estimated at the time, after making some easting that he was 3 leagues or 9 miles inland. On the 27th, after making more easting cautiously because of this, he estimated his position to be 7½ leagues inside Australia, near Dirk Hartog's landing, which shows how far out the charts then were. Later on the morning of the 28th, he estimated he was at a position which was 21 leagues or nearly seventy miles inland on his chart.

However by then land was believed to be close by. He sighted some *"goismon"* (rock fish) and some *"bunches of grapes"* (grappe de raissans) which in fact is a local type of sea weed which looks like bunches of grapes. These indicated to him that land was close by. At this stage they felt adverse currents might be holding them out to sea, and so explain their apparent incorrect eastward position which placed them inland.

In the afternoon on 28 March at 2.30, land was sighted from the masthead at a reported distance of 9 leagues or 27 miles. The ship continued to make an easting shoreward in a fresh south to south-east breeze. At 3.30 the coastline could be seen stretching along the east horizon. At 6 in the evening they sounded and found the bottom, consisting of sand and shell at 60 fathoms. They continued sounding towards the coast, finding 35 fathoms at 7 p.m. and finally anchored at 10 p.m., on a long line of 180 fathoms.

The first anchorage made by St. Allouarn which he estimated to be at 25° 52' S and 107° 20' E, judging by a comparison of the map made by Rosily and the present Admiralty chart of the area, was off the coast of Dirk Hartog Island about five miles north of South Passage at the southern end of the island. The ship was so far out from the coast that the crew did not observe that Dirk Hartog Island is an island. That is understandable. An observer has to be very close in to see that an entrance exists south of the island. South Passage is a very tortuous channel which from the sea does not have the appearance of a passage.

St. Allouarn's latitude calculations at that time were out only by approximately 1 minute, so it is possible to predict with a good degree of accuracy, his position in the area.

At daylight the explorers observed that the coast was bordered by reefs and not accessible by boat. Sails were therefore set at 7.00 a.m. and the vessel continued north, at approximately two and a half leagues out from the coast. The land was observed to be low-lying sandhills covered with scrub, and the protective reef was seen to be continuous.

At sunset they came to the northern end of the island and observed a further island, later called Dorre Island, and at the same time saw Turtle Bay, the place where Dirk Hartog landed, and which is a good safe anchorage and landing place for vessels entering Shark Bay by this middle passage.

The boat was edged towards Turtle Bay, with continuous soundings being made. At 6.30 p.m. on 29 March the anchor was dropped in 30 fathoms of water, about three miles north of Turtle Bay. Unfortunately, the anchor fouled, and a troublesome night was spent in a fresh south-west wind with a current from the same direction pushing against the anchored ship.

At daybreak the vessel was better secured and two parties were prepared to land. Ensign Mingault or Mingau was detailed at 8.00 a.m. to take the large dinghy to survey the north of Dirk Hartog Island. Smoke had been sighted inland the day before and it was felt it might be inhabited.

Western Australia annexed for France

The party under Ensign Mingault landed at Turtle Bay on the morning of 30 March and took possession of the country in the name of the French King. A bottle containing a parchment recording this event was buried at the foot of a tree, together with two French coins. The party then continued with their survey. (see map p.260)

No signs of human beings were found. The explorers reported the land was sandy and covered with bushes. They found evidence of fires. There were burnt shrubs and charcoal. But it was considered this was probably the result of the hot sun drying the wood. They discovered what they believed to be more tangible evidence of human life than fire. Inland, in the bush, they discovered a cleared circular area which they believed was possibly used for dances by natives. But the only living things they saw were animals. They sighted what they thought was a small, tailed animal *(maque)*. This no doubt was a wallaby. Their efforts to catch one failed. They also sighted a dog-like animal, scratching for turtle eggs on the beach. This undoubtedly was a dingo.

A search party was sent ashore at midday on the 30th under the command of Saint Allouarn's cousin, Charles de Boisquenneux, accompanied by Ensign Sausmenil and Ensign Rosily and a crew of rowers. They surveyed a small area along the coast. Like the other party they found no sign of habitation.

The parties were impressed by the large number of turtles on the beaches, and collected a quantity of eggs.

The large dinghy or "chaloupe" returned to the vessel at 7.00 p.m. Others in the small dinghy went ashore in the night to catch turtles, without success.

The boats were taken aboard the *Gros Ventre* early the next morning. Anchor was weighed at 9.00 a.m. and the ship sailed eastwards in a slight south-east breeze, aiming to round Cape Levillain at the eastern end of Turtle Bay. The wind dropped and the ship was anchored off Cape Levillain at 10.00 a.m. The day was spent in fishing, and they reported catching great quantities of fish including pilchards *(sarde)* and sharks.

While at anchor here they ferried ashore the body of a sailor who had died from scurvy, and buried it. This presumably was interred somewhere in the vicinity of Cape Levillain.

Although this incident is mentioned only in passing in the log, it is a significant one for the historian of French maritime explorations. French vessels at that time were more unhealthy than British ones and French sailors suffered more from scurvy and other illnesses than their British counterparts in the eighteenth and early nineteenth centuries.[22] The log books of French vessels and accounts of travellers contain frequent references to illness, especially scurvy and dysentery.

The main cause of illness, no doubt, was the food and water which decayed and putrified as the voyage wore on. The water and wine became sour with the passage of time, and the biscuits became weevily. The bad water, which was often added to with stocks from possibly tainted sources in the tropics no doubt caused some of the stomach upsets and dysentery. But the main cause of the more serious complaints was the diet. The men apparently lacked vitamin C and as a result suffered from scurvy. Normally a French ship would call at a refreshment port to benefit the men. But on this trip there was nothing in the way of fruit or vegetables available for them either at Kerguelen Island or at western Australia.

Added to the diet were the unhealthy condition of the ships. Livestock was carried to provide fresh supplies, and these offered sources of infection.

Besides this, the French preferred to bury their dead ashore, and not commit them to the deep like the British. Those who died at sea were, therefore, usually kept in the ballast until the ship came to land. This no doubt provided a further source of infection. The burial of the sailor at Shark Bay is indicative of the French method of their disposing of those who died at sea.

French ship's doctors could do little to improve the men's health and the general situation. They tended to use old techniques such as blood letting, and lacked the medicines and knowledge to treat diet deficiencies and what were for them new tropical and other diseases. St. Allouarn himself died as a result of disease just after his return to Mauritius.

The ship moved off on the same evening. It made into Shark Bay, seeking to discover the so-called Dampier's River. The broad openings of Freycinet Estuary and Hopeless Reach, which extend southwards each side of the Peron Peninsula were seen and noted and marked on the map as "appearances of rivers", which had led others to erroneously conclude that Shark Bay contained the mouth of a large river.

The *Gros Ventre* did not enter the reaches of these bays. Instead St. Allouarn sailed north surveying the waters between Bernier and Dorre Islands and the coast, which Baudin later attempted to do. Here they lost two of the ship's anchors in the swift currents and winds they encountered in the open anchorages.

While there Rosily drew a most accurate map of the regions that had been visited in the bay, further indicating his qualities as a map maker.[23]

Departure for Timor and Mauritius

The *Gros Ventre* then made southwards again, and left Shark Bay by the same way it came in, by way of Naturaliste Channel instead of going north before the southerly winds and leaving by the way of the Geographe Channel which they had seen.

After leaving Shark Bay, St. Allouarn sailed along the north coast of western Australia as far as Melville Island. This whole section of the coastline was then something of a mystery. It had only been partially explored by the Dutch in 1616 and 1619, and by Dampier in 1699. In 1772 little was still known about it. St. Allouarn was therefore setting himself to perform an important task for his country by exploring these unknown waters instead of making straight for Timor to refresh his crew. But unfortunately for France he made little contribution to knowledge. He mainly discovered the dangerous tidal races which no doubt kept him away from the coast, and sailed through and charted some of the numerous shoals and archipelagos which abound in those waters. Nevertheless, he made an important impact not only by bringing the Bourbon flag to the region but by specifically paving the way for the great scientific expedition sent in 1801 under Baudin who was given the task of verifying and completing St. Allouarn's work in that area.

Sailing northwards, St. Allouarn doubled Northwest Cape on 9 April 1772 and then bore eastward passing through the shoals which he found north of King Sound, on 20 April. The following week, on 26 April, the *Gros Ventre* was among the islands of the Bonaparte Archipelago which was later surveyed and named by Baudin, who made the wry comment in his log book on 7 August 1801 (19 Thermidor Year 9) that the islands were the "number of dots" which St. Allouarn put on his chart to show the archipelago. From there, St. Allouarn sailed across Joseph Bonaparte Gulf towards Melville Island which was left astern on 2 May as the ship headed for Portuguese Timor to get vital supplies of fresh food and sweet water.

These were urgently needed on the *Gros Ventre* by this time. No supplies had been found and taken on board at inhospitable Kerguelen Island, and no rivers to provide sweet water, and no grains and fruits had been found by him in western Australia as has been indicated. There was only fresh sea food to be had there, and this did not provide a healthy diet. The rigours which the crew of the *Gros Ventre* suffered on the voyage, and St. Allouarn's tenacity as an explorer can be judged by the log entries after arriving near Dili on 7 May. These record that sixty of the ship's crew, that is half of the ship's company, were down with scurvy. Consequently a considerable period was spent refreshing before leaving for home in Mauritius, via Batavia.

The vessel finally arrived at Port Louis, Mauritius on 5 September 1772, with a sick and exhausted crew. St. Allouarn and Ensign Mingault, the man who took possession of western Australia for France, did not last long after landing.

St. Allouarn died on 27 September at the age of 35 years and took the story of his life and voyage to the grave, which is unfortunate. A book by him would have significance. His log book is replete with information of scientific interest about climate; currents; magnetism and natural history, as well as providing the hydrographic and cartographic data which the French later sent Baudin to check.

St. Allouarn's arrival and his reports put an end to the wrongful impressions Kerguelen had created about the land he discovered, and the idea that a rich France Australe could be established in that part of the southern Indian Ocean. This information did not arrive in France in time to prevent Kerguelen going on his second fruitless mission. But already the French naval authorities had their own hard-headed views about the matter. There were by then sufficient reports to show that the southern hemisphere did not have the same weather as the north, and that it was pointless to look for salubrious, temperate lands there in latitudes equivalent to France. In the southern ocean where rich France Australe was held to be, was only an area of fog, snow and bad weather.

Kerguelen's expeditions in consequence had an effect he little dreamed of. His failures and findings, together with those of others, resulted in France turning away from explorations of the southern Indian Ocean in general, to surveys of western Australia.

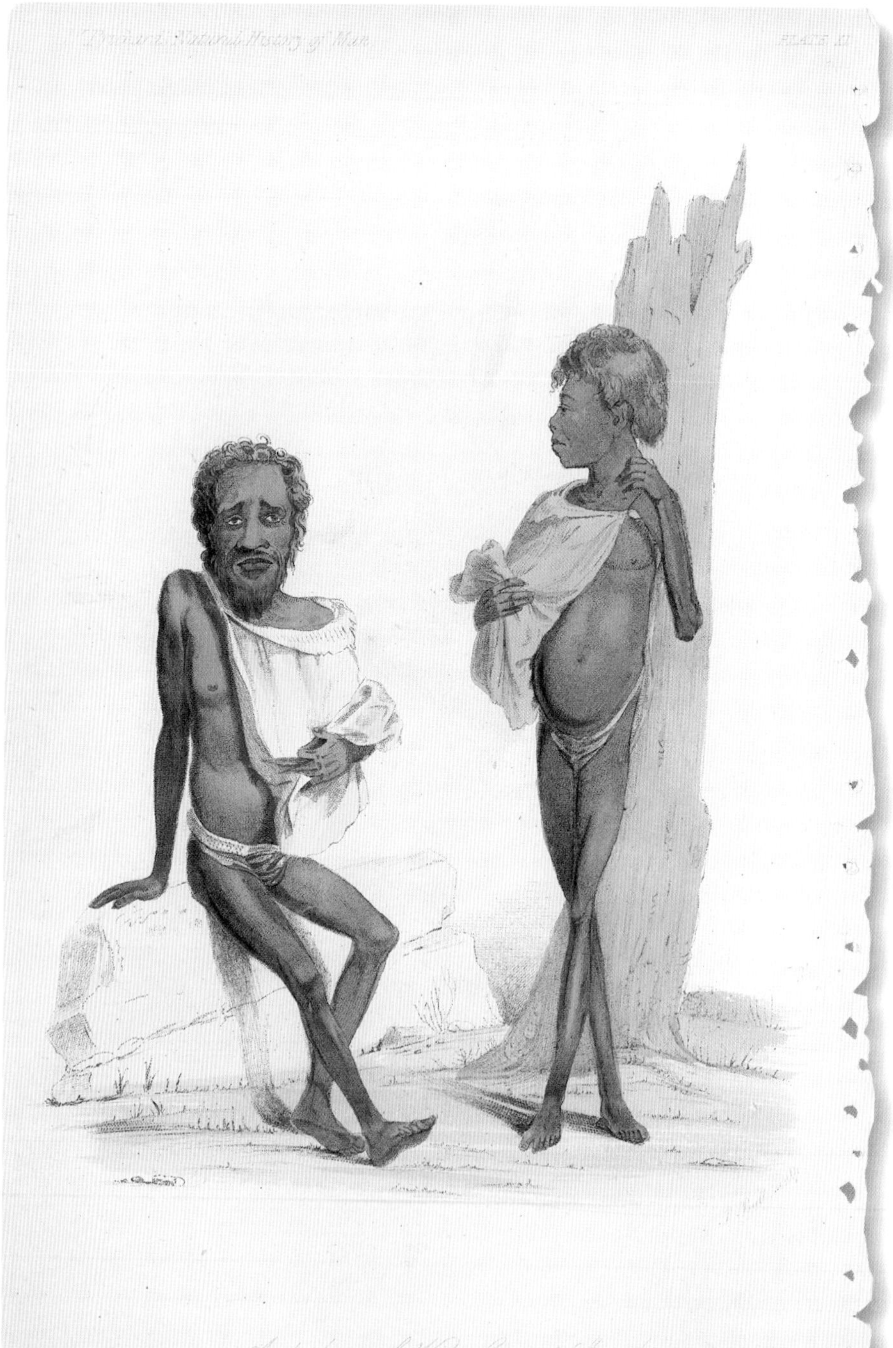

Sketches of Aborigines at King George Sound, by Louis de Sainson, 1826.

PART II
Revolutionary France and western Australia

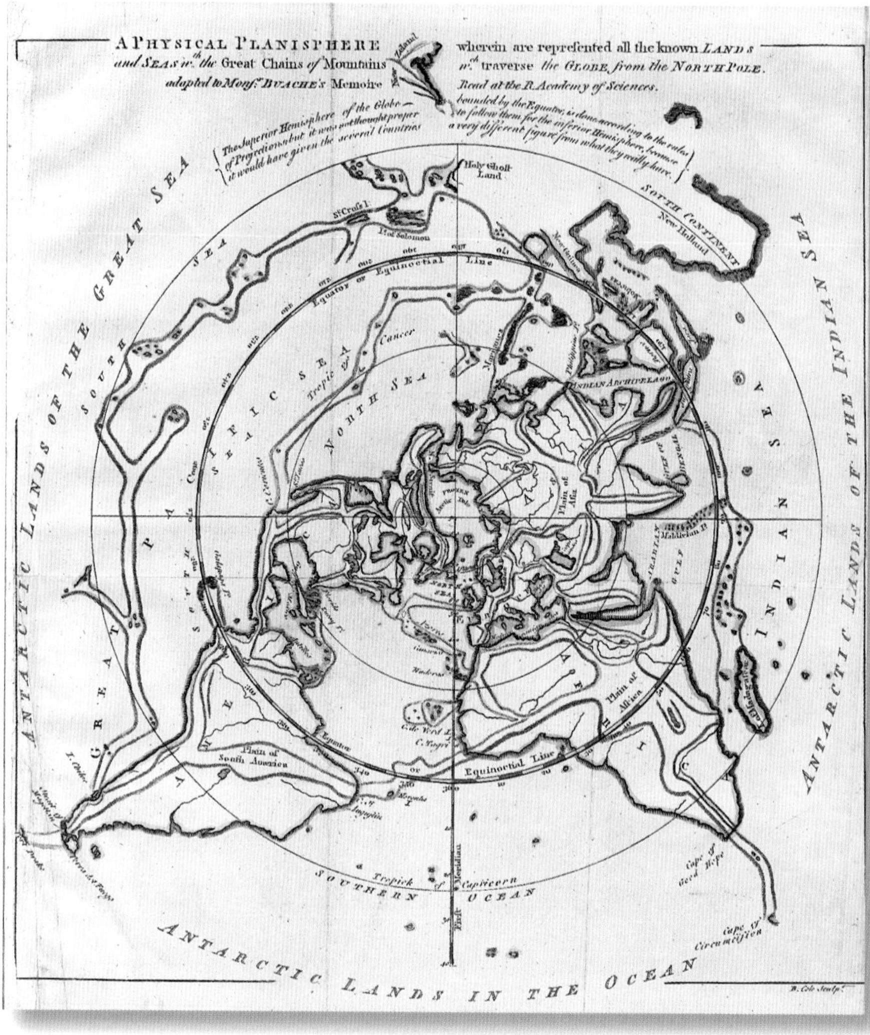

This physical map of the world by the French geographer Philippe Buache (1700-1773), based on explorers' discoveries, shows the continents linked by mountain chains which form the separate oceanic basins. Buache's theory led him to predict the existence of the Alaskan Peninsula. The French named a large island off Fremantle, now Garden Island, in his honour. The island, now a naval base, was renamed by the British. A bay on the Island now bears the name Buache.

CHAPTER 4

D'Entrecasteaux's survey of the south coast of western Australia, 1792

The French did not immediately exploit St. Allouarn's early survey work in Western Australia, as Britain did in the eastern part of the continent where Cook's early work and his annexation of the territory made not long before St. Allouarn claimed territory in the west for France, was followed up by the establishment of a colony in New South Wales.

Twenty years passed before the next direct contact with western Australia was made by France. And the expedition which came then under D'Entrecasteaux made only a cursory survey of the south coast of western Australia, eastward from Cape Leeuwin. The west coast was not visited and examined.

This state of affairs was more the result of bad luck and accident than design. Two earlier French expeditions were proposed to visit the region and chart it.

Latouche-Treville

Not long after Kerguelen returned from his second, and equally abortive visit to Kerguelen Island, a distinguished naval officer, Louis René Madelene le Vassor de Latouche-Treville, submitted to the French naval ministry in 1774 a proposal to further explore the Pacific, and then survey the little known southern coast of Australia.[1] The plan was of interest. The actual shape of the southern and western part of Australia at that time was one of the remaining mysteries in the southern hemisphere which needed to be cleared up. It was then not known if, in fact, Australia consisted of two islands like New Zealand. There was a possibility that the Gulf of Carpentaria in the north, and the southern coast which turned north at the Great Australian Bight, were joined by some sort of inland sea. If this was the case then the French would have more reason to consider the place for settlement and feel secure about the security of their part of Australia if it was a separate island. Canada had proved to them that it was difficult to share a continent with Britain. Later on

LATOUCHE – TRÉVILLE ALTERNATIVE ROUTES PROPOSED FOR EXPLORATION 1774

in the 1840s, French views in this regard were clearly demonstrated when France established a colony at Akaroa in the South Island of New Zealand, leaving Britain to dominate the North Island which was separated by waters like the English Channel.

Although France still lacked a half-way house on the way to the Pacific and desperately needed a place which could supply suitable wood and other supplies,[2] Latouche's proposals came to nothing. Part of the trouble was that there existed in French naval circles and among savants, a division of opinion about where France should explore. Latouche emphasized the Pacific. Savants and some naval men in Paris preferred to consolidate work in the still unknown Indian Ocean area. Latouche's mistake was that he left the Indian Ocean to be surveyed later in his voyage after the Pacific had been explored. If he had elected to journey via the Cape of Good Hope instead of Cape Horn, he might have received some support for his proposal, if not success. The best base for supporting French explorations eastward at the time continued to be Mauritius which was in close proximity to the large unknown territory of western Australia which continued to intrigue savants. Latouche-Treville's voyage, unfortunately for France's links with Australia, was not approved by the naval ministry.

La Pérouse and the French plan to survey western Australia

By 1783 more definite proposals were made. This time the initiative was from high officials and not from a potential explorer seeking approval and favours.

A large scale comprehensive scientific survey mission designed to visit the Pacific and Indian ocean was planned and prepared in that year. The difference about priority areas to be searched first continued to exist. Here King Louis XVI played a direct role. He was greatly interested in the Pacific. Consequently, he personally helped plan the expedition. Thus when La Pérouse[3] was selected to lead it he was sent to explore the Pacific in the first instance.

This French survey was primarily designed to clear up all of the remaining great mysteries of the "south seas" including Australia. The French planned to chart the still unknown areas west of Cape Horn, then chart the then unknown areas of America in the north Pacific and Bering Sea, then chart the then little-known coast of Siberia and Korea, and finally chart the then unknown coast of west and south Australia.

La Pérouse did not keep to this plan. He was given a lot of leeway in his orders. Once in the Pacific he used his discretion and completed his north Pacific survey. After this he sailed through the Pacific to east Australia, landing there about the same time as Captain Phillip and his convict settlers. La Pérouse sailed from there to complete his survey plan, but disappeared at sea. Western Australia consequently remained unexplored until a mission was sent specifically to complete some of the Australian section of La Pérouse's work, in particular part of the western Australian coast.

D'Entrecasteaux

The next French maritime expedition to come to western Australia, following St. Allouarn's visit, arrived primarily as the result of good fortune, and conducted a hasty reconnaissance of the south coast rather than making a comprehensive survey of the unknown parts of the area, as was intended. Consequently, the mission led by D'Entrecasteaux made little contribution to knowledge about the region, which savants in France were waiting for. D'Entrecasteaux's scientific mission merely proved that the chart of the south coast of western Australia, drawn by Pieter Nuyts,[4] in 1627, was surprisingly accurate. The only addition D'Entrecasteaux made to this was a

LA PEROUSE EXPEDITION 1787

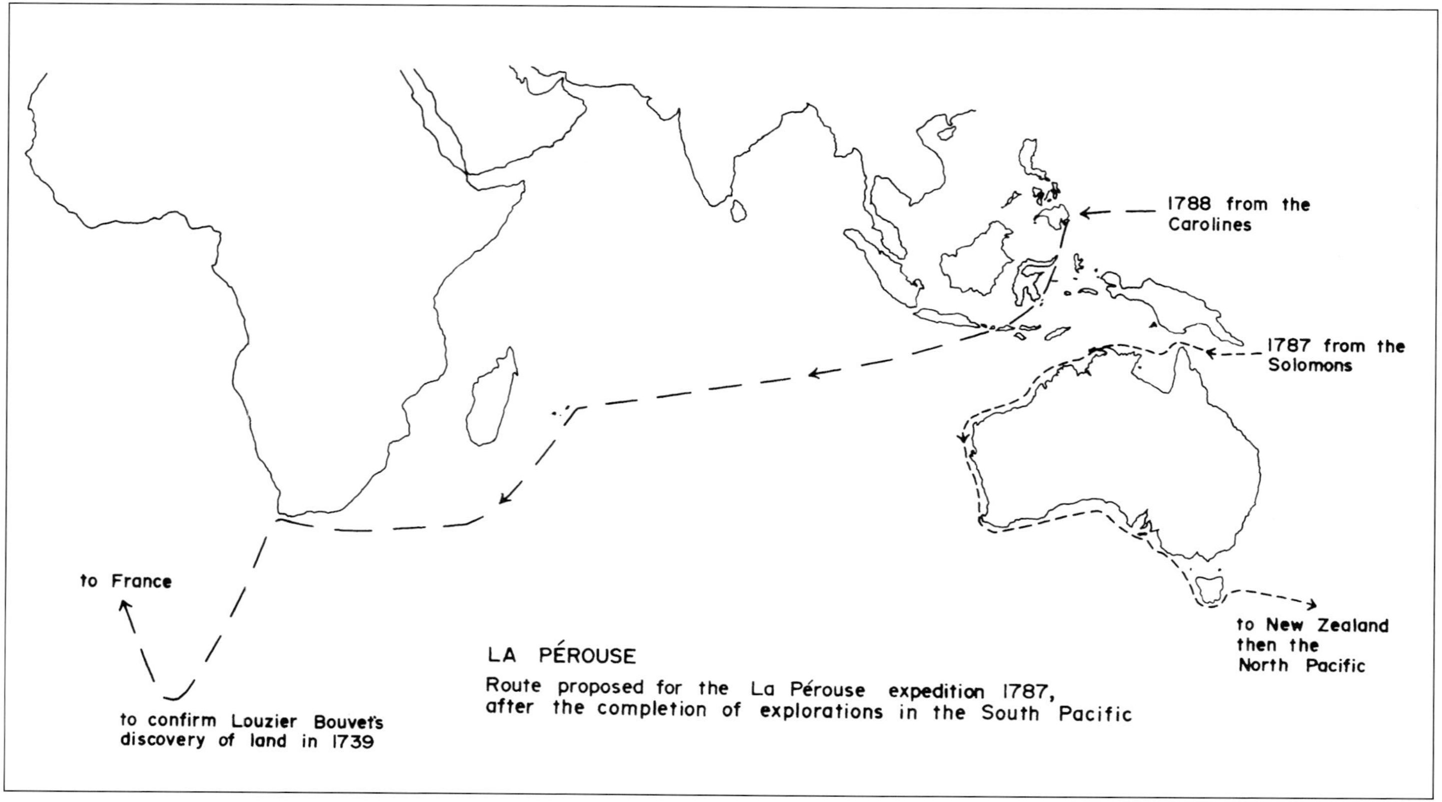

detailed but not comprehensive map of the western part of the Recherche Archipelago which had been located and charted by Nuyts, but not closely surveyed. Besides this D'Entrecasteaux con-

NUYTS' MAP OF THE SOUTH COAST
OF WESTERN AUSTRALIA 1627

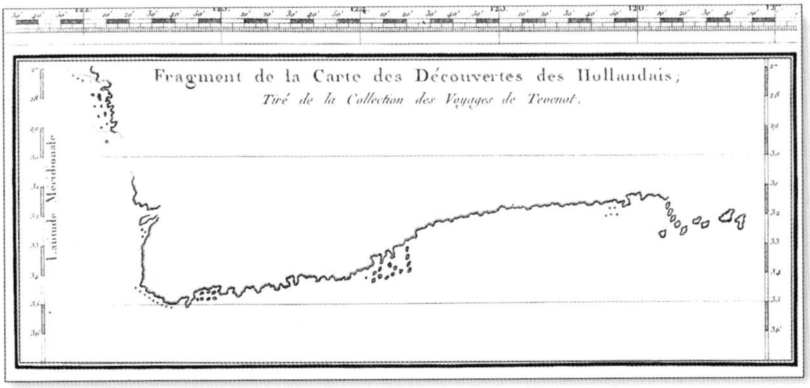

was "arid", although he landed only once, seeing the coast elsewhere from a distance. This assessment was incorrect. The area he reconnoitred is among the most productive agricultural land in the state and is not as arid and waterless as he believed.

Whether or nor La Pérouse would have offered different information and assessments if he had come to western Australia is an interesting question for the historian. La Pérouse was well-equipped to make surveys and was well-intentioned in that regard. He was a careful investigator who aimed at perfection, and was prepared to spend time to achieve this. More significantly he was prepared to aid the development of knowledge in the field of natural history. His scientists went ashore and conducted surveys in a detailed manner. D'Entrecasteaux, in contrast, preferred to have dinghies make hydrographic surveys rather than have them used as ferries for scientists wanting to get to the shore. And when they did get ashore he preferred to have them keep close to the boats and not search far. When D'Entrecasteaux was at anchor at Port Esperance in the Recherche Archipelago, for example, the main effort was made in hydrographic surveys and not in natural history research.

D'Entrecasteaux's hasty judgement resulted in the persistence of the myth about the aridity of western Australia, and the displacement of the idea that it had a salubrious productive temperate environment, as described in French utopian socialist novels set in that region. Instead Dampier's view,[5] and that of the Dutch, that the land in western Australia was unproductive, arid, and mostly lifeless, and that the few savages who lived there were "unique human beings" who seemed to be able to exist where others could not, gained ground, creating the impression that western Australia should be avoided by those seeking refreshment and an attractive environment.

The firm conclusions about the land and life in western Australia, made by D'Entrecasteaux as a result of his hasty survey resulted in the propagation of Dampier's dismal view about the area. The early encyclopedia produced by French intellectuals in the enlightenment provides a very unenthusiastic view of western Australia in the brief entry for that area. The 1765 edition which referred to the country as having the "most miserable people on earth, who closest resemble brutes, and who had no bread, grain or vegetables at their disposal",[6] was still characteristic of later editions which still relied heavily on Dampier's assessment.

The critical views of western Australia, in contrast to the enthusiastic accounts of eastern Australia which was praised by European maritime visitors, raises the question of the values they used to make their judgements, which were reached without scientific analysis. The journals and log books of the early French navigators indicate that they were judging the lands they saw by colour, and by land form. Like most Europeans they liked their land green, and equated bright green with productivity. The bushes they saw from the decks of their ships in the distance from the sea looked very dark green or brown, and they concluded the land was waterless, arid and non-productive. This was reinforced for them by their failure to find mountains and valleys which indicated to them the presence of water-courses, and by their failure to find river mouths or deltas. In western Australia, these signs do not necessarily mean the land is waterless and arid, as later settlers found. Nor does the dark colouring of the vegetation, often enhanced for distant observers by the atmosphere, mean that it is unproductive. It mainly did not appeal to Europeans who were used to lush green fields. This is why D'Entrecasteaux and others were more attracted to Tasmania, where the French always preferred to go for refreshments and rest.

This attitude towards western Australia, that it was arid, remained until after settlement, although it was diminished to some degree by later French scientific explorers who were not all convinced that it was a silent desert, with no birds and little game.

The disappearance of La Pérouse

That D'Entrecasteaux came at all was very much a fortune of history. La Pérouse disappeared at an awkward time for France. When he last reported home from Botany Bay in March 1788, France was in the throes of an economic and political crisis that was to culminate in revolution and the end of the *ancien régime* which had sent La Pérouse on his expedition.

La Pérouse in his last report home, estimated he would arrive at Mauritius in December 1788. By then, at home, the States General had been called, the economic crisis had worsened, and the intellectual and popular forces Voltaire and his colleagues had stimulated, had revealed their purpose and power.

La Pérouse was not missed for some time after his estimated time of arrival had passed. It was then not unusual for ships to be delayed for considerable times, especially if they were in monsoonal areas, or if they were on survey missions in strange waters. Communications were bad. Survey ships mostly relied on chance encounters with passing ships to send messages home. In distant waters such as the Pacific from where La Pérouse sailed, heading for western Australia, these chances were few.

Consequently, it was not until the northern summer of 1789 that fears began to be held for La Pérouse's safety. Shortly afterwards the Paris mobs stormed the Bastille and the *ancien régime* fell. The various revolutionaries who emerged to power in France had other more serious priorities to deal with than allocating time and resources to search for a lost explorer sent abroad by royalists. There was the matter of a constitution to consider, and there were economic and political reforms to be undertaken. There was a need to propagate "the rights of man", and to make quite clear what was meant by this campaign; and there were foreign policy problems. The revolution in France resulted in the emergence of a combination of outside non-revolutionary powers who threatened war, and confrontations intensified as the emigrés who had fled from France added to the threats to the revolution coming from abroad. As a result France feared invasion and intervention.

Nevertheless, despite the awkward situation and the political changes and uncertainties, proposals were made to send an expedition to search for the missing mission which by then was believed to be in trouble.

Proposals to search for La Pérouse

Pressures and initiatives to organize a mission to search for La Pérouse stemmed from three sources. Scientific intellectuals pressed for action to have their lost colleague found. Merchant traders proposed a search mission which would be combined with a commercial venture designed to help the failing French economy. And the naval minister Fleurieu and his colleagues made official moves to select a leader and equip an expedition.[7]

The initial and most noticeable pressures for a search mission came from scientific intellectuals. The main driving force in this regard was the impressive Society of Natural History. At the beginning of 1791 this Society urged the new National Assembly to vote funds for a rescue expedition.

Fortunately for the proposal, Charles Pierre Claret, Comte de Fleurieu[8] came to the Navy Department as Minister on 25 October, 1790, just as the pressures were mounting for an expedition. Fleurieu was himself a scientific intellectual as well as a distinguished seaman. During his career he had worked with Berthoud to develop the marine chronometer, and had conducted tests on this. Subsequently, he further distinguished himself as an author, writing about the history of French discoveries in southeast New Guinea, and later compiling sailing guides for pilots. He personally supported the idea of a mission.

The submission by the Society of Natural History was therefore swiftly acted upon. Guided by the recommendations of a specialist sub-committee, the National Assembly approved the plan for a rescue mission, granted monies and this recommendation was approved by King Louis XVI, who was still in power, on 25 February, 1791.[9]

Joseph Antoine Bruny, Chevalier D'Entrecasteaux who was selected to lead the mission, was then regarded as one of the most knowledgeable seaman in France, in regard to eastern waters. After a distinguished early career, where like Fleurieu, he demonstrated an aptitude for science as well as seamanship qualities, he served in eastern seas, leading an expedition which showed the French flag in China and in Southeast Asia after the American War of Independence. Subsequently, he served as an administrator in the Indian Ocean region, returning to France just before it was decided to despatch a mission.[10] He seemed an obvious choice.

Dupetit-Thouars' attempt to find La Pérouse

However, D'Entrecasteaux did not lead the only mission sent. Private merchants in France, once it was realised that La Pérouse was missing, took independent action to send help. One of these proposals, made by Aristide Aubert Dupetit-Thouars,[11] was got under way as a result of help from financial subscribers, and from the King. His plan was to sail westwards from France via Cape Horn to the Pacific and search from that direction, while D'Entrecasteaux searched east from the direction of Africa.

Dupetit-Thouars' mission proved to be abortive. He departed from France on 22 August 1792. Suffering from sickness among the crew, and with his ship carrying a large number of ship-wrecked sailors whom he picked up in the Atlantic, he put into Latin America for

Bibliothèque Nationale

supplies. There he was interned and his ship seized as a result of the growing threat of the French and their revolution to Europe and its colonies.

D'Entrecasteaux's departure from France

D'Entrecasteaux left France before Dupetit-Thouars, departing from Brest on 29 September, 1791. The months before were spent in preparing the two ships selected, stocking them with supplies and choosing crewmen.

Unlike previous explorers from France, D'Entrecasteaux was not involved in the determination of the aims of the expedition. These were discussed and set down by the government and scientific intellectuals. The aims they determined for the expedition were two-fold. D'Entrecasteaux was to search for La Pérouse along the course he proposed to take from the west Pacific. And he was to take scientists and artists to make further scientific studies for France.[12]

Pressures by revolutionary scientific intellectuals

This was a significant decision with far-reaching consequences. It indicated the commitment of the new revolutionary regime to science and discovery overseas. The decision to make the journey into a scientific one was not taken lightly. There were a series of discussions about the specific aims. In these France's scientific intellectuals played a major part. The later documents and petitions of the Society of Natural History, produced after it had been decided to send a mission, indicate the place of scientific missions overseas in the French revolutionary period, and the continuing interest of France in the South Seas.

A printed communique written for the naturalists being sent to the South Seas, printed on 29 July, 1791, made the new revolutionary government's attitude to exploration overseas quite clear. While condemning the "old explorers in the time of the *ancien régime*", as "rapers of the world" who sought to get precious metals, it indicated that the new alternative government would not stop this effort, but would change the aims of exploration to promote "the better gold of science", to aid "truth" and "enlightenment" on all manner of topics.[13]

It was for this stated attitude and other similar reasons that French intellectuals thus saw the revolution as a process of liberation for them and their work, and they became closely identified with the revolutionary movement. But these views were not shared by all, nor was the principle of involvement by scientific intellectuals in politics accepted by all. D'Entrecasteaux's mission, like the later Baudin's was consequently marred and adversely affected by political confrontations and involvements by committed and non-committed scientists and seamen who held different views of the revolution. The basic fault was that respect was given to researchers from then on not necessarily for scientific attainment, but for political alignment. Scientists committed politically to the revolution had no inclination to let those who were not committed, make contributions. In their view the appearance of the new regime in France heralded the emergence of the "new science" and "the new scientist". In their view these alone had the ability and right to make discoveries and achieve notice and fame.

D'Entrecasteaux's specific aims were to make comprehensive surveys on the land and sea, to look for natural resources in all their varieties, to assess the productivity of lands, their potential for commercial use, and to describe the types of men found and their livelihood and cultural characteristics.

These orders contained the seeds of conflict, especially in regard to the study of man. D'Entrecasteaux's orders to describe man in his varieties represented an attack on the church. Up to then the church had been responsible for teaching about man, and these teachings emphasized his common origin with Adam and Eve and his universality. D'Entrecasteaux's orders meant he was to reject existing assumptions and research in previously forbidden areas of knowledge with state sanction.

KNOWLEDGE OF THE SOUTHERN INDIAN OCEAN
ON THE EVE OF BAUDIN'S EXPEDITION

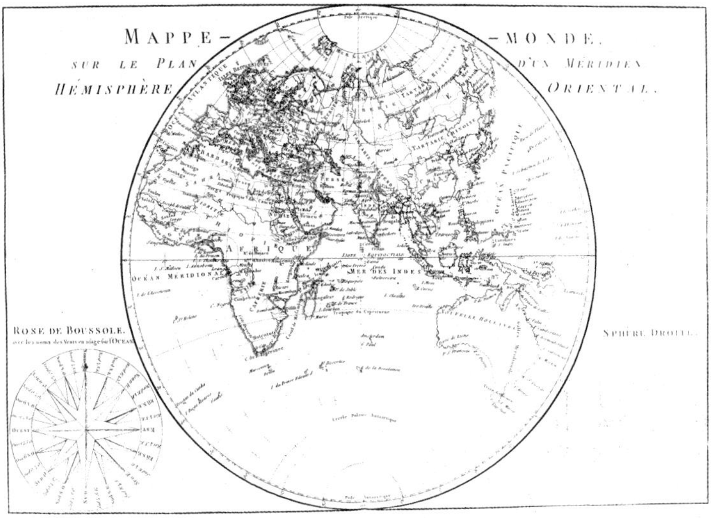

The last major land masses left to be mapped and made known at the time of the French revolution, were Australia, New Guinea and the Antarctic. The Baudin mission was sent out to complete the mapping of Australia.
One of the main discoveries made by Baudin in western Australia was that the coast does not go north in a straight line from Cape Leeuwin to North West Cape. He found that it goes east at Cape Naturaliste to form the large expanse of Geographe Bay.
At the left hand side of the map, below southern Africa, the two mythical islands of Marseveen and Denia are shown.

Although the question of the nature of man and human society was a main point of attention and a cause for division, there was a general interest in a variety of scientific fields by the mission. La Pérouse's unfinished work left a lot of unsolved problems, especially in the Indian Ocean area. There the coastline and physical environment of western Australia was a major mystery to be solved. Not only was western Australia still a mystery. Also, to the south of South Africa there still appeared two mythical islands, Marseveen and Denia, which were reputed to exist there. These and other incorrect locations of land missed by Cook had to be cleared up.

D'Entrecasteaux's planned itinerary

D'Entrecasteaux was given specific orders by the King on 16 September 1791, to sail to Mauritius by way of Finchal in Madeira, and Cape Town.[14] From Mauritius he was to go to Botany Bay, and from there to the New Guinea area to search for La Pérouse. After this he was to sail through the East Indies then to the western Australian coast and sail along it to Tasmania. From there he was to go back to Mauritius on the way looking at Rottnest and the Swan River area, and the so-called William River which was believed to be near North West Cape.

While in western Australia, D'Entrecasteaux was requested to plant grains. This was for a two-fold purpose. The planting of grain was planned as a type of French "foreign aid" for what was believed to be the most unfortunate depressed people existing. It was also planned to make the area more useful for European visitors, as a place of refreshment. At the time western Australia appeared to have no food supplies available for ships calling there, apart from fish.

D'Entrecasteaux did not plant grains in western Australia. Where his crew landed seemed arid. But when in Amboina, on the way down, he secured some deer to land. However, these died before they could be landed and so this plan "to aid the natives and posterity" failed.

These particular plans to cultivate western Australia raises the question about whether France at that time had designs on the region which was already annexed to France.[15] Britain had acquired New South Wales, in eastern Australia, as a colony in the year before the revolution. The "humanitarian" idea of transportation for criminals, to a new crime-free environment away from the corrupting influences of civilization was not lost on the French. The new criminal code passed by the National Assembly in Paris on 26 September, 1791, three days before D'Entrecasteaux departed to examine the South Seas, provided for deportation or transportation as one of the eight approved punishments. Article 29 left open the determination of the place to which the prisoners were to be transported.

Nothing was done about this. There was some confusion in particular about the suitability of Australia. In March, 1791, a French translation of a book about Governor Phillip and his experience was on sale in Busson's bookshop in Paris, at the prohibitive price of four pounds, four sous. This did not present a rosy picture of the struggling colony. Articles about Botany Bay had earlier appeared in the *Moniteur*, indicating the food and other problems Phillip experienced, painting a dismal picture.

Although some revolutionaries were favourably inclined to the idea, it was not until after the revolutionary wars that France took up seriously the idea of establishing a convict colony in western Australia. In the meantime France was involved in revolutionary wars which prevented that type of expansion abroad.

The decision to send the mission and the quick concerted effort at preparations, allowed D'Entrecasteaux to leave before the "Reign of Terror" commenced in France. If delays had been made, or the matter left until then for a decision, it is doubtful if a mission would have been sent. Fleurieu, the Minister who played such a part in organizing the mission, for instance, was arrested in the Terror, and others involved fled to escape the guillotine.

D'Entrecasteaux's expedition did not hear of these events until they reached the Dutch East Indies in 1793. There the news of what was happening on the political front at home caused the expedition to break up, and led to investigations of the mission by the "Committee of Public Safety" in Mauritius and France, and to accusations against the leaders.

Departure from France

The mission left France with two ships of doubtful sailing quality; the former storeships *Truite* and *Durance* which were renamed the *Recherche* and the *Espérance*. The *Recherche* was the lead ship of the expedition, and had a crew of 113, including the scientists, under the command of D'Entrecasteaux. The *Espérance* was under the command of Huon de Kermadec, and had a crew of 106 including the scientists.

Among the crew were a number of young officers such as Elisabeth Paul Edouard de Rossel,[16] who later joined Rosily as a naval administrator, both of whom later became responsible for organizing surveys of western Australia which resulted in specific proposals for France to colonize the region in the restoration period.

Also included was Jean Baptiste Willaumez,[17] who made the hydrographic survey of Esperance Port, and who went on to achieve fame in the naval administration in France.

The scientists and artists included Charles François Beautemps Beaupré,[18] an hydrographic engineer who later rose to prominence; the botanist Jacques Julien de la Billardière,[19] who incidentally was deeply committed to the revolution; and Charles Riche, a doctor and botanist. The official artist was Piron,[20] who left a series of impressive art works of the voyage.

The two vessels left Brest deeply laden with stores, gifts and armaments to face the unknown. The gifts included substantial quantities of red materials, the colour of the revolution. Each ship carried provisions for eighteen months.

The idea of the trip was attractive. There was no shortage of recruits offering themselves. The first part of the trip was in fact delayed by the discoveries of three stowaways in the *Recherche* who were put ashore. Three others were later found in the *Espérance,* but had to be permitted to sail. Nevertheless, despite the apparent attractions, D'Entrecasteaux's voyage was one of hardship, danger and discomfort. Of the 219 men who sailed, 89 died by the time the expedition ended, including both D'Entrecasteaux and Kermadec, and a clerk who was found to be a woman.

The day after leaving Brest, D'Entrecasteaux opened his sealed orders which indicated the route he should take, and was informed in the same document that he had been promoted to rear admiral.

On the way south they took astronomical sightings for scientific reasons, but left the precise observations until they were ashore, indicating that it was only there that they could achieve accurate results. At Teneriffe they anchored near a British vessel, with whom they established good relations despite the deteriorating political situation, indicating the neutrality of science.

The trip in the south Atlantic was slow. There the ships revealed their poor sailing qualities which was to affect the expedition's close shore surveys later made at places such as Esperance. The slow passage of the ships resulted in water shortages. The ration of water was reduced, but even so stocks were so low in the Atlantic Ocean that it was feared the ships would have to reach with the trade winds across to Brazil instead of proceeding to Cape Town. However, Cape Town was reached with difficulty on January 17.[21]

The water that was left by this time was foul. But there were no sick men reported. Constant fumigation of the ship, and the serving of anti-scorbutics such as vinegar and sugar appeared to have had the desired effect.

D'Entrecasteaux's change of plans

At Cape Town, the whole plan of the trip was altered. There, there were reports waiting for D'Entrecasteaux informing him that traces of La Pérouse's expedition had been found in the Pacific.

The evidence for this was very flimsy. The Commander of the French forces at Mauritius, Saint Felix, took depositions from two

French sea captains, Préaudet of the ship *Jason* and Lepinay of the ship *Marie Hélène*. These reported that a British seaman, Captain Hunter, reported that he saw men dressed in European clothes in the Admiralty Islands north of New Guinea. He believed these could be from the La Pérouse expedition.[22]

This was a strange incident. Captain Hunter in fact was in Cape Town at the time. He sailed to there, heading for Britain, not long after D'Entrecasteaux's expedition arrived. Yet he not only did not approach D'Entrecasteaux, when he was ashore in Cape Town, but also did not mention the incident to highly-placed administrators when he met them, although it was known that a search for La Pérouse was under way and searchers had arrived. He reported the matter only at Mauritius.

Nevertheless, D'Entrecasteaux acted on the depositions even though they were unverified second-hand reports, and determined to go to the Admiralty Island group as quickly as possible.

Easting run to the Pacific

D'Entrecasteaux left Cape Town on 16 February 1792, making directly for the East Indies. The ships' progress became increasingly slow after it came to the trade winds in the lower latitudes towards the equator. He therefore changed course and went southwards to pick up the westerlies and go south of Australia to the Pacific.

Both bad luck and good fortune struck at this time. The carpenter Louis Gargan was found dead "from excessive living in Cape Town". He was a valuable crew member. He was essential if the ships got into shoal waters and were damaged.

Fortunately, two stowaways were found, one of whom proved of signal use. He was a German instrument maker who escaped from a British convict ship on the way to Botany Bay. He and a soldier crept aboard the French ships at Cape Town, just before they sailed. He later on served as ship's blacksmith, repairing delicate rudder equipment at Esperance Port where the *Espérance* suffered damage.

The passage via the southern route was slow. St. Paul was passed on 28 March, and was seen to be burning. There was much speculation about this, with theories ranging from the English lighting it to spite the French expedition, to volcanic action. However, D'Entrecasteaux saw no sign of signals for help and believed no one was ashore and in danger, and so went on to Tasmania to refresh and to replenish his water supplies. From there he made for the Admiralty group, arriving at the end of July 1792.

His mission proved abortive. The natives there were more conspicuous for the lack of clothes than for their uniforms. There were no signs at all that Europeans had visited the area. A mass of wood near the coast, which had been reported as a wreck of a ship, was seen to be an uprooted tree.

In actual fact D'Entrecasteaux was not far away from the site of the wrecks of La Pérouse's ships. La Pérouse was lost at Vanikoro in the Santa Cruz group. This was not discovered until after the French revolutionary wars ended. The mystery was finally solved by an Irish Captain Dillon in 1827.[23] He gathered information that there had been a wreck and survivors had landed at Vanikoro, and he sailed in that year to collect information. He secured relics and real evidences of a shipwreck which was identified in France as belonging to the La Pérouse expedition. Dillon was subsequently rewarded. In the meantime Dumont D'Urville, who had been sent on an expedition to the Pacific in 1826 had been instructed to look for evidence of La Pérouse. He learned of Dillon's discoveries when he was in Tasmania and he sailed to Vanikoro where he recovered more relics and evidence, and left a memorial on the spot of the shipwreck. However, what happened to La Pérouse and his crew remained a mystery. It is not clear if they were all marooned, or if some managed to escape.

Arrival at Amboina for stores

D'Entrecasteaux called at Amboina to replenish supplies and refresh the crew after he came from the Pacific intending to explore the western and southern Australian coasts. However, he was not able to get sufficient supplies. There was little meat available and that which he purchased when the barrels were opened, was found to be made up of a lot of bone. Flour was also short and the poultry was unimpressively lean. The supplies he received consisted of yams, sweet potatoes, pumpkin, pigs and goats. The local buffalo could not be taken to sea live. They were considered too violent to take to sea to slaughter for fresh meat. But spices were taken, mainly pickled bamboo shoots, cloves and nutmeg and considerable quantites of sago. Three deer were also taken to land in western Australia.

The trip from Amboina, therefore, did not start out as an attractive venture for the men. The food taken on was commented on as "more suited to the Indian palate" than the French. In the circumstances it was difficult to keep the men healthy. Sago was the main fresh food supplied, but it was reported, "in spite of all the arguments of our surgeon, they [the crew] conceived such a disgust for it, at the end of a few months, that they preferred to it salt meat even of the worst quality".

Water was also an immediate problem. The barrels were apparently not properly scoured at Amboina and the fresh water taken on soon went foul, emitting such a noxious gas that it was difficult to enter the water storage area. The problem was overcome to a certain extent by the use of an agitator. The water was agitated before use, and this caused the gas to rise, leaving the water potable.

The ships left Amboina on 14 October, 1792, and sailed north of Timor into the Indian Ocean. The men had been permitted to buy their own supplies, and the ship was cluttered with poultry and pigs, which added to the health risks. The poultry proved difficult to preserve and supplies were rapidly depleted. Apart from the difficulty of food and water supplies, the ship was infected with cockroaches which consumed everything left lying about, even eating the vitriol.

Abandonment of plan to visit north western Australia

Running into southerly headwinds, the ships were steered out into the Indian Ocean after leaving the East Indies. Consequently, as these winds persisted, it was decided to abandon the plan to visit William River and the west coast of western Australia. The expedition therefore headed south west far out into the Indian Ocean. On 23 November the course was changed and they steered eastward by south to reach Cape Leeuwin.

D'Entrecasteaux, in doing this, revealed a difficulty that was later to cost France dearly. It is not easy to approach the western Australian coast from the north and survey it from there southwards. Strong southerlies blow in those waters and make progress difficult. The best course that can be followed by a sailing survey vessel in the region is from south to north.

Failure to realize this in 1824 prevented Duperrey from coming to western Australia as planned, when he was sent to investigate the possibility of a French settlement there. By the time France could take action to correct this, the British were alerted and took possession of the continent, thus forestalling the French in 1826.

Arrival in south western Australia

At 4.30 a.m. on 5 December, 1792, the west Australian coast was sighted to the north east, at a distance of 4 or 5 leagues (12 or 15 miles). The landfall was reckoned by calculations to be just to the north of Cape Leeuwin. The ship's course was therefore shifted to east south east to better observe it.

The expedition was not impressed by what they saw as they closed

D'ENTRECASTEAUX EXPEDITION, 1792

ROUTE FOLLOWED BY BRUNY D'ENTRECASTEAUX
(RECHERCHE AND ESPÉRANCE) 1792

Location of anchorages and area surveyed: Esperance, 9-17 December, 1792.

in on the coast. The land appeared bare and arid, with dark vegetation and no sign of human habitation. It was also impossible to make a closer survey. High seas were seen to batter the rugged coastline.

Journey eastward along the south coast

This sight stood in sharp contrast to their previous landfall, the lush tropical green, inviting islands of the East Indies. D'Entrecasteaux's log book, from 5 December, contains nothing but critical accounts of the land he viewed. The further east he sailed, and the lower his water supplies became, the more he desired to visit the green, lush forested territory of Tasmania, which attracted him like the East Indies.

Hastiness combined with bad weather led D'Entrecasteaux and his hydrographic engineer and chart-drawer, Beautemps-Beaupré, to make a significant mistake about their landfall, which resulted in subsequent confusion about the shape of south-west Australia. D'Entrecasteaux believed he had discovered a large island off the

MAP OF SOUTH-WEST AUSTRALIA
DRAWN BY BEAUTEMP-BEAUPRE,
D'ENTRECASTEAUX EXPEDITION 1792

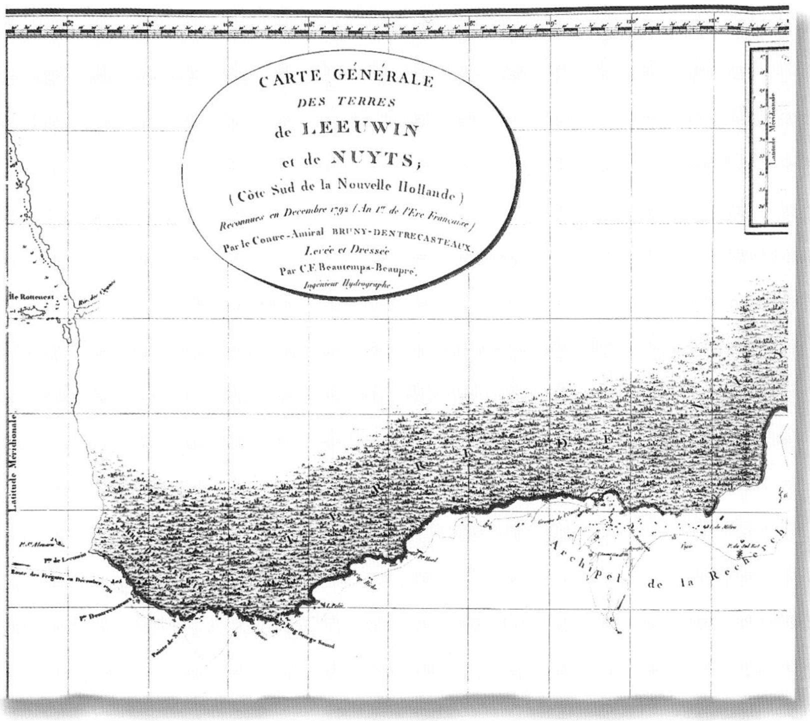

mainland, where Cape Leeuwin is situated. On his map he left a gap of approximately 40 minutes of longitude between what he called St. Allouarn's island and the mainland coast. In fact what he first saw was Cape Leeuwin itself and the highlands stretching north from it, which, like other parts there, when seen from the distance at sea, appear to be an isolated island. The low lands about Hardy Inlet, which stretch between the high land at Cape Leeuwin and the high land eastward near Point D'Entrecasteaux, can only be seen from close in. The peculiar thing is that this area was quite clearly and more accurately mapped by Rosily on St. Allouarn's expedition in 1772. The islands St. Allouarn saw were the small rocky outcrops lying south eastwards out to sea from Cape Leeuwin, which are passed to port as a ship enters Flinders Bay.

On D'Entrecasteaux's map, therefore, Flinders Bay which was mapped by Rosily, is shown as a strait of water between St. Allouarn Island and the mainland, and the coastline of the mainland is consequently made to seem more eastward than it is, running in a nearly straight line from there up to North West Cape instead of coming out to make an extensive wide promontory between Cape Leeuwin and Cape Naturaliste. This misconception helps explain why Baudin was

D'ENTRECASTEAUX'S
INCORRECT MAP OF CAPE LEEUWIN, 1792

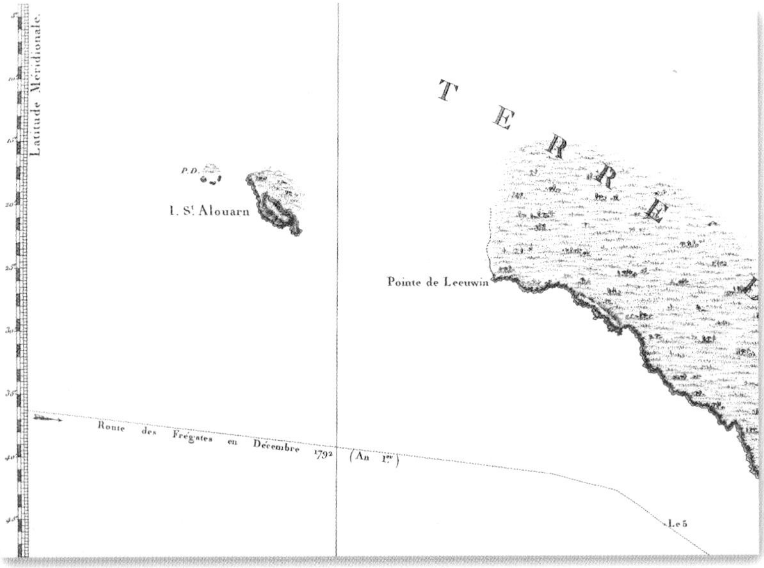

When he approached Cape Leeuwin, D'Entrecasteaux kept too far off the coast to make clear and accurate observations and maps. He consequently shows an island lying west of Cape Leeuwin with a possible passage in between. In fact the land D'Entrecasteaux called St. Alouarn Island (I. St. Alouarn) is Cape Leeuwin, and the place he calls Pointe de Leeuwin is most likely Cape Beaufort which is at the eastern end of Flinders Bay.
The information given by D'Entrecasteaux led Baudin to the region to solve the problem of the location of Cape Leeuwin, which he attempted to do in 1801 and finally in 1803. (See Appendix 17, page 291).

surprised to find an immense bay, Geographe Bay, on the north side of the promontory when he followed D'Entrecasteaux after Napoleon assumed control of France and the revolution.

The ships continued eastward in boisterous weather, with the winds blowing fresh from the west north west and with the seas heavy. Standing right out to sea, away from the apparently dangerous coast, the expedition passed the opening of King George Sound at 6.00 p.m. on 6 December. They clearly observed "the bay opening", and although they wanted to anchor there, were too late to turn and make their way up wind in the boisterous weather. They consequently passed it one league off and left it for the British explorer George Vancouver to make known the fine sheltered deep water port, with its fresh watering points at King George Sound.

On 7 December the ships followed the coast at a distance, observing it to turn to the east north east as it started to form the Great Australian Bight. Cape Riche, named after one of the botanists on the expedition, was passed in the course of the day. The following day, 8 December the ships met with the coastal reefs and islands that lie off the south coast near the present site of Hopetown. The ships passed inside these islands and shortened sail at night to slacken the pace of the ships till the sun rose.

On the morning of 9 December the level of the barometer dropped and a fresh wind came from the west north west. With the barometer continuing to fall the wind rose and shifted to the west, accompanied by heavy broken seas.

Arrival at Port Esperance

This sudden gale put the ships in a dangerous position. They were on an eastward course in the vicinity of the Recherche Archipelago which had been located and charted by Nuyts. By rights, as soon as the barometer level fell, putting the reefs and islands directly on the lee, D'Entrecasteaux should have gone on a starboard tack out to sea to find more room. But driven by the wind, continuing along the coast, the expedition sighted the islands at 9.00 a.m., and by 11.00 a.m. were among them. The wind by this time had further shifted to the south west and the position was perilous. It was now impossible to make a tack seaward safely. The seas were big and the wind high, so anchoring was impossible. If they did that and their anchors dragged, they would be driven into the islands with no chance of manouevring. Before them stretched what appeared to be "an uninterrupted chain of reefs and islands". To make matters worse the boisterous wind churned the waves so much that it was difficult to separate turbulent surf on reefs from the churned-up deep water. To further add to the troubles of the expedition, the strong wind forced D'Entrecasteaux to furl the topsails, to save his masts, although these were needed to give as much headway to the ship as possible, and to provide them with a good steering capacity in following winds and seas. The furling of the topsails also increased the motion of the ship in the rough seas. The seas there, driven by west winds, when they come to the shallower part of the coast near the Archipelago, are short and exceedingly uncomfortable.

The *Espérance* at the time was ahead of the *Recherche*, and appeared to be standing directly into danger. In fact Huon de Kermadec had the situation well in hand. A very competent young seaman, Le Grand,[24] was sent to the mast head of the *Espérance* as the ship was moving east. From there he steered the ship through the deep water passage, in West Channel which lies between the westernmost group of islands in the Recherche Archipelago and Butty Head. The ship then made into the calmer waters in the lee of Observatory Island where anchor was dropped and the sails furled.[25]

D'Entrecasteaux observed the manouevre, saw that the masts were all still standing which indicated clearly that the vessel had not struck bottom, and having no alternative, followed. Rounding up near the *Espérance*, he cast anchor, but was apprehensive about the bottom, and the ability to hold the ship at anchor. His first anchor commenced to drag. Men were consequently stood by the masts with axes to chop them down so as to reduce windage, but the second anchor dropped pulled up the vessel and a third anchor made it secure, although the anchorage was not a perfectly sheltered one.

In honour of the *Espérance*, and the fine manoeuvres it had performed, the area behind Observatory Island was named Port Esperance. This is the name now given to the township and port several miles east of this, behind Dempster Head. The name of the officer, Ensign Le Grand, who saved the ships in the gale was immortalized by D'Entrecasteaux naming the prominent cape at the mouth of Esperance Bay after him (Cape Le Grand).

No communications could be made between the ships that day because of the weather. But twelve islands were counted. However, D'Entrecasteaux had no intention of making a close detailed survey of the area which appeared to be as rocky, sandy and as uninhabitable as the rest of the coast he had seen. The expedition by then suffered a serious water supply problem. They needed fresh supplies urgently. But an immediate departure for a place with known supplies was not possible until the wind became favourable. At anchor in the gale, tossed by the waves and wash behind Observatory Island, the rudder bars on the *Espérance* snapped which meant a further delay. The expedition thus prepared to stay on to fix the rudder after the gale moderated and to use the occasion to survey the islands to find passages through them, and to search for water supplies.

On 10 December the wind slackened and the seas flattened and communication between the closely anchored ships became possible.

The water situation was found to be acute. The water ration was further cut and D'Entrecasteaux decided "with pain" that the proposal to survey the south coast of Australia would have to be abandoned and haste made to get to Tasmania which had water supplies in plenty.

Surveys of the Esperance area

Soundings were taken around the ships, and a landing was made on Observatory Island where a forge was set up to repair the damaged rudder bars. Landing was not easy. In the course of attempting to get on to the island, which has no landing beach, the ships' chaplain fell into the water as he was jumping from the dinghy to the rocky shelf. The danger to the men was increased by the sight of large sharks following the boats.

On the island seals were killed for food, and La Billardière, one of the botanists, made a brief study of the island which he found to be covered with shrubs, some of which he collected for research in Paris. His main scientific finding of note was that the seals, previously drawn by the French scientist Buffon,[26] had been drawn wrongly. La Billardière found that in fact they had no neck as pictured before, but were more like fish, with their heads being part

BLACK SWAN BY PIRON, 1792

of their bodies. He assumed Buffon was mislead by using an incorrectly stuffed model made from a skin for his drawing.

An observatory tent was set up on the island, and sights were

AN OBSERVATORY CAMP SET UP BY D'ENTRECASTEAUX 1793
A PAINTING BY PIRON

A camp similar to this one at Recherche Bay in Tasmania, was set up at Observatory Island near Esperance, Western Australia.

taken. Observatory work, however, was difficult. It was hard to land men and equipment on the island and the terrain was rocky and uncomfortable. The weather also continued to be problematic. Waves and surge continued to beat on to the island. Consequently some men had to spend the first night there, with only a few biscuits and little water, being unable to get back to the ships. Pools of water were found, but they were all salt. La Billardière, however, found himself a small watercourse southwards on Observatory Island, and got enough water to quench his thirst. For food he and his companions roasted penguins on the coals, but even this attempt at comfort was adversely affected when the wind shifted to the east and blew the smoke into their small cavern, blowing the fire out.

Two more extensive expeditions were sent in dinghies while the ships were at anchor, to make brief surveys of the islands of the archipelago and of the mainland.

The major survey undertaken was an hydrographic study made under the direction of Willaumez and Beautemps Beaupré.[27] This expedition departed at 3.00 a.m. on 13 December, to draw a plan of the

ENTERING ESPERANCE BAY IN A GALE

A painting by Piron, 1792.

islands and passages in Esperance Bay, and to search for wood and water. Supplies were taken to last several days, when it was estimated the rudder would be fixed and the ships could leave.

Only the western part of the archipelago was examined by this expedition. An east wind came in on the 11th and blew consistently until 14 December. Making eastward in the small boat was therefore difficult. In any case the archipelago was extensive, and a proper survey of all of it would have taken much longer than the time estimated.

No wood supplies and no water streams of use to the ships were found on the islands. Willaumez consequently sailed north into

MAP OF D'ENTRECASTEAUX'S EXPEDITION
IN ESPERANCE BAY WITH RECHERCHE ARCHIPELAGO

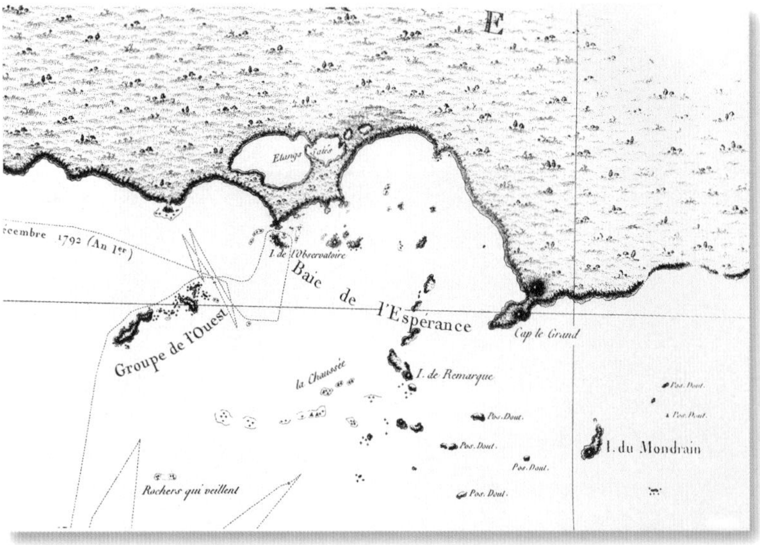

Esperance Bay and landed on the coast, somewhere in the vicinity of the present course of Bandy Creek, east of Esperance Town, where the explorers walked inland. What they saw did not impress them. They found merely "a sliver of a stream" of fresh water half a league from the coast which could not possibly replenish the ship supplies. The land they saw was mostly sand dunes, which they believed quickly absorbed any rain that fell, and the surrounding rocks also meant to them that the water was quickly shed into the sand.

No sign of human habitation was seen. There were no camp sites, no remnants of fish or shellfish were to be seen to indicate the presence of humans as was the case on the coast of Tasmania. Smoke was seen in the distance, but it was concluded that the inhabitants of Esperance had no knowledge of boats and hull construction, that they did not venture on the sea, and therefore were more "savage" than the Tasmanians who knew how to build and use canoes.

Willaumez and his party returned on the evening of 15 December with his unimpressive reports. By then the repairs to the *Esperance* had been completed and preparations were made to leave.

Inland survey of the mainland at Port Esperance

On 15 December a second party went to the mainland from Huon de Kermadec's ship, the *Espérance*. This group, accompanied by the botanist Riche, had strict orders to return at 1 p.m. They were meant to make only a brief survey of the coast near the anchored ships, north of Observatory Island.

This expedition after sailing north, went eastward along the coast to what is now Blue Haven Beach, which provides an excellent landing place. The beaches further westward, nearer the ships, are washed by waves, have rocky ledges and are difficult to land on. Blue Haven Beach is tucked between high Dempster Head and a high-reaching bald head of rock. There is plenty of room in this bay, which is about a mile long, and would have been attractive to the explorers not only because of the safety for their boat but also because from the hill tops there it was possible to see the ships at anchor, and to see a way inland to take stock of the situation.

From the hill above the landing beach, which is now an attractive picnic spot, the view inland is not immediately helpful and enlightening. From there it is possible to see only series of scrub-covered sand dunes and little beyond. The shore party divided into two, to traverse the hills and explore. One party went due north, the other north west.

Once across the sand dunes, which are not difficult to traverse, but which make walking hot going and therefore a hardship for explorers on low water rations, it is possible to see Spencer or Pink Lake and Lake Warden and the flats surrounding them, which lie beyond the line of dunes. The party going to the north reached these and drew maps of the lakes, noted that unlike salt lakes in Tasmania, they contained no fish or living things. They noted the marshy nature of the country, noted that smoke could be seen further inland, and then returned to their dinghy. The party which went north west traversed only dunes and scrub.

The botanist Riche becomes lost

In the meantime the botanist Riche walked off alone to look for specimens. He became so absorbed collecting materials that he lost his way. Thus when the party was ready to depart Riche was missing. His disappearance resulted in upsetting delays for D'Entrecasteaux.

Riche was not only attracted by the new specimens for science which he found, he also sought to make contact with the Aborigines. He saw camp fires in the distance and made for them. But the smoke he saw always seemed to be no closer then when he first set out.

It is not suprising Riche got lost. The scrub was quite thick in the area, and there are no outstanding landmarks that can be used for bearings unless a careful note is made before walking through and beyond the dunes.

However, Riche acted sensibly, doing the only thing that can be done in the Australian bush. He found a small fresh water hole and kept close to it.He left this on the 16th after he became panic-stricken at the thought of being left behind by the ships. Taking some water in his specimen bottles, he made for the coast and managed to sight the ships, probably from Dempster Head. He made his way along the beach and met a shore party which had been sent to search for him. He arrived back on the ship the same day.

The surprising thing is that Riche and the others were not bitten by snakes. He saw only one, but the territory he traversed is one of the most snake-infested areas in western Australia. However, the men were heavily clothed and armed and Riche carried a lot of equipment and must have made a noise and this could have cleared everything out of the path ahead.

When Riche was discovered missing, De la Grandière, who was in charge of the shore party, returned to the ship at 7.00 p.m and reported the matter. Although he had been ordered to return early, De la Grandière did not become apprehensive until after 5.00 p.m. Riche had been lost before on the voyage. He was an avid collector of specimens and time seemed to matter little to him. He had to be searched for in the Azores and elsewhere.

De la Grandière, before he returned to the ship, left a supply of arms, a warm coat and a note on the beach for Riche, promising he would return the next day.

The search for Riche

Early the next morning a boat was sent ashore to see if Riche had returned. But the supplies left for him were untouched, and there was no sign of the missing botanist. The search party made a brief survey, but saw no sign of him. They saw only some Aborigines who, by this time, had appeared. This raised their fears. They felt he had been taken and perhaps "eaten" by savages.

The party returned at lunch time as ordered. D'Entrecasteaux, by this time, was worried about other matters. The shortage of water still worried him. He wanted to depart and had no desire to imperil the health and lives of the rest of the expedition. His next guaranteed water supply was in Tasmania, over a thousand miles to the east.

A conference was consequently held. At this the remaining scientists prevailed upon D'Entrecasteaux and the others to delay the departure for 48 hours to provide an opportunity to search for their colleague. They pointed out that a similar situation had once occurred on Cook's ship, and he delayed for 48 hours during which time his missing scientist was found by a shore party.

Two parties, therefore, were sent ashore, one from each ship. One searched to the east and to the southeast from the landing cove. The other searched to the north and to the northeast. These left the ship on the afternoon of 16 December, taking supplies sufficient for four days.

Fortunately for the search parties, the fine weather they enjoyed on the dinghy trip gave way to cloudy overcast conditions, making their task a cooler one. They did not have to traverse the dunes under the boiling sun, as they had to before.

They were fortunate to find traces of Riche inland. Sets of footprints, the place where he lay down, a piece of paper with writing, and a pistol were discovered near the water hole Riche camped by, which was not far from the present town of Esperance. But by then Riche was heading towards the beach on his own, this time on a proper course. When the shore parties returned they found Riche waiting for them. But his collected specimens were all lost. He was emptyhanded. However, the reports he made indicated that the land was sterile and unproductive. He reported locating only one plant which produced edible materials. The country, he indicated in support of the other observers, was arid. Partly as a result of this the French thought that the Aborigines in western Australia were a unique species of man who could live on salt water and on special types of nourishment not suited to others. This was specifically commented on and investigated by the subsequent French exploratory missions which were charged with making a special survey of the characteristics and livelihood of the Aborigines.

Departure from Port Esperance

D'Entrecasteaux departed from Port Esperance on 17 December, and continued to reconnoitre the coast until he reached just past the head of the Great Australian Bight, where he turned south at 131° 38′ east longitude from Greenwich, making directly for Tasmania.

The natural history researches his scientists conducted at that place, among the Tasmanian natives in particular, revealed that D'Entrecasteaux's men had the capacity to make studies of use to science.

Their performance in western Australia had not been impressive, and raised doubts about their capacity to achieve the aims set for them. The only useful work done, as the scientists themselves said in complaint, was in the field of hydrography.

D'Entrecasteaux's expedition therefore did little to solve the mysteries of western Australia and its native inhabitants. The savants waiting in France and elsewhere in Europe for information about the last great unknown land area in the Indian Ocean and Pacific area, had to bide their time until a further expedition could be sent.

D'Entrecasteaux himself was to play no further part in this quest. He died on the way from Tasmania to the East Indies and Huon de Kermadec died soon after. Command of the expedition then went to the royalist officer, D'Auribeau. From then on the expedition was continually in trouble, stemming from the tide of revolution.

The expedition reached the Dutch East Indies in September, 1793. By then, unknown to the members of the expedition, the French king had been executed by the revolutionaries; a republic proclaimed and war declared on most of Europe. The expedition consequently found no welcome or help from the Dutch, despite the illness of the crew.

Revolutionary sentiments soon divided the expedition. D'Auribeau declared himself to be an anti-republican and declared in favour of the royalist emigrés in February, 1794. Republican sympathisers were arrested by the Dutch and the ships raised the royalist flag and were handed over to the Dutch by the French royalist officers. The expedition consequently broke up and did not complete its survey.

The revolutionary Committee of Public Safety consequently gathered evidence on "the traitors" to the revolution.[28] But D'Auribeau died in August. The rest of the personnel gradually made their way back to France. Rossel and the scientific collection were taken by the British on the way back home. But eventually he and the collection returned with others to Paris to play a further part in the quest for knowledge about western Australia.

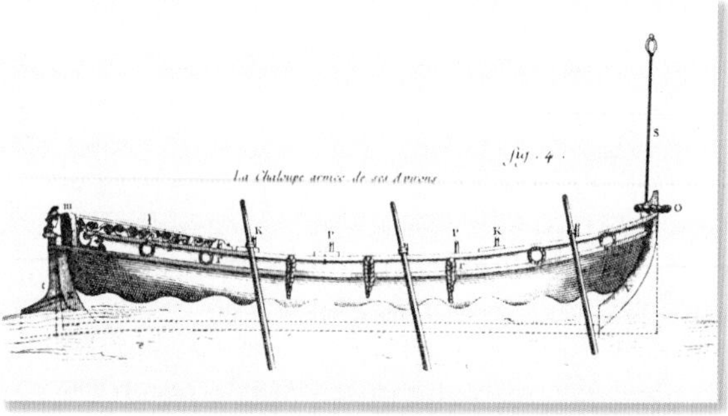

French chaloupe (long boat). Baudin's chaloupe, used for survey work and ferrying scientists ashore, was wrecked in a gale in Geographe Bay in June 1801.

CHAPTER 5

Baudin and Western Australia. Section 1: origins of the mission and the survey of the west coast by Hamelin and Baudin

State of the survey work

The greater part of the Australian coastline still remained unsurveyed and the interior of the continent unknown as the eighteenth century drew to a close. The only stretch of coast that was known with some certainty was the Pacific seaboard. Bass and Flinders had examined the area around the south east corner of the continent, to the south of where Cook had commenced his survey northwards at Hicks Point, discovering Bass Strait in 1798. But details about the discoveries there were not universally known until the early years of the following century. Thus at the time there was no certainty in Europe about the exact shape of the greater part of the Australian continent. Nor was there much information about the dangers to navigation which existed in Australian waters; nor about the state of the local Australian winds and tides which could help, hinder or endanger ships. Little was also known about the resources of the continent and the surrounding seas, and about the human and other types of life which existed there.

Desire for further scientific knowledge

A variety of reasons existed to prompt the French to complete D'Entrecasteaux's unfinished work after the mission broke up. Foremost among these was the continued desire in France to contribute to scientific knowledge. The urge to be in the forefront of the drive for scientific enlightenment in Europe, which was evident in the old regime, if anything was intensified by the revolution. There was a feeling among scientists and other intellectuals after the fall of the old regime, that scientists and the mind had been "liberated" from the influence of traditional religious and mystical beliefs, and that the new influence of rationalism would produce unprecedented advances in knowledge. The new mood in regard to science was evidenced not only in the drift of young intellectuals from church schools and courses for the priesthood to the new secular institutes of scientific learning in revolutionary France, but also in the creation of a new "rational" revolutionary calendar, devised by the mathematician and physicist Gilbert Romme[1] and his colleagues, and based on decades, to replace the septenary system and the dating style of the "vulgar Gregorian calendar" based on religious revelation.

The one area French revolutionary scientists could make an international name for themselves, for France and for the revolution, was in Australian exploration. The continent stood out as the last great mystery area in the South Seas ready to be made known, and if French scientists could now reveal the secrets, France would gain the honours.

French Revolutionary research scientists disadvantaged

Added to this incentive was the need to quickly supply French scientific research institutes with new collections of specimens of scientific interest and significance.[2] War and political chaos at home had had an adverse effect on French scientific work. There was a lack of new specimens essential for work in the scientific classification of types which was then the most prominent and popular field of scientific research enquiry. The last collection of specimens of significance to arrive in France had come from La Pérouse who was sent out in the pre-revolutionary period by the *ancien régime*, and this was only of Pacific materials. D'Entrecasteaux had not added much to this from the Indian Ocean-Australia region. He was reluctant to let the natural scientists with him use his longboats as ferries to get ashore. He invariably used his small boats for hydrographic research work and mapping. French scientists in consequence suffered shortages of research materials.

In contrast, scientists in competing nations and anti-revolutionary states in Europe were well-served and equipped. British scientists were still sifting through the extensive world-wide collection of specimens made by Captain Cook, and this was being added to continually by local collectors such as Bass[3] in the new British colony in New South Wales, and elsewhere in the British Empire and its environs. This put French revolutionary scientists at a disadvantage, and adversely reflected on revolutionary science which the French believed was in advance of science in non-revolutionary societies where, it was maintained, non-rational views prevailed.

On the other side of France, scientists in Austria were as well, if not better off, than the French. A series of scientific specimen collecting missions were despatched abroad by the emperors Joseph II and Leopold II from the mid-1780's, to improve in particular the holdings at the acclimitization garden at "the palace of enlightenment" at Schönbrunn, just outside Vienna. These culminated in the extensive mission despatched in 1792 under the guidance of a young French naval officer, Thomas Nicolas Baudin, who accepted service with the Austrian monarchy, to collect botanical and other specimens from the Indian Ocean, Australia and East Asia for the benefit of European science being promoted by Austria.

Wars at home and abroad and political chaos made it difficult for France to make up the lack. The French government from 1792 had more important priorities than equipping scientific expeditions. Scientists such as Lacépède therefore had real reasons to believe that French science and scientific institutes were in danger of slipping down from the paramount position they held in the world of European science unless something was done. Constant pressure was consequently exerted by scientists on the government to get help in opening new important fields of research.

Added to this was a further cause prompting quick action by France. There was a fear among officials in France that Britain would step in, solve the mysteries of Australia, and gain glory and recognition for Britain in that area of exploration as Cook had done in the case of Pacific discovery and the southern Indian Ocean.

Britain was well-situated to explore. There was evidence of trouble in the navy at home, but this was offset by the existence of British establishments abroad, such as at Sydney, which could be used as local bases for exploration. Bass and Flinders revealed the utility of Sydney in regard to Australian exploration with their discovery of Bass Strait. Flinders, moreover, was rapidly acquiring a reputation as an Australian explorer. Acclaim for his work preceded him, before his return to Britain in 1800. Shortly after this, French fears of British action proved well grounded. The British Admiralty appointed Flinders to investigate the southern part of Australia in particular.[4] His expedition left nine months after the French mission departed.

Absence of political motives for the Baudin expedition

It has been suggested by writers using British historical records, that France also had political motives in sending out the mission. Although there has been speculation and controversy among historians for some decades on the question of whether or not Napoleon had designs on Australia when he despatched the Baudin mission, the extensive collection of records of the expedition makes it clear that the mission was despatched as a result of pressure by scientists, and was sent for scientific purposes.

A variety of points of evidence contained in the manuscript records of the mission reveal that the theory that Napoleon sent the mission because he had designs on Australia, is a tenuous one. Firstly, the records reveal that the Baudin mission was not Napoleon's brain-child, and was not obviously part of his expansionist schemes. The mission was planned by the Directory which ruled France before Napoleon, and was done with the advice of scientists. Napoleon's role, as far as the documents are concerned, was merely to approve the already determined plan.[5]

Certainly the plan approved by Napoleon was changed from the earlier one to give more emphasis to Australian exploration, and he had a bust of Dampier sculptured to place among the busts of the notables who had made the world known, which adorned his Gallery of Consuls, which has been made much of to show his interest in Australia. But these facts, without further evidence, cannot be used to conclude that Napoleon was pre-occupied with Australia and had designs on it. In fact it was French scientists who changed the plan he approved. For some time past they had wanted to clear up the mysteries of the fifth continent and decided it would be best to concentrate their efforts on this area in 1800, rather than diversify French efforts. The bust of Dampier, incidentally, did not symbolize Australia for Napoleon and the French. Dampier was a well-travelled explorer and amateur scientist. He had helped make a lot known, and not just Australia. His bust was in the Gallery of Consuls for his general contribution to knowledge. In any case the adverse description Dampier gave of Australia in his works would hardly attract Napoleon and the French expansionists to have designs on the country. New South Wales incidently also then was not obviously proving to be a success.

Secondly in the proposals for a mission there is only one reference to political motives, and that was made in the earlier proposal by the Directory and does not relate to the possible acquisition of Australia by France. Those drafting the plan after outlining the scientific benefits of the voyage, then indicated that the mission could be of political use if a comprehensive study was made of newly-discovered Bass Strait and the strait to the north of Australia between Cape York and New Guinea. Baudin was ordered to survey these, but in the end examined only Bass Strait.[6] Again it cannot be inferred from this evidence that France wanted to acquire territorial control of the vicinity of the straits. France required a knowledge of the waters in the straits for strategic and commerical reasons. They represented new access ways to and from the Pacific, one in the west wind system area, the other in the monsoonal wind area, both of which were used by French naval and commercial vessels en route to and from the Pacific region which was of interest to them.

Thirdly, in the records of accounts of the mission, there is a further reference to political motives which adds weight to the point that the Baudin mission was sent for only scientific reasons. When Baudin arrived at Mauritius on 16 March, 1801, on the way to Australia, he strongly objected to the reception given to him, claiming it did not suit a naval officer of his rank. This gave rise to a series of differences between himself and the scientists who accompanied him which served to mar the expedition. For this outburst Baudin later incurred the displeasure of French officials[7] who wanted to stress the neutrality of the scientific voyage, and he incurred the displeasure of scientists who held to the principle that science was international and neutral and should be devoid of national and other politics.

There is one other brief reference to political motives in the private journal kept by sub-lieutenant Saint Cricq who sailed on the *Naturaliste* with Hamelin.[8] When this vessel was anchored off the western Australian coast between Rottnest Island and Cottesloe, while surveys were being made of the Swan River and the islands outside, he observed it was a useless and dangerous place; that the river could make a good harbour but was closed by a bar, and that he could see no reason to propose an establishment there. Certainly this raises doubts about the intention of the mission in regard to secret political motives for the voyage. But Saint Cricq was a junior officer. He would not have had a knowledge of important secret policy documents of the Napoleon naval ministry. Baudin certainly was ordered to look for suitable harbours, but this could be viewed as a French endeavour to find and keep a record of places of refuge and refreshment in the Indian Ocean, as it had done before in the Pacific, and not a basis for political expansionist designs.

Coming across such comments it is easy to understand why critical scholars such as Ernst Scott[9] tackled the question about political motives for the mission, and gave rise to controversies. France had a tradition of expansionism which did not die with the revolution, and it continued to desire a base in temperate lands in the southern oceans during the *ancien régime*, throughout the revolutionary period, and in the period of the restored Bourbon monarchy. But older scholars have committed an error by basing their theories on information contained in published works such as Péron's history of the Baudin expedition, and on inferences about Napoleon. Péron upheld in his book that Napoleon sent the mission. This information, as indicated above, is not correct. The correct account is contained in the manuscript records which the older scholars who have written on the subject, did not find and use.

These records indicate clearly that the French plan to colonize Australia was left to be made in the restoration period, after the fall of Napoleon, by Bourbon and not revolutionary French ministers.

Proposals for a scientific mission of exploration

It would be unfair and playing favourites for the historian to give to any single individual or institution in France the credit for creating and sending the Baudin mission. Early proposals for revolutionary France to send a scientific mission abroad came from a variety of interested individuals and organizations. Baudin himself submitted an early proposal for a scientific voyage around the world to emulate Cook's, to bring glory to revolutionary France. This proposal is undated, but seems to have been composed in 1796. By then law and order was restored in France, and not only was peace made with Spain by the Treaty of Basel (1795), but also Spain, which had joined the anti-French revolutionary forces in Europe, changed sides in the war and moved towards a French alliance, later actively joining France in the war against Britain as a result of the friendly Treaty of San Ildefonso (1796).

This new Franco-Spanish alliance opened for France a world-wide series of bases and refreshment points and opportunities for overseas voyages. Such political changes must have had an influence on Baudin when he drew up his proposal and submitted it. His plan was to have an expedition sent to the Pacific by way of Cape Horn, where Spain dominated, and collect "tropical specimens" from Chile to Peru, after which he would go to the Society Islands and then to the south west part of Australia. The main aim of the proposed expedition was to collect plants and animals for the under-stocked gardens of acclimatization in France.

However, it could not be claimed from this that the later mission of 1800 was the brainchild of Baudin as has been suggested before. A scientific mission abroad was being pressed for at the same time by influential scientists such as Lacépède and Jussieu. Besides this the Minister of the Navy had developed a new interest in exploiting for France, which was experiencing economic difficulties, the rich commercial resources of the Pacific, which it was believed Britain was gaining.

Baudin's scientific voyage to the West Indies to collect specimens for scientific research

But instead of being despatched on a grandiose voyage as proposed, Baudin, with the support of the natural history scientists in France, in the end was sent on a short voyage to the Spanish West Indies to bring back an existing collection of specimens which were stored there.[10] Permission was given from then friendly Spain for this mission to go and Baudin set out in the summer of 1796, taking with him the Austrian-born botanist Anselme Riedlé and the zoologist

Rene Maugé who were to add to the collection by searching in the West Indies. These two scientists later accompanied Baudin to Australia.

The impressive collection brought back to France from the West Indies by Baudin was not wholly satisfying for the waiting scientists at home. Many of the specimens had been stored for years and were dried and the fresh ones were all tropical and contained little that was new. A new proposal was therefore sent by the Institute, by natural historians in particular, to the Directory on 24 July, 1798 to have a more long-ranging mission sent. The Directory took the matter up and approved a voyage to be made by Baudin who by then had the support of the French scientists. The original plan was to have

FRENCH KNOWLEDGE OF WESTERN AUSTRALIA AFTER D'ENTRECASTEAUX'S EXPLORATION, 1792

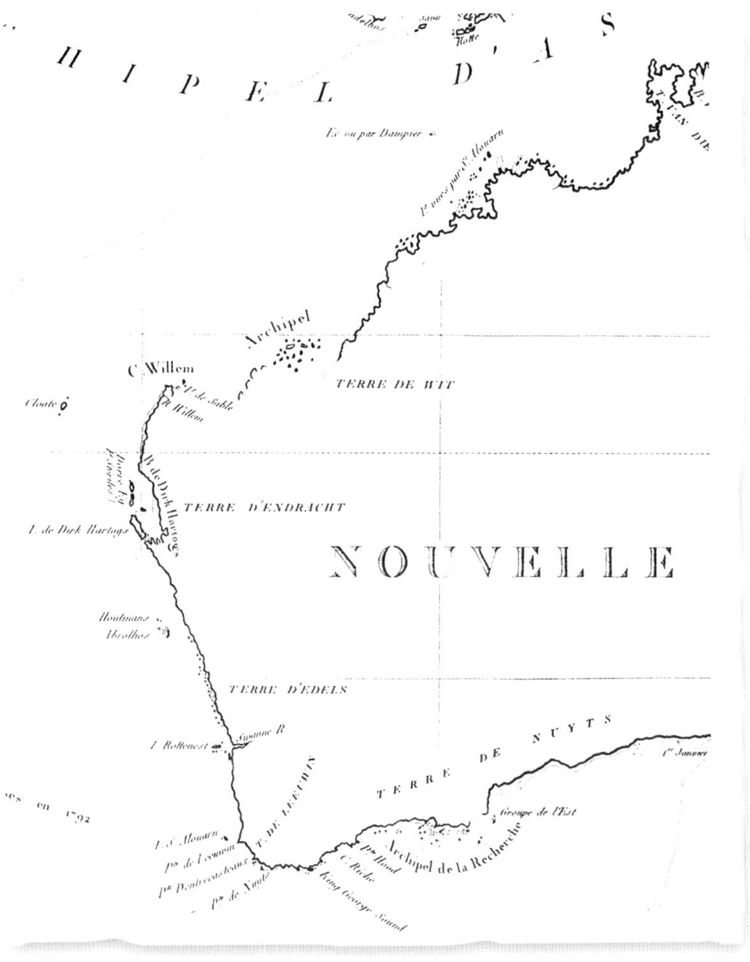

Baudin follow Cook's tracks, and also to have him thoroughly examine the south west part of Australia. The difficulty at that time was there was no satisfactory research boat available and in view of the wars in which France was engaged it did not have time to build one especially for the task. However, the planned invasion of Britain by France by then had been deferred, and two spare boats were found for the round the world voyage. But before the expedition was further equipped it was shelved due to economic and political difficulties which gave rise to Napoleon emerging as consul. After Napoleon was established in power a final and less grandiose plan to explore only the western part of Australia was submitted to him by scientists at the Institute and approved.

Plans for Baudin to explore western Australia and collect specimens for scientists

The new plan called for Baudin to thoroughly explore the north, west and south parts of the last great unknown continent in the southern oceans. He was not only to map the coast, but was to look for good harbours, journey up rivers, examine the products of both the land and sea and report on native life and customs.

His specific orders were to take the eastward route via the Cape of Good Hope to Tasmania to explore that island first. Then he was to explore the south coast westwards from where Cook had left off his survey at Point Hicks, to where D'Entrecasteaux had left off his survey at the head of the Great Australian Bight. After this Baudin was to sail west to examine the Swan River and Rottnest Island, the supposed William River which was believed to be somewhere near North West Cape, which was probably Exmouth Gulf and then the west and north coast as far as Endeavour Strait at Cape York.

The confirmation of Thomas Nicolas Baudin as leader was natural. He had the support of the French scientists and was approved by the naval administration and government. He was both an accomplished seaman and an experienced leader of scientific missions despatched to collect specimens.

Baudin's earlier leadership of the Austrian scientific voyage to Asia and Australia

Thomas Nicolas Baudin was born on 19 February, 1754 at the Ile de Ré on the wind and tide swept coast of Britanny which is a good nursery for seamen. He commenced his sea career in merchant ships, later joining the navy as a cadet in 1774. As he was not of noble birth he could not look forward to early promotion and command during the *ancien régime*, except in war. Following the naval reforms made by Castries from 1780, new opportunities were offered to officers such as Baudin. He was soon promoted to sub-lieutenant. But further advance proved slow in peacetime. Baudin consequently accepted service abroad with the Austrian monarchy, which was then not an unusual thing to do. His selection to take an expedition to collect specimens from the Indian ocean, Australia and Far East regions for the garden of acclimatization at Schönbrunn,[11] after leading earlier missions, shows the high regard in which he was held in Austria.

He was leading the Austrian expedition to the Indian Ocean when war broke out between France and the power he was serving. War was declared by France on 20 April, 1792, four days before Baudin left Italy for the Cape of Good Hope. He was intercepted by a French naval vessel near Gibraltar. He consequently offered his services to his country, as a patriot, but accepted advice given to him and proceeded with the mission to benefit science and knowledge.

THOMAS NICOLAS BAUDIN

On his return he went back into the service of France and as has been indicated was selected, because of his experience and reputation, to lead the mission sent to collect specimens from the West Indies. Good reports of his work there resulted in his selection to lead the great mission to Australia.

Preparation for the voyage to Australia

His second in command was a professional naval seaman Jacques Felix Emmanuel, Baron Hamelin. Hamelin was a younger man than Baudin. He was born at Honfleur in 1768, the son of a pharmacist. Like Baudin he commenced his career in merchant ships, later joining the navy, serving in the revolutionary wars. Hamelin was a good seaman, and strict disciplinarian. He commanded the *Naturaliste*, which was slower and a worse sailing vessel than the *Géographe* commanded by Baudin. But despite this, he and his crew made the most distinctive contribution to the study of the western part of Australia, outshining Baudin. He seemed more prepared to stand into the coast and make close inshore surveys than his commander who appeared to be often over-cautious which led to dissatisfactions among the scientists who were eager to get ashore, thus making Baudin increasingly

BARON HAMELIN

THE GÉOGRAPHE AND THE NATURALISTE UNDER SAIL

unpopular with them. Hamelin's qualities and the recognition of these in France are demonstrated by his being selected to take charge of the naval force being assembled on the French Atlantic coast to invade Britain, after his early return to France from Australia. He was created baron in 1811, after further service in the Indian Ocean, and subsequently held high office as Director-General of the Charts and Plans Office in the French naval ministry.

Despite the hardships of shipboard life, and the danger of going to explore an unknown coast which had a reputation for shipwrecks, there was no lack of recruits to make up the rest of the crew of officers, scientists and seamen required. The opportunity of being a member of the greatest exploring mission since La Pérouse was a popular stimulus to join. The attractiveness of the voyage for adventurous and patriotic Frenchmen at the time is demonstrated by the names of the recruits. Young Bougainville,[12] son of the eminent explorer, joined as midshipman. His name in the list was added to by other young and old officers who were to achieve prominence in the navy and in exploration. Among the latter was Louis de Freycinet who commanded the next French scientific mission to the western part of Australia, in the reign of the restored Bourbons.

The 23 scientists and skilled draftsmen sent to collect or make drawings of fresh specimens for the benefit of posterity were all distinguished men in their own fields. They represented a wide variety of disciplines. Bissy and Bernier were sent as the astronomers. Leschenault and Michaud and Denisse were sent as botanists. Riedlé was sent to join them as head gardener. Depuch and Bailly were sent as mineralogists. Boulanger was appointed engineer-geographer and Faure as geographer. The zoologists consisted of Maugé, Péron, Levillain, Bory St. Vincent and Dumont.

Péron and the origins of the differences which marked the voyage

The man who was to achieve most prominent popular notice among the scientists was François Péron, a product of the revolution. Péron was born at Cerilly, Allier on 22 August 1775. His widowed mother endured hardships to have him educated in theology at Cerilly College. The principal there was impressed by Péron's ability and trained him for the priesthood. War broke out and interrupted his studies. He joined the revolutionary army as a volunteer patriot and served on the Rhine where he was wounded, losing an eye and was captured. He was kept a prisoner of war in Magdeburg Fortress and spent his time reading about voyages of exploration. He was repatriated home to France in 1794, because of his wounds, and was inval-

FRANÇOIS PÉRON

ided out of the army. In Paris he attended the medical school, and studied at the museum. As a result he developed an interest in natural history. His health was not good, and this together with an unhappy love affair led him to give up his studies and seek a place on Baudin's mission. He already had demonstrated his devotion to work, his drive and his abilities. He was an indefatigable research worker. Con-

BAUDIN EXPEDITION, 1801

ROUTES FOLLOWED BY THE FIRST BAUDIN EXPEDITION TO WESTERN AUSTRALIA 1801

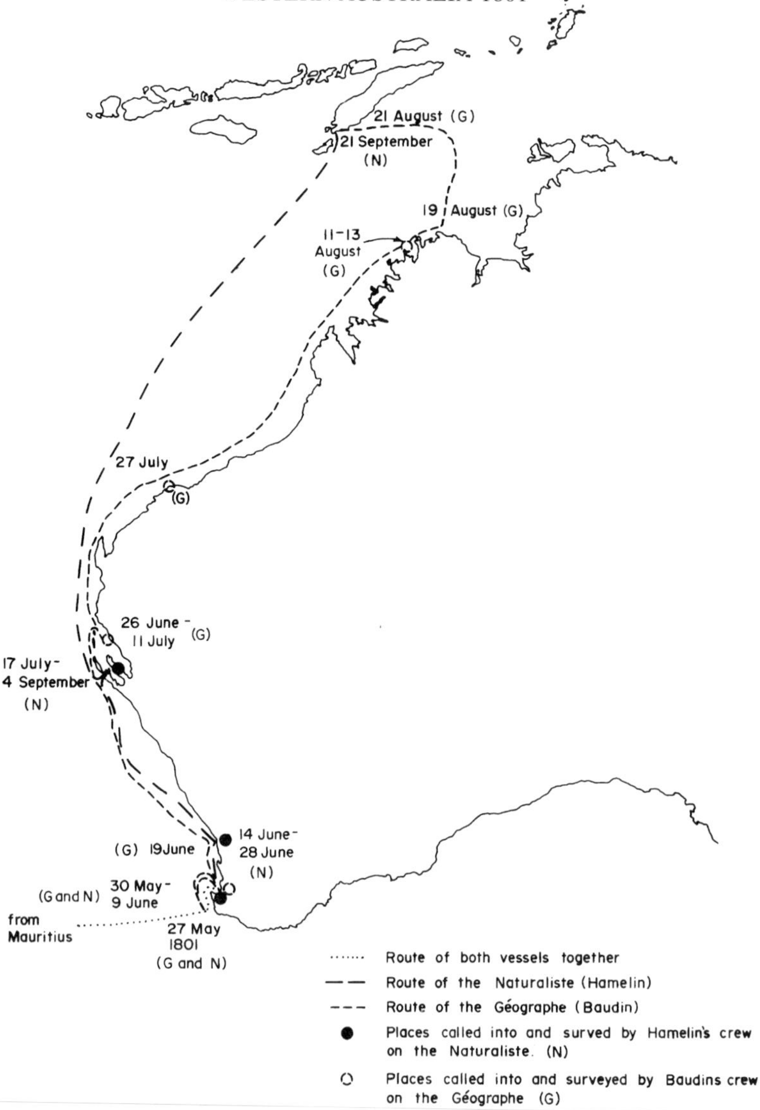

Location of anchorages and areas surveyed:

Baudin (Géographe) and Hamelin (Naturaliste):
Geographe Bay, 30 May–9 June 1801.

Baudin (Géographe) independently:
Shark Bay, 26 June–11 July 1801. Dampier Archipelago, 27 July 1801.
Bonaparte Archipelago, 11-13 August 1801.

Hamelin (naturaliste) independently:
Swan River and approaches, 14–28 June 1801. Shark Bay, 17 July–4 September 1801.

sequently, he was supported by the Institute and was given the position of zoologist-anthropologist for the mission. Péron subsequently wrote the history of the mission, which was completed by Louis de Freycinet after Péron's early death. This written work clearly reveals Péron's passionate nature, his devotion as a scientist and the difficulties and weaknesses of the Baudin mission. Péron with other scientists, was soon at odds with Baudin. Not long after they left France differences developed, and Baudin rapidly found himself isolated from the scientists and artists and young officers grouped about Péron. Feelings ran deep for the whole trip. A number of the scientists in consequence left the expedition at Mauritius before Australia was reached.[13] The trouble was caused not only by personal differences which emerged in close living in confined spaces on the small ships. The scientists objected to being subjected to naval discipline, and being told when to go to bed, when to put their lights out, when to go ashore and when to come back, and that they had to join in physical exercises. They sometimes worked late into the night, and found it difficult to fit into the strict system of naval shipboard life.

Added to this were the political differences that continued to affect every aspect of life in Paris. Péron was a revolutionary. Baudin was an officer of the old school. Péron had fought for the revolution and his beliefs. Baudin at the time Péron was fighting and was wounded, was serving in the ships of the enemy of the French revolution, Austria, whose armies had captured Péron. Harmony between these two was thus not readily possible. The bitter differences which marred the trip have been revealed for posterity by the fact that Baudin's name is not mentioned in Péron's history of the voyage, although Baudin was commander. Nor, on the Australian coast, is there to be found a prominent place bearing the name of Baudin, which reveals how those who sailed with him felt about him, which must have hurt Baudin.

The scientists were expected to make researches, keep records and collect samples in a wide variety of different fields, ranging from astronomy, geography, mineralogy and botany to branches of natural history.

Proposal to scientifically study man in Australia

Prominent among the latter was the study of man in Australia.[14] Baudin's naturalists were ordered to make a most comprehensive study of the physical and moral conditions of the indigenous people reported to be living in the western part of Australia, but who had never before been investigated, and seldom seen.

The comprehensive orders given to the naturalists in this regard, represents a turning point in the study of man which occurred in

France as a result of the revolution, and resulted in the foundation of scientific anthropology, which soon took a place alongside archaeology in the general effort being made to explain the origin and nature of the human race.

There was a long existing interest in man by thinkers and others in France. But before the revolution concepts about man tended to be made in the context of Christian beliefs, which insisted that man was universally created in God's image. This was added to by the accepted idea that Christianity was an agent of and synonymous with civilization. However, evidences coming from China especially, before the outbreak of the revolution, raised questions about the role of Christian religion in civilization, and about whether or not man was a universal creation. Evidence indicated he existed in different forms in different parts of the world, and there was no clear indication whether or not this was because of evolution.

The revolution which swept aside the clergy and their institutes of learning in France, opened the way for new insights and new avenues to be explored and valued. A variety of books, summarising knowledge and seeking explanations about man apeared in the Republic, among which was François Péron's *Observations on Anthropology*, (Paris, 1802). It would be tangential to describe this literature here. What is more significant to note is that special organizations of intellectuals, such as the *Société des Observations de l'Homme* (The Society for the Study of Man) were established in France at the same time as the interest in literature and research on man in different parts of the world appeared.

Péron in particular applied himself to research in the field of anthropology in Australia. One of his more important published research papers in that area, is his study of the comparative strength of "natural" and "civilized" man, where he indicates that civilized man in fact is stronger, thus destroying the romantic idea that "civilized" man is degenerate and that men living in natural conditions in a natural environment are "healthy noble savages".[15]

Unfortunately, as a result of a number of scientists leaving the expedition at Mauritius, Péron had to undertake many scientific duties which prevented him from following his specialist interests. Moreover, as Péron sailed in Baudin's ship he was seldom given the opportunity of getting ashore. Information gleaned about the Australian Aborigines, on the voyage, is therefore not extensive. In any case, Aborigines proved hard to locate and accost. And when they were met with, the language barrier was insurmountable. The French landed in western Australia equipped to talk to Polynesians. Their greetings and other comments were therefore not comprehended. Nevertheless, the records of the voyage contain a lot of useful information about Aboriginal life, their weapons, their housing, their apparent diet and life style.

Departure from France

The two ships attempted to leave Le Havre on 18 October 1800, but contrary winds kept them harbour bound. After a further night ashore for all, they left the following morning accompanied by the cheers of the populace who put on "a memorable display" to mark the occasion.[16]

A favourable east wind took the ships seaward to the English frigate which was blockading the harbour. Once the purpose of the expedition was explained, the English joined the other well wishers offering success to the scientific mission, and the vessels headed for Teneriffe to complete loading stores.

Little of note occurred on the way there. Some stowaways who wanted to go on the great voyage of exploration were discovered. The masters of the ships' crews also revealed that several had deserted. The only other incident of note occurred near Teneriffe. The *Naturaliste* was chased by what was believed to be an English privateer, but this made off after some manouevres by the *Géographe*. Otherwise the passage was a normal one.

Troubles on the voyage

The troubles which thereafter marred the expedition emerged at and after Teneriffe, although a hint of them appeared as the vessels neared the Canaries. When land was first sighted the scientists and many officers who had never before experienced a landfall of a ship, acted excitedly. Baudin not only remained aloof, he also demonstrated his disapproval of this by recording in the official journal of the voyage that the scientists and officers behaved like madmen, creating bedlam. This is a petty observation to enter into a ship's journal, and indicates at least part of the source of later conflicts. Baudin had a very paternal and condescending attitude to those accompanying him on the expedition. He did not hesitate to reprimand his officers and scientists or to confine them to their cabins or banish them from his presence and his table when he felt this was warranted. And throughout the journey he developed a habit of writing curt notes to those who appeared to challenge his authority and by not carrying out or departing slightly from his orders. These notes were always entered in the official log and did little to win Baudin friends or respect. Indeed, the notes became a source of amusement to those who found Baudin's demonstrations of authoritarianism unpleasant.

Baudin's task admittedly was not easy. Looking after a number of independent-minded scientists with different interests and backgrounds is never an easy task, and those accompanying Baudin were no exception to the rule. He often had cause to complain at their behaviour. The French naval ministry realised this could occur. Consequently, when they appointed Baudin as commander, they stressed that he was leader, that his word was law, and that he had the responsibility to impose discipline. Baudin's trouble was he did this with little feeling or thought. Throughout the mission he appeared to take more note and care of his ships and equipment than his men. He seldom took advice even about when and where to collect scientific specimens, which was the main purpose of his mission.

This seems to be out of character for Baudin. He was not unpopular on the previous voyages he made. In fact he gained the respect and confidence of scientists so much that not only did they push forward his candidature as mission leader, but also two scientists who went with him to the West Indies in 1796 joined him again to go to Australia.

Several possible explanations can be made for Baudin's apparent change of character. Baudin died of tuberculosis on 16 September, 1803, before the mission ended. He could have already been a sick man and wasting before he sailed from France. He certainly suffered from dysentery and other complaints early on in the trip. Sickness therefore could have affected his judgement-making ability, which would not have been helped by the unhealthy diet.

Alternatively or correlatively, Baudin could have been inclined to authoritarianism. He was an officer trained by the *ancien régime*. He never fought for the revolution. He served the Austrian monarchy in the years of revolutionary extremism. He served the Directory quietly and well. His trouble might have been in part that he was later placed in a difficult position by the rise to power of Napoleon, and Napoleon's emphasis on discipline, law and order, with which Baudin seemed to agree, but which others did not accept as part of the revolution. When at sea in the Consulate period, Baudin could have been acting according to what he believed was the spirit of the times. There is evidence to suggest this is true. Those with whom he clashed most severely and persistently on the voyage were dedicated revolutionaries such as François Péron, for whom discipline, law and order carried out in an authoritarian manner were not acceptable.

Upsets at Teneriffe

Once the ships were anchored at Teneriffe a further side of the story of the conflict between groups on the mission was revealed. The excited scientists and officers left for shore as soon as anchor was dropped and few were seen again until the ships sailed twelve days later. To add to the displeasure this "neglect of duties" gave Baudin, some returned suffering from venereal disease. This surprised Baudin. The quality and reputation of the street girls at Teneriffe, which was then an international port of call, was well known at the time, and all but the most indiscriminating and promiscuous seamen avoided them. Baudin complained that men of the quality of his officers and scientists should have known better. He subsequently confined some of them to their quarters for their behaviour. Others were reprimanded for bickering and insolence and for neglecting their duties.

Unfortunately for Baudin the quick departure of the expedition from Teneriffe was prevented by his failure to purchase supplies. Wine had not been supplied to the expedition in France. It was decided that Baudin should purchase Canary wine which was of good quality and travelled better. But when Baudin reached Teneriffe prices were high and supplies were low because of recent English purchases. A search of the outlying islands arranged by the French consul failed to find supplies so the expedition departed on 14 November for Cape Town, lacking wine supplies, intending to make up the deficiencies in Mauritius.

After leaving Teneriffe, Baudin made a serious error of judgement, taking a gamble which did not pay off, which cost him dearly in regard to his reputation. He set a course southwards, sailing near the west coast of Africa, hoping to make a quick journey to Mauritius by taking the shortest route. This route was seldom used at that time. The winds and currents close to the African coast are contrary for ships sailing south. The most favourable and often-used course was to head out into the middle of the Atlantic, and then head south, or go further west to near the east coast of Latin America, and then go south there to pick up the west winds and currents to make an easting to Mauritius. Baudin's expedition took nearly five months to reach Mauritius. Supplies were already short and complaints were being made. Baudin's error of judgement thus provided his growing number of enemies with good ammunition. His authority could be tolerated if it worked for the benefit of those amenable to it. But when the authoritarian figure proves inefficient then a real basis for trouble exists.

Already Baudin had run into further trouble well before his error brought physical suffering as a result of short supplies. On the way down he strictly enforced the lights out rule, and ensured that all on board had adequate physical exercise, social recreation and rest which was not acceptable to scientists who had neither the inclination to exercise with the crew nor go to bed early. Added to this, whenever a specimen was caught, each of the different scientific groups wanted it, and Baudin had to adjudicate which did little to win him friends. For instance, Baudin records on 1 December, 1800 a porpoise was caught. A shipboard row immediately broke out. The zoologists wanted to dissect it straight away, and turned it upside down. The artists wanted to first draw it, and wanted it the right way up. This matter was settled quickly when the artists went to complain to Baudin and ask for his intervention. When they got back it was already dissected. Baudin from then had to give priorities which did not please all. There was continued bickering.

Further troubles and bitter differences at Mauritius

It is not surprising to note in the circumstances that a number of crewmen deserted and some of the scientists left the expedition after it arrived at Mauritius. The five week stay there, from 16 March to 25 April, in fact was so marred by troubles that Baudin later regretted calling there for stores.

Trouble for the mission there already began as the ships moved in to anchorage at St. Louis. Baudin himself sowed the seeds of discontent by objecting to the low-key reception given to him and the expedition by the port authority. He objected to the immediate close political scrutiny of the expedition members by the colonial administration who feared revolutionary ideas about liberation for slaves, and he objected to the censorship of the mails and documents he had brought. He and the administration fell out and relations soured for the rest of the stay. Fresh supplies were immediately denied Baudin. The scientists and crew, faced with a continued diet of salt meat in sight of land, consequently drifted ashore to live there. This was followed by differences with the authorities about the arrest of deserters. Baudin felt he was getting no co-operation from the authorities. Even those arrested as a result of pressure, escaped from the guard which Baudin complained was "made up of publicans and tavern-keepers". All the expedition received from the authorities in Mauritius was a supply of forced labourers from among the local native population to help with the refitting of the ships, and even these did not impress Baudin. He complained ten Europeans could have done in one week what twenty-six natives found difficult to do in three weeks.

The initial upsets and differences evident at the beginning of the stay were followed by bitter and more deep divisions. Baudin soon felt there was a plot against him and the expedition, carried out by discontented scientists and officers and the local colonial administration who wanted to see the mission fail. For example some of the scientists and officers who left the ships, to live ashore, did so officially by getting sick leave certificates to go to hospital. Baudin made personal visits to see them on several occasions, out of sympathy, but could never find them. Whenever he visited the hospital he was told by the authorities there that his men were away somewhere in town, and it was never certain when they would return. More disturbing, after several days in the colony, placards appeared around the city calling the mission "the expedition that failed".[17]

Various reasons can be offered at least to explain the differences between Baudin and the government authorities on Mauritius. Mauritius at the time was not only short of finances and fresh supplies, but was also threatened with attack. There were rumours that Britain had grouped a fleet in India to capture the island. Baudin's vessels in fact were believed to be invaders as they approached the island and were intercepted. The administrators of the colony thus had no desire to use their powder for unnecessary ceremonies of welcome, and had neither the money nor surplus supplies to give Baudin in view of a possible seige. They had received no orders from France to do this. Baudin was consequently expected by them to purchase what he wanted in the market place.

Added to this was the fact that there was a renewed suspicion of France and the political regime that had arisen under Napoleon, by colonialists in Mauritius. In particular they were apprehensive about whether the French government intended the declaration of the rights of man to be applied to local natives as well as colonialists. Indentured and forced labour existed in Mauritius and other French colonies. Colonialists had no desire to see this abolished by an idealistic government in distant Paris. By the time Baudin arrived the matter had become a very emotional local political issue. The cold war and ideological confrontation as well as hot war with England persisted abroad. In the mid 1790s, Britain added a new dimension to this when protestant missionaries from Britain went overseas to work among native peoples for the first time. This occurred at the same time the anti-slavery campaign rose and developed. With this Britain had a new cold war weapon which was difficult to combat. It promised real "liberation", "equality" and "fraternity" to oppressed native peoples in revolutionary French, Dutch and other territories in contrast to theoretical "liberation" by France, thus promising the colonialists in the Indian Ocean and Asia a real new form of subversion effected by the agents of God and Britain. France's only answer was to proclaim for the rights of man. But this could mean releasing slaves in French colonies which raised fears there. It was for this reason Baudin and his men were closely questioned about the political situation in France and about their own views when they arrived at Mauritius.

Mauritius thus was a fertile ground for Baudin's enemies and detractors to work against him. Discontented scientists and seamen soon found refuge and support.

No definitive evidence exists to support Baudin's contention that there was a widespread plot there to delay or stop the expedition leaving so that its men and equipment could be used to help with the defences of the island against the expected British invasion. Baudin was certainly quietly warned about this by contacts in Mauritius. But some of the deserters appear to have left not because of this but because of the poor food on the ships, and because of the prospect of earning very high wages which were at the time being paid in Mauritius.

Nor is there evidence to suggest that the colonial administration was responsible for putting up the placards about the "expedition that failed", as Baudin believed they did. This could have come from discontented scientists, or disaffected officers or crewmen whom Baudin had upset.

Plan for the voyage altered after leaving Mauritius

Whatever the cause, Baudin left Mauritius with a bitter taste in his mouth, a depleted crew of seamen and scientists, and with short supplies which affected his plans. He managed to purchase some food in the free market with monies advanced by Scandinavian traders who wanted to demonstrate their support for scientific enquiry and the international nature of it. But the stores were insufficient for the planned voyage. After the ships put to sea the expedition was immediately put on salt rations and a sparse ration of bread. Baudin, as a result of this and the delays, decided to alter the planned itinerary. He determined to make for Cape Leeuwin first, instead of Tasmania, to coast northwards on the western coast of Australia, following the sun, and then go to Timor to get his fresh supplies.[18]

He sailed from Mauritius with 15 men and 11 scientists short in the crew of the expedition. Among others left behind was Bissy the astronomer.

Only two scientists remained on the *Naturaliste* when the expedition put to sea on 25 April, 1801. On the 26th the system of watches was rearranged, and transfers were made to better balance the complements of the two vessels. Among other changes Charles Lesueur, a trained topographical painter who had to join the expedition as a volunteer gunner, was appointed a designer or sketch artist. He subsequently worked closely with Péron and left a rich collection of high quality sketches, many of which are preserved in the Museum of Natural History at Le Havre, France. The indefatigable Péron, who had wide interests and a broad education, undertook the bulk of the scientific research work, making up the deficiencies left by desertions.

The voyagers were not in a good frame of mind when they left. It soon became apparent that their main diet would consist of biscuits, which already were affected by weavils, and salt rations. Nevertheless, by the time they reached the western coast of Australia the replacements for the deserted seamen who were a worry, had settled down and became skilled in manouevring the sails and the ship; and the scientists, especially Péron, worked with dedication and enthusiasm.

Arrival in south western Australia

A landfall was made near Cape Leeuwin on 27 May, after the expedition witnessed a most colourful, spectacular sunset as they neared the coast. This impressive display of red glowing clouds can sometimes be seen off the coast of western Australia, especially after

ERRONEOUS MAP OF CAPE LEEUWIN TAKEN BY BAUDIN FROM D'ENTRECASTEAUX

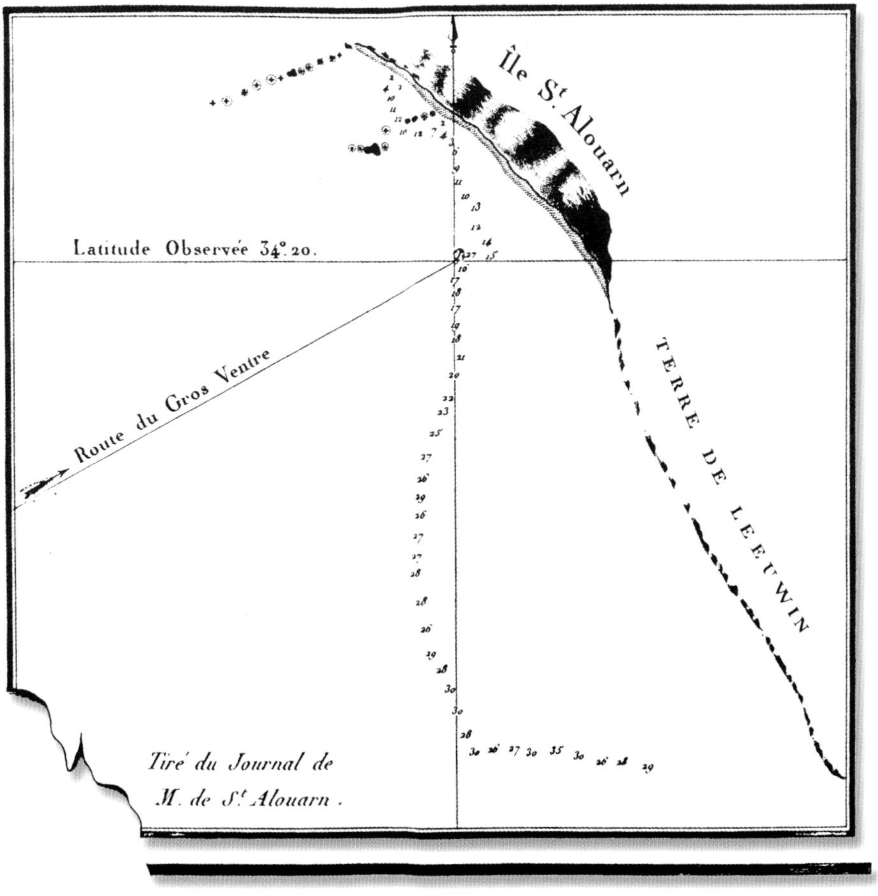

severe volcanic activity in the close-by islands of the East Indies. Sometimes the dust fragments carried by jet streams result in magnificent displays of colour as the sun sinks below the horizon.

Cape Leeuwin at the south west tip of Australia, near where the expedition made its landfall, was one of the few positions fixed on the chart of that region at the time. The expedition consequently

FRENCH KNOWLEDGE OF SOUTH WESTERN AUSTRALIA
BEFORE BAUDIN'S SURVEY

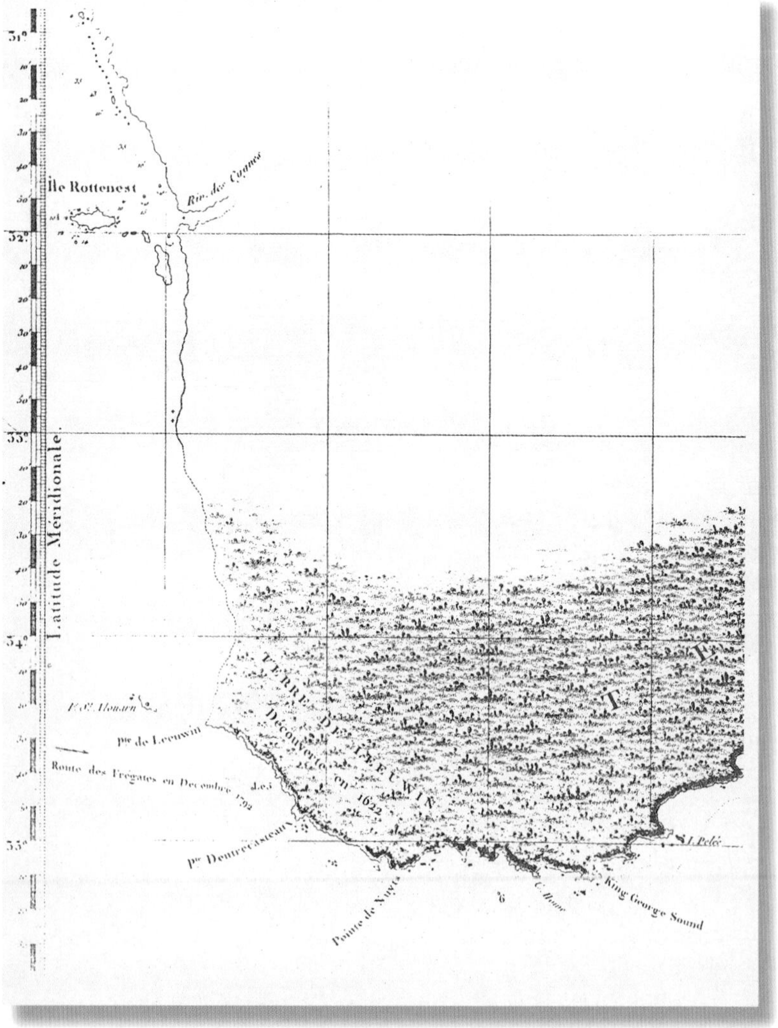

Both ships arrived at the coast together as planned which meant that they did not have to put their alternative plan into effect. This arranged for a first rendezvous at Rottnest and a second rendezvous in Shark Bay if and when the ships became separated.[19]

It is not clear from the journals where the expedition actually made its first landfall. The coast of western Australia was sighted 10 leagues away when the *Géographe* was positioned at an estimated 34° 25′ 55″ south and 112° 51′ east of Paris (115° 31′ east of Greenwich). Cape Leeuwin is situated at 34° 22′ south and 115° 08′ east of Greenwich. A later observation made on the ships at noonday was 34° 35′ 38″ south and 112° 57′ 42″ east (115° 17′ 42″). The chronometers however indicated 111° 42′ 7″ (114° 2′ 7″) for one and 111° 47′ 19″ (114° 7′ 19″) for the other. Both the latitude and longitude assessments made by the explorers in fact were wrong and remained so throughout the survey of western Australia, making it very difficult to pinpoint exact anchorages and landing places on the western Australian coast. Baudin himself realised this later, in 1803, when he returned and observed Cape Leeuwin at close quarters.

From the evidence it seems the ships came on the western Australian coast to the north of Cape Leeuwin near Cape Hamelin, and then turned north for Rottnest to commence their survey.

Discovery of Geographe Bay

On the morning of 30 May the ships observed a previously unknown cape jutting outwards into the sea well north of Cape Leeuwin. This was named Cape Naturaliste. On rounding it the expedition members found themselves in an immense bay "15 leagues from one side to the other". This bay, Geographe Bay, in fact is some seventy kilometres across, reaching from Cape Naturaliste to Bunbury. The high points of both extremities can be seen from sea and make an inviting sight for those seeking a sheltered anchorage, although the bay is open to the north and dangerous during the winter gale periods when winds come from that direction in south western Australia.

But south winds were blowing at the time the expedition arrived so the vessel turned into the bay and anchored east of Wright Bank about four miles out to sea north of Bunker Bay near the tip of Cape Naturaliste. The anchorage position was determined to be 33° 30′ 50″ south and 111° 56′ 56″ east of Paris (114° 16′ 56″ east of Greenwich). The assessment of latitude was reasonably accurate. Wright Bank is 33° 30′ to 33° ′ south and the ships were anchored behind this. But their longitude assessment was too far west. Their proper position was more likely 115° 2′ east, giving an error of approximately 45′. This error was not wholly Baudin's fault. Cape Leeuwin was wrongly fixed on his outdated charts. Correction of this by observation by him was difficult. The horizon was mostly hazy when he reached the coast and in any case he was using a system of calculations which did not offer a high degree of accuracy. His astronomers calculated position by means of the observed distance between the sun and the moon which is liable to produce errors and different results by different observers. These heavenly bodies are too big to achieve accuracy, in particular when the instruments Baudin's astronomers had at their disposal, are used. Consequently, his exact anchorages can only be determined by making corrections and by informed guesswork. This consistent error of position, incidentally, explains the incorrect shape Baudin gave to both Geographe Bay and to the whole of the western coast of Australia in his draft map.

Exploration of Geographe Bay

The expedition stayed in Geographe Bay from 30 May until the night of 8 June when a winter gale from the north caused them to make a hasty departure. In the meantime the ships were at four separate anchorages, moving from Cape Naturaliste gradually eastward to Bunbury, making an effort at consistent and thorough survey of the whole of the bay and its environs; providing opportunities for the scientists to examine the land from east to west.

Because of the need to move in with care it was not until after dark that the ships were finally securely anchored in 23 fathoms at the first anchorage. Landing in small boats was consequently left to be made on the following day, much to the disappointment of the eager scientists.

On the way in the newly-discovered bay looked attractive and promising. The high brown rocky, heavily-wooded hill peninsular which rises out of the coastal plain near Busselton to form Cape Naturaliste offered shelter from the west and south winds and seas which seemed to prevail. It was open to the north, but there seemed no danger from that direction. The bay therefore seemed safe, in particular because of the sandy bottom which was good ground for holding the anchors firm.

At the north end of the hilly peninsula, punctuating the brown cliffs which went to the water, were two sandy coves which seemed ideal for small boat landings. More significant and intriguing, the hills behind the eastern-most cove appeared to be cut by a T-shaped watercourse which ended at the cove. This promised fresh water which, up to that time, had not been found in quantities by visitors to western Australia.

MAP FROM CAPE LEEUWIN TO BUNBURY WITH ANCHORAGES IN GEOGRAPHE BAY, 1891 AND 1803

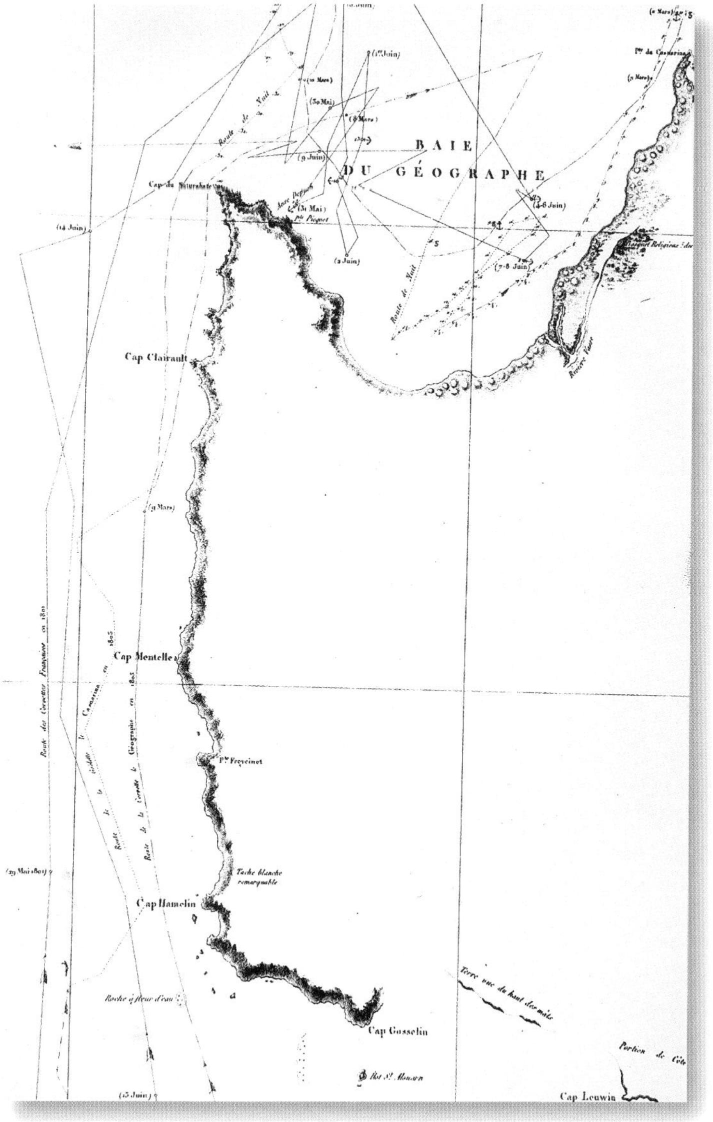

This chart shows the patchy quality of Baudin's survey work. The coast from Cape Hamelin north to Bunbury, including Geographe Bay is sketched in detail although not accurately. The south coast was incompletely surveyed. Baudin believed Cape Beaufort was Cape Leeuwin which he named Cape Gosselin. Baudin could have solved the mystery of the location of Cape Leeuwin and the shape of the coast near it if he had followed St. Allouarn's track into Flinders Bay. Instead he stood well out to sea to round the Cape and consequently sighted only a few pieces of land between Cape Leeuwin and Cape Beaufort. He marked this on the map with the note "land seen from the masthead".

Baudin decided not to waste the night at anchor. He employed the crew to make soundings all around the vessels to assess the quality of the anchorage as a whole, and put out the dredge to collect specimens from the bottom so the eager scientists could get to work. The catch in the dredge was disappointingly small and unexciting. But samples of the bottom which contained clay, indicated to the waiting men that not only was it an ideal anchorage, but also that there was in fact a river nearby. The presence of clay in the ocean at that time was regarded as a sign of a nearby river mouth or delta. This heightened the expectation of finding supplies of fresh water and game.

But further disappointments lay ahead for the scientists and men. Baudin had no intention of staying for long and making a thorough survey of the land at this stage. He decided to send only two very small parties ashore on 31 May, each having only two scientists. The others had to remain on board and were kept there until 4 June, which did not add to Baudin's popularity.

One small boat under the command of Sub Lieutenant Picquet[20] of the *Naturaliste* was ordered to take Bernier[21], the astronomer, and Boullanger[22], the geographer, to examine Cape Naturaliste and determine its exact position by taking sights from the shore. Another small boat under the command of Henri de Freycinet[23] was ordered to take the head gardener, Riedlé[24], and the mineralogist, Depuch[25], and make a landing in one of the sheltered coves so the scientists could get an idea of the country and test the soils.

Picquet's mission was a failure. He left the ship at 10 a.m. in fine weather, but by the time he reached the Cape the weather had become rough and cloudy and he had to return without landing. If he had set out earlier he might have been more successful. Winds and waves tend to increase in intensity as afternoon approaches in that area. But where he went, at the tip of Cape Naturaliste, in any case is a difficult place to make a landing in a small boat because of the surge from the large swell coming from the west, which continue around the Cape.

Freycinet was more successful, although he also had difficulty in landing because of the same surge which similarly affects the waters in the northern coves close to Cape Naturaliste. Nevertheless he got his party ashore. It is not clear from the log where this party actually landed. The ships' anchorage is not clearly known. Judging from the evidence they were lying almost north of Rocky Point, between Bunker Bay and Eagle Bay which are two sandy coves that can be clearly seen from a distance out to sea there. Behind Eagle Bay the hills are cut by a watercourse which can also be seen a way out to sea. It seems certain, therefore, that it was at Eagle Bay that Freycinet landed, and that it was along the dry bed of Eagle Bay Creek that the scientists walked to seek information about supplies and the land.

In this sense Freycinet's mission failed. There was no water in the creek. Eagle Bay Creek flows only after heavy winter rains. Baudin arrived before these fell. This was therefore a big disappointment. It meant also there was little opportunity to find game. Indeed the scientists found little evidence of life. There seemed to be no traces at all of human beings. They found what seemed to be the marks of "cloven hoofs" indicating the presence of small quadrupeds, and found animal droppings like horse dung. But their one prize was a small lizard. Even the birds seemed to be few, adding to the impression that the land there was "sterile", and all about was silent.

The general disappointment of those who went ashore was offset to some degree by Depuch finding "stratified granite". This was quite a momentous find. Just before Baudin's expedition left Le Havre there was a controversy in Europe about "granite". An eminent Swiss naturalist, Horace Benedict de Saussure[26] as a result of researches in the alps there, put forward the theory that "granite" could exist in stratified form. This was not generally accepted. Depuch's find in the Eagle Bay region of Geographe Bay provided evidence to support Saussure's theory. This alone made the expedition ashore scientifically worthwhile.

But for the bulk of the scientists — those who remained aboard the ships — there was only continued disappointment and envy. Baudin decided not to stay and let the others land although there were mysteries intriguing for the zoologists and others. There was the mysterious "quadruped" to be identified, and the natural history of the region, which was briefly observed by the gardener and the minerologist in their very brief stay ashore, to be properly surveyed and described. Despite this Baudin decided to weigh anchor. He was held up by Picquet which upset Baudin. Like Freycinet, Picquet had been given strict orders to return before dark on 31 May. But because of the weather he had to spend the night at sea. Immediately he returned the next morning, to be reprimanded by the impatient Baudin, anchor was raised and the ships beat about the bay, making eastward. In so doing, in sympathy with the reprimanded officer Picquet, a small cape near Dunsborough was named after him by the crew. Anchor was again dropped after dark that evening, about three miles to sea, a short way east of the previous anchorage. As in the case of the other anchorage, no definite location exists in the log books. The chronometer gave varied readings and Baudin at that stage gave up observations. But from the evidence in the reports, and judging by the compass bearings made in the survey boat, it is fairly certain that the vessels anchored about 3 miles off the coast just east of north of the present town of Dunsborough.

None of the scientists went ashore from this anchorage, although ravines and small water courses were seen, and fires were observed burning in the woods behind the beach, inviting investigation. Only one long boat was launched. This was placed under the command of Midshipman Bonnefoi.[27] He moved shorewards towards the south west corner of Geographe Bay, into the fine sheltered cove at Dunsborough. But he did not land, he merely made an hydrographic survey, noting that the water was shallow, that there were turtles and that there seemed to be no fresh water. He then returned to the ship as ordered.

In actual fact there are a number of small streams in the region, which is a well-watered area that has now become a rich grazing land. The trouble is these streams are difficult to detect from the sea. They normally do not flow down into the ocean, as streams do in Europe. In Geographe Bay they tend to end at the vegetation line and then disappear into the sand dunes that make up the beach. This makes it difficult to see the mouths of creeks from the sea.

The scientists on board managed to get a close look at the land. Baudin weighed anchor after Bonnefoi left and followed the long boat in towards the cove near Dunsborough, anchoring there for the night.

The next morning, 3 June, anchor was again weighed and the ships again ran out to sea although more fires were observed, and despite the fact Bonnefoi had reported the existence of a good safe well-protected small boat landing place. That area in fact is one of the safest places for landing for small boats in Geographe Bay. It is well-protected from the swells coming in from the Indian Ocean. The gradually shallowing waters there soon reduce wave action.

The ships anchored that night for the fourth time, well out from shore. The position of the anchorage was reckoned to be two leagues — approximately 7 miles — from the coast. The position recorded for this was 33° 26′ 44″ south and various longitudes ranging from 112° 8′ 29″ east of Paris (114° 28′ 29″ east of Greenwich) to 112° 29′ 4″ east of Paris (114° 49′ 4″ east of Greenwich). Correction of these, correlated with information gleaned from bearings taken in the survey boats indicates that the vessels were anchored a little to the east of north of Wonnerup Inlet, and almost to the west from Minninup, that is approximately 33° 26′ 49″ south and 115° 25′ or 115° 26′ east of Greenwich.

It is clear Baudin did not intend to stay at this anchorage any longer than at the others. But he was forced by events to stay, and it was from this fourth anchorage that the scientists went ashore *en masse* and made their much delayed and detailed study of the Geographe Bay region, which they had been waiting impatiently to do.

Relations between Baudin, the scientists and many of his officers had worsened since the arrival in Geographe Bay. He set strict time limits for survey trips made in the small boats, and severely reprimanded those who did not keep to the schedules he set. He obviously displeased the scientists by continually keeping off the coast, and continually keeping them aboard although the weather was good. He also seems to have disturbed Hamelin. There is an interesting little note in Hamelin's diary, where he says he did not go to the scientific lectures Baudin arranged on board the ships as he did not "play at being a savant". Baudin did. He endeavoured to make a scientific contribution, and used his position of authority to express his theories which often irked the professional dedicated scientists such as Péron who had little time for such amateurs.

Baudin's main trouble was that he again miscalculated. The scientists wanted to complete the task they were sent to do. The weather was good when the ships first came into the bay, and remained good for some time although it was winter and the time for dangerous seasonal gales. Baudin should have taken the opportunity to let the scientists go ashore while the weather remained fine, and the barometer steady. Instead he left this until his fourth anchorage on 4 June. By then the wind had shifted to the north east which is a bad sign in those waters at that time of the year. He was soon caught in a severe northerly gale which brought death, scientific loss and near disaster.

Survey of the eastern part of Geographe Bay

A number of separate expeditions, designed to make brief surveys, were sent out from both ships on 4 June as the wind set in. Midshipman Bougainville was despatched by Baudin from the *Géographe* to examine the bay and the coast to the north east. He followed the coast almost reaching Casuarina Point, the high land at the east of the bay which seems to form the end of Geographe Bay in that direction. But the rising north winds held him up and he had to turn back in order to keep to Baudin's schedule, thus missing discovering Koombana Bay, which was left to be found two years later, when the expedition returned.

Bougainville did not land anywhere along the coast to examine the hinterland not only because of the pressure of time imposed by Baudin, but also because that is not an inviting part of Geographe Bay for small boats to land at that time of the year. If he had done so and climbed the high dunes he would have seen the wetlands and the wooded area in the vicinity of the Capel River which is now a rich agricultural, dairying and wine producing district with attractive water bird life. A close examination of the area beyond the sand dunes might have corrected Baudin's dismal impression.

It was unfortunate for the French expedition and for French colonial planners that Bougainville did not do this. That region has good supplies of fresh water. If he and the others had discovered this, then Geographe Bay might have become of greater interest to the French. Instead they turned later to concentrate on the Albany region where fresh water was found.

Two other sorties were made from the *Géographe.* Midshipman Breton[28] was sent away in a small boat to take the impatient zoologist Péron and the botanist Leschenault ashore. They had been left on the ship until then. Péron in particular was keen to get ashore. He had been charged with making studies on anthropology and wanted to contact the natives whose fires had been seen from the ships. His mission was unsuccessful. His eagerness and disappointments are reflected in his reminiscences. "As soon as the boat landed us," he records, "I ran towards the interior in search of the natives with whom I had a strong desire to be acquainted. In vain I explored the forests, following the print of their footsteps . . . all my endeavours were useless and after three hours fatiguing walk I returned to the sea shore where I found my companions waiting for me, and rather alarmed by my absence."[29]

Breton returned to the ship arriving there with his party at 6 p.m. There is no indication where he had landed the group. But his hard trip back against the currents indicates that he travelled somewhere to the leeward of the *Géographe,* that is he landed a few miles south of the Capel River, between there and the dead-water at Wonnerup Estuary.

Baudin himself took a party of scientists, made up of Bernier the astronomer, Riedlé the head gardener, Depuch the mineralogist and Maugé the zoologist. Bernier determined the location of their landing place as 33° 30′ south but provides no longitude. This latitude cuts through the east coast of Geographe Bay just south of Capel River. This and other evidence, in particular the references to the big deposits of shells which still exist there, indicates the party landed just south of the reefs which extend along the coast at the concealed mouth of the Capel River.

There the party had more success than Péron and Bougainville. As the party landed they saw a native standing in the water spearing fish. He went into the sand dunes when the party left the boat. Depuch went after him, offering him beads and other presents. But then Baudin records "he began to shout violently, waving at the ships". He then picked up three types of spears and ran off. But they saw enough to make a description of him. He was described as very black and naked from head to foot except for a skin or a piece of bark on his back. He seemed old, with a grey beard, and carried, besides his three spears, a fire brand which Baudin concluded was for cooking the fish as soon as it was caught.

The expedition was ashore only for three hours. In that time Riedlé collected some specimens, and the prints of quadrupeds were seen, and excrement like that of a dog was found which raised interest. Baudin examined among other things the shore line, and noted the tides and wave actions. He made a disturbing find. He found a line of seaweed 15 feet higher than the rest and concluded correctly that this was the level to which the water rose in storms. This was a frightening observation. Geographe Bay is a huge natural trap for the northerly gales which blow in the winter, and at that time is a very dangerous anchorage. Ships anchored there in winter are well-advised to move into Bunbury or sail around to Flinders Bay, behind Cape Leeuwin as soon as the north wind sets in, if they want a safe place. Already the wind had shifted towards the north, which is the first sign of an impending gale. Baudin was therefore quite right in being cautious and in being impatient to leave.

Before they rowed away from the shore Riedlé planted maize, apple and pear seeds and apricot and peach stones and a variety of vegetable seeds on the plain between the coast and the forest to supply "additional food to the natives whose existence seemed very miserable". But the planting also had a scientific and economic motive. Baudin had been requested to test the soils, and plant European products with the idea that this would help visiting ships get supplies. D'Entrecasteaux had tried to do the same. Before they headed back for the ship the party also left a bottle, some biscuits, a necklace, a nail and some other trifles as presents for the natives.

Baudin's party brought back specimens. Of particular interest to them was the "wild celery". This was the only fresh food they found and was used to make a fresh vegetable soup for the men. Apart from this the land seemed to offer little in the way of supplies.

Discovery of Wonnerup Estuary

In the meantime several boats departed from the *Naturaliste,* to make surveys further to the south and west. One of these, under the command of Sub Lieutenant Heirisson,[30] landed some six miles to the leeward of the ships near Wonnerup Inlet and on walking over the dunes found a large expanse of water like a river or lake. His party followed this for some miles and found no end to it. Heirisson was most impressed by what he saw. Among other things it was rich with birdlife.

Keeping to orders, Heirisson returned to his ship before dark. At dinner time that evening Hamelin went to visit Baudin and reported the find. Baudin was by then all set to sail early the next morning, but on being informed of the find and on being pressed, decided to stay and have a more detailed survey made of the river. There was no reason not to, the weather still appeared to be fine with the wind blowing from the land, coming from the north east and not from the sea.

Baudin therefore prepared his long boat to go ashore, and placed it under the orders of his second-in-command, Commander Le Bas,[31] with Sub Lieutenant Ronsard[32] appointed to assist him. But the main shore party was led by Hamelin himself. This group left at 3 a.m. on 5 June, guided by Heirisson. Weatherwise they were already in trouble before they left. The wind by then had shifted more to the north and had freshened. From the debris Baudin had seen on the beach, that year's winter gales had already started and another one was approaching, and the small boats had gone ashore southwards, to the leeward. This would make it difficult for them to get back to the ship, if they could ever manage to get off the shore which is subject to a big surf in north storms.

Le Bas left the *Géographe* at 3 a.m. on 5 June and made his way to the nearby anchored *Naturaliste,* with his boatload of scientists — all but the astronomer Bernier and the zoologist Maugé[33] went with Le

HEIRISSON'S MAP OF WONNERUP INLET AND ESTUARY

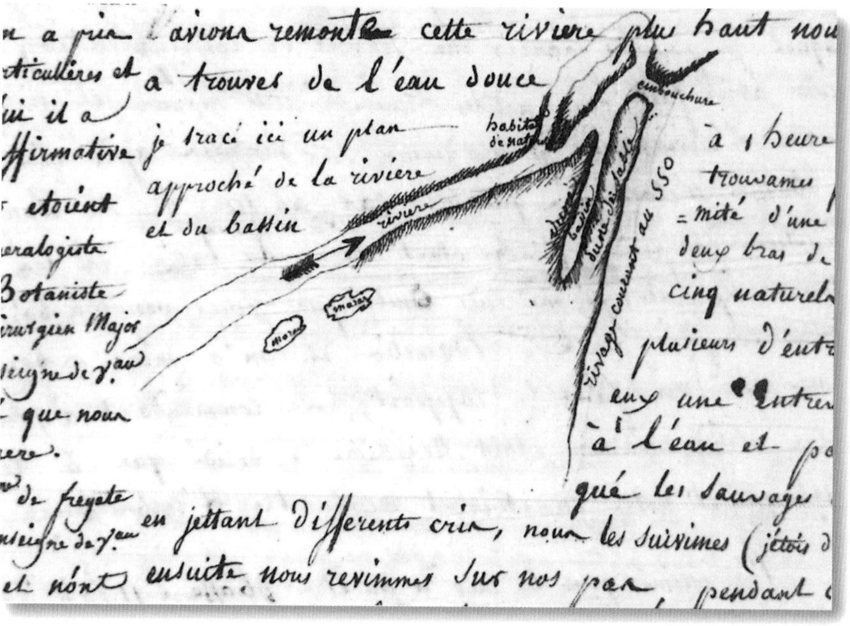

Bas. There he was joined by the two boats from the *Naturaliste,* one under Hamelin who had decided to join the expedition, and one under St. Cricq.[34] Running before the wind the boats made shore

quickly, at a position nearly due south of the anchored ships which was not far east of Wonnerup Inlet. Back on the *Géographe* after the boats departed, Baudin noted with concern that the wind was rising, and became concerned about whether or not they would be able to return early that night as ordered. Later in the afternoon the wind shifted to the north and the seas rose, giving those left on board an uncomfortable time which added to their anxiety.

Those in the shore party were less anxious. Once they set their feet on the ground they seemed to forget the wind and waves, and thought only of making a scientific expedition into the interior of the country, which would mark the first occasion Europeans had visited the region. An entrance to the river was found along the coast south of the anchored ships. This was Wonnerup Inlet which provides access to Wonnerup Estuary, the Vasse River, the Sabina River, the Abba River and the Ludlow River and to the Deadwater, which all have a common sea entrance. The river bar was blocked off by a native fish trap made of stakes, but room was found between these for the two boats from the *Naturaliste* to get over the shallows into the deep water that forms the inlet. Baudin's long boat had too deep a draft to get across the bar which was then not scoured out by winter rains. This boat therefore was anchored beyond the surf line in the care of some sailors. The remainder of the party separated into two groups. Captain Hamelin took one group to row as far along the river as he could. Another pary of scientists, consisting of Péron, Depuch, Leschenault, Riedlé, Lasueur[35] and the ship's doctor, Laridon, went to walk along the river bank into the interior.

The latter group immediately began to collect specimens, showing great excitement at finding "extraordinary animals that had only previously been found in a fossilized state".

Investigation of Aboriginal life

Péron, the "anthropologist", after a short while decided to cross the river and go further into the interior to look for natives. The others refused to go, and tried to prevent him. But Péron must have been a very brave man and keen scientist. At eleven in the morning he removed his clothes, plunged into the river and went into the forest. This was across the Deadwater near the present Ludlow River. Inland, where he was, there was no wind and the weather was fine. It is not possible to say from the journal exactly where Péron walked. From the evidence he seems to have headed into the tuart forest inland beyond the Deadwater, the arm of water extending for some way north of the inlet. Péron saw no Aborigines, but came across interesting evidence of occupation. He left a long report on his finds. Briefly, he believed what he found to be a ceremonial ground in a

MAP OF GEOGRAPHE BAY AND THE VASSE RIVER WITH A PLAN OF A SUPPOSED NATIVE SACRED SITE ON WONNERUP ESTUARY

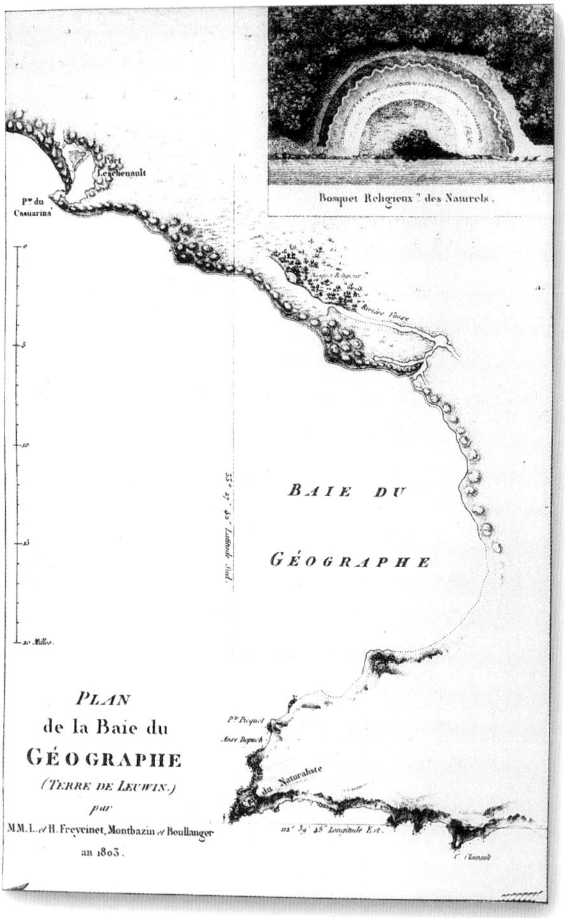

Wonnerup Inlet and Estuary, discovered by Herrisson and explored by Hamelin and others are shown on the right-hand side of the bay. The entrance to Wonnerup Inlet is shown to be closed by a bar. Inside, lying parallel with the coast, is the stretch of water now called Deadwater. Beyond this is a large stretch of water also lying parallel to the coast. This was believed to be a river and was given the name Vasse River in honour of the sailor who was drowned in the surf near the inlet. The French-named Vasse River is now called Wonnerup Estuary into which the Ludlow River flows. The present Vasse River flows into the Vasse Estuary, the mouth of which is shown on the left bank of Wonnerup Inlet near the bar. The French did not explore the present named Vasse River.

The "religious grove" or supposed sacred site was found by François Péron who believed from this that the natives of western Australia had a high civilization like the Egyptians. He found the site marked out with mathematical precision. His sketch of what he found is a birdseye view of what appears to be an amphitheatre.

Further along the coast, the bay which bears the name Anse Depuch (Depuch Cove) is where the French found granite which resolved a scientific argument then going on in Europe. The bay is now called Eagle Bay.

thicket of large trees. In the thicket were twelve white (paper-bark) trees formed in a series of semi-circles. Inside the semi-circles, rushes appeared to have been cut and placed in geometric patterns consisting of polygons, parallelograms and the like. Péron decided it was a place of worship, and in consequence concluded that he had discovered a new race of Egyptians who blessed the stream with their geometric figures. He also saw in tracings in the sand, the evidence of Runic figures and Mexican hieroglyphs which indicated the presence of a high civilization.[36]

It is unfortunate for anthropology and our knowledge of the Aborigines that Péron did not meet the Aborigines and observe them in their habitat during his walk. Not much can be deduced from his word pictures. If Lesueur or another artist had accompanied him and made sketches for posterity to examine, then more would be known. As it is, what Péron wrote is imaginative. He seems to have used a lot of his acquired knowledge to interpret what he saw in the paper-bark tree grove on the banks of the creek he had visited. In his early years he was religiously trained. In his Christian education he was exposed to lessons on man and his origins, which had come to be highlighted in church schools in France as a result of the finds of the encyclopedists and the attack by scientists on the Christian theory of origins of the human species. Péron, like Christians at the time, was a universalist. He believed there was one type of man, and consequently thought in terms of correlative links between men. He thus readily compared the Aborigines with the Mexicans and others he read about after he rejected his religious upbringing and became a revolutionary. Other more scientific revolutionaries who later visited Australia rejected the concept of universalism, and endeavoured to look at the Aborigines solely in the context of their own habitat, rejecting beliefs and pre-conceptions, thus leaving more scientifically valuable accounts.

After making this find Péron made his way back to the coast, observing that there seemed to be nothing much but trees and vegetation on the way and that these gave no fruits, which, he concluded, explained why there was so little game. In fact he was wrong, there is a lot of game but it is nocturnal. When he arrived on the beach he met Lesueur and Ronsard, and together they went to find the long boat. Péron was pleased at suddenly meeting the others. He feared the boat might have gone back without him.

But other bits of news was not so pleasing. Firstly Lesueur and Ronsard told him that they had actually met an Aboriginal woman. These two had gone back to the beach after a walk inland. There they met the second boat from the *Naturaliste,* under the command of St. Cricq who was about to return to his ship. While they were on the beach they saw two people approaching them. At first they thought they were sailors. But when they got close they saw they were an Aboriginal couple. The French gave chase. The Aboriginal man ran and disappeared over the dunes into the bush. The woman, who was found to be pregnant, could not move as fast and was caught up with. She then stopped, "sat on her heels and hid her face with her hands, remaining as one stupefied, perfectly without motion and insensible to all that went on around her".

The picture they presented of the Aboriginal they observed contrasts markedly with the impressions of Péron. They gave a most unenthusiastic report. Far from being representative of a high civilization, "the wretched woman was entirely naked", they recounted. All she wore was a small bag made of kangaroo skin, tied round her with strings of rushes. This contained a few orchideous bulbs, which they concluded formed a good part of the native diet. Her personal appearance did not impress the men. In colour of skin and hair, and in proportion of the body, she resembled Aborigines seen in other parts of Australia. She was viewed as "horribly ugly and disgusting". She was reported to be uncommonly lean and scraggy, and her breasts hung down almost to her thighs. Her extreme dirtiness, it was reported, added to her natural deformity, and "was enough to disgust even the most depraved among the sailors". To make matters worse, when she was caught, apparently out of fear, she evacuated her bowels, according to the French "like a frightened animal".[37]

Lesueur also came across and disturbed an Aboriginal settlement. This also did not impress him. "The settlement consisted of several huts built on the humid banks of the salt marsh. They were roughly constructed of slender branches of trees stuck in the ground and fastened together at the points, somewhat like an arbour, and covered on the outside with the useful sort of bark which I have before noticed. They are about three feet in height, three feet in breadth and about six feet long."

What particularly struck the observers was that the natives who apparently ate fish, had no canoes.[38] The scientists in D'Entrecasteaux's mission also specially mentioned this point, noting that unlike the Tasmanian natives, the Aborigines in western Australia did not seem to have discovered boats and floating hulls.

There were thus two views of the Aborigines given by the French explorers. One is a roseate one by Péron the anthropologist, and the other not very attractive. The latter impression did not have a lasting effect. After all, the observers had seen only one Aboriginal at close quarters, and nothing conclusive could be drawn from this by scientists. Consequently in later expeditions a more concerted effort was made to research into Aboriginal life and customs.

Wreck of the long boat

If the first piece of news told to Péron made him envious, the second piece of news made him disturbed. While the scientists were in the interior, Baudin's long boat was blown onto the beach. There it had filled with sand, and could not be moved and was threatened with destruction by the pounding surf. Some stores and rifles were rescued from it and preparations were made to camp on the lee side of the dunes until help arrived.

Exploration of Wonnerup Estuary

In the meantime Hamelin was inland following along the river by boat while other scientists walked along the bank. It is not clear exactly where he rowed. No clear directions are given in the sketch maps left, which was also the case when Heirisson mapped the Swan River. It seems he went straight inland from Wonnerup Inlet. There one branch of the Vasse goes suddenly south. But this is a difficult entrance to see. It bends towards the coast on flat ground and does not have the appearance of a river. Hamelin seems to have taken the more inland branch which goes to the Ludlow River. On the way there is another branch of the Vasse Estuary. It seems that Hamelin rowed along this for some way. On the way back they sighted some Aborigines on the shore. This was not far inland, probably the piece of land on Wonnerup Inlet, north of the old tramway line, on the south of the main road that goes from Wonnerup Inlet to Bussell Highway, on a piece of land registered as Crown Grant 4 which is now a farm.

The scientists who had been walking along the river bank were by then in the boat. Some of these and Freycinet landed when they saw the Aborigines, intending to meet them. However the Aborigines disappeared, but they were heard calling in the forest. As on other occasions the French left presents of beads, looking glasses, knives and other trifles to show their friendly intentions. While they were doing this the Aborigines moved between the shore party and the river, cutting them off from the boats, and adopted a seemingly threatening attitude waving spears. Not wishing to be cut off, the French opened fire and retreated amidst "savage howls" from the Aborigines.

Continually "threatened by spears and clubs" the French party moved down into the river and were picked up by Hamelin who came to their rescue while offering them cover.

Hamelin had no desire to establish bad relations. In fact the expedition had been ordered by the government to make only friendly approaches. To do this Hamelin's men called out "friend" in

Polynesian, and spoke some Malabar and some clearly-pronounced French words. They were answered by the Aborigines. The French tried to record some of the shouts they heard, but these were not very clear.

Exploration party marooned

The party eventually returned to the beach, convinced that what they had discovered was in fact a big swamp and not a river. A number of bird and other specimens of animal life in the region were collected. When they got back to the coast they also found the long boat firmly grounded and impossible to launch. As the sea was by then bad, the whole party and their specimens were therefore marooned. To make matters worse, they had expected to stay only a short while and they had only a few supplies. To get a little more comfort they made a shelter from the sails behind the sand dunes, lit a fire and cooked some of the birds they caught as specimens, making a stew out of these with "brackish water", a little rice and some vegetable matter which they found on the bank of the "swamp". Thus fortified, they then spent an uncomfortable night in the wind and rain, with guards posted in case of attack by the natives.

Out on the *Géographe* the worsening weather gave the already worried Baudin good reason to fear for the safety of his ships and men. As night fell the badly pitching ships swung north with the wind shift which created a dangerous lee shore, and made a run out of the bay difficult. Baudin had not thoroughly surveyed the waters around Cape Naturaliste. He realised there was a dangerous reef north of this, which was seen on the way in. This meant the ships would have to make well north of the area to clear it. With the current and wind now set from the north, making past that danger area would be difficult. But Baudin was a very good proven seaman. His knowledge of the sea and its moods and dangers gave him a respect for it which the scientists on the expedition regarded as excessive caution. This time Baudin's judgement was sound. He rightly concluded he would be better off in the deep waters out in the open sea. However, there was no chance to leave. The scientists and boatmen were still ashore and he could not desert them. They had few provisions and from all reports the coast they were on seemed waterless and sterile.

Baudin's concern

A little before sunset on 5 June, Baudin caught a glimpse of the two small boats off the coast beating into the wind. But there was no sign of the long boat which "was always easily distinguishable because of its sails". Baudin sent up a light to the masthead and spent an uncomfortable and fruitless night waiting for news. "Nothing happened at all" he recorded. All that was seen was a light which appeared on the mast of the nearby *Naturaliste* at 2 a.m., which Baudin took to mean that its boats had returned. This was correct, one of the small boats from the *Naturaliste* returned in the night. This was observed from the *Géographe* at daylight on 6 June, but by then the wind was too strong and the seas too rough for the ships to communicate. So nothing about the bad situation on the beach was known on the *Géographe*. The only good news was that the anchors of the big ships were holding, although only 70 fathoms of cable were out. However, the pitching and rolling had worsened. Consequently the topmasts were lowered in case they split. This was a necessary precaution at that time. Although rigging ropes then were shroud laid, they were made of material that stretched and gave and caused topmasts to shift and split. This bad rigging system was not rectified until shroud laid wire rope was manufactured and used for rigging later in the century. The main point in regard to the rigging of the *Géographe* and *Naturaliste* is that the decision to lower the topmasts meant they would not be able to raise sail and leave.

Baudin's nervousness increased as the day wore on. In his journal he commenced to reprimand the shore party for not returning the day before. He considered that they had had plenty of time to see if there was or was not a river mouth and come back, thus revealing that he did not really understand the minds and purposes of the scientists he had let go ashore.

In the afternoon of 6 June the wind moderated, and the small boat which had returned to the *Naturaliste* the night before came over to the *Géographe*. Bailly the mineralogist was in the boat. He made an immediate report to Baudin, but this was about Geographe Bay and not the situation of the shore party. He had left the beach in the small boat under St. Cricq before the long boat was wrecked. He informed Baudin that the stretch of water Heirisson had discovered was not a river but a swamp or lagoon, which was disappointing, and he gave an unenthusiastic report about the Aboriginal woman who had been seen at close quarters, and reported that from what he had seen of the place it seemed of so little importance and interest that he would be better off returning to the ship. This is why he returned.

Baudin from this assumed that his officers had been beguiled by the scientists to stay, and waxed wrath in his journal, and vowed "they shall not go ashore again, unless there is absolutely no risk".

Leschenault and others favourably impressed by the countryside near Wonnerup

Other members of the party who stayed on to survey the estuary, had a different impression. Leschenault in particular was impressed. He wrote about the area at great length in his journal. He recorded that he looked with admiration on the flat country covered with trees which formed a magnificent forest, and appeared fertile with black soil formed of vegetable matter.

His botanical survey indicated that there was a wide variety of trees and plants. A great part of the journal, however, is taken up with a description of the Aborigines the party encountered inland, and the unsuccessful attempts the party made to establish friendly relations with them on the river trip. Leschenault has left in particular a most valuable and detailed report on Aboriginal artefacts, weapons and houses which he saw, and on their environment.

Like others ashore at the time, he was not impressed by the Aborigines. He wrote in his journal that the Aborigines in Geographe Bay were the people the furthest from civilization that he had seen. They lacked instruction, had a fear like ferocious beasts, and had no clothes or canoes like other people.

There was still no sign of the men or boats on the beach from the ships, although the visibility from the ships was good. The scientists in fact, at the time, were still behind the dunes which provided some shelter from the wind and the blown sand and spume which speeds across the beach from the surf line there when a gale blows.

After sunset lights were again hoisted to the mast head, and a cannon was fired each two hours as an order for the parties to return. "This should ensure that they will be back aboard by ten o'clock the next morning at the latest" recorded Baudin.

At 9.30 that night Hamelin came bearing the bad news about the shore party. He had left the shore after returning from his river trip, in an endeavour to get back to the ships to get help for the stranded party. The trip must have been a nasty one. It took Hamelin and his four rowers twenty-two hours to make the six miles from the shore to the anchored vessels, toiling all the way in the heavy seas. Baudin was appalled at the news, "It was a terrible blow to me", he recorded. He feared in particular, as a result of Hamelin's report that the men would be attacked by hostile savages, while relatively defenceless. He was not over-worried about their food and water supplies. He believed the party had enough supplies for a further two days, although in fact much of these were lost or spoilt and the stranded men were already famished. What also disturbed him was that his survey boat needed for inshore work seemed to have been lost.

Operations to rescue the marooned scientists commenced

The wind dropped in the night, and Baudin had two long boats prepared, one from each ship, to go to the shore to see what could be done to help. At 4 a.m. on 7 June, when the boats were to leave, the storm again rose. A change of plan therefore was made. Hamelin was ordered to sail inshore close to Wonnerup Inlet, and make the rescue from there. Unfortunately Hamelin's ship suffered some damage in the storm and the anchor was fouled. He had to delay his departure. This was unfortunate. Manpower was needed ashore if the survey boat and castaways were to be rescued that day.

Baudin hoisted his anchors and sails and made shoreward alone in a fine display of seamanship. But unfortunately for his public relations he had already written a reprimanding letter to Le Bas about losing his long boat which he entered in the official ship's log.

Baudin anchored off Wonnerup Inlet at midday, in rough conditions, in shallow water in 7 fathoms. He immediately sent off his largest dinghy under the command of Midshipman Bonnefoi who was ordered to search near the river mouth for the missing long boat and for the scientists, none of which could be seen from the anchored ship. Bonnefoi was given strict orders not to set foot ashore. Baudin had had enough of shore parties, and had no desire to lose another survey boat. He saw Bonnefoi anchor off the coast, saw no one come to the beach, and went down to his cabin, "not being able to hold back his tears".

Fortunately Baudin had dropped anchor just near the wrecked longboat. Unfortunately for posterity he took no bearings and its location can only be surmised. Parts of it, and the thirty muskets and other items must still be under the sand on the beach near Wonnerup Inlet.

From the ship Baudin saw what he believed was a line sent ashore to pull off the longboat and observed men on the beach. But he found out little until Bonnefoi got back at 9 o'clock that evening. He brought Péron back with him. It was Péron and not the longboat who had been pulled out from the beach through the surf to the boat, "nearly drowning the scientist in the process". He came aboard the *Géographe* absolutely exhausted and indicated the plight of the famished men on the beach. Bonnefoi's report gave Baudin little reason for hope for the longboat. It seemed to him to be beyond salvage. It was deeply embedded in the sand and was being pounded by waves.

Hamelin brought the *Naturaliste* in to anchor near the *Géographe* by mid-afternoon. He then joined in the preparations being made to effect a rescue. The major effort to do this was made on the morning of 8 June. By then the wind had shifted back to the north east, improving the weather situation, although the seas were still rough and the surf on the beach bad.

Two boats were sent off that day, one from each ship. The one from the *Géographe* was under the command of Midshipman Bougainville and contained the chief carpenter who was sent to assess the situation and supervise the salvage. Bougainville took with him a letter from Baudin to Commander Le Bas, written in the most intemperate terms accusing him of wilful disobedience, and telling him that the master-carpenter was now in charge.

Le Bas reprimanded for losing the longboat

Baudin was most apprehensive about the rescue. From the ship he could see the big surf and undertow on the beach and was now concerned about the men being able to reach the ships' boats. However they made it with difficulty, although the scientists lost their valuable collection of specimens. Riedlé, the chief gardener, in particular suffered a loss reporting that his valuable tin box was carried away in the undertow. By four o'clock all the scientists were safely on board, and Le Bas had apologized, which Baudin did not accept, recording in the log "I heard him without replying. His appearance was such that I did not then reprimand him as he deserved".

The loss of Vasse during the rescue operation

Baudin hoisted his boat aboard and prepared to sail out of the Bay which he later termed "that fated place". But his troubles were not yet over, two of the *Naturaliste's* boats were still near the shore helping with the rescue and attempting the salvage. All of the salvage equipment Baudin had sent had been landed and was on the beach. Baudin and Hamelin had no desire to leave this, but the weather once again deteriorated. Fires were lit on the ship and rockets sent up to recall the boats. These eventually arrived back at the *Géographe* between 8.30 and 9 at night. Commander Milius[39] who was in charge of the small dinghy brought more bad news. He told Baudin that one of the rowers, second class helmsman Vasse[40], had been swept overboard near the surf line and had been drowned in the undertow while trying to get the salvage equipment. The river near where he was reported to have drowned was consequently named after him.

There is a local myth in Australia that Vasse in fact did not drown, but reached the beach and lived with the Aborigines. This has been featured in a number of books and stories about the Busselton region and about western Australia. The origin of the story is not clear. It appeared quite early. It was brought to the notice of the French government shortly after the remnants of the expedition returned to France, and was investigated. The conclusion then reached was that Vasse in fact had drowned, and that stories to the contrary were the invention of sensationalist news makers. The stories also could have been circulated by anti-French activists in Britain who spread bad stories about French ships and conditions, implying that seamen would rather desert in wild places than sail in them. The story later ended up as a romantic tale about the mythical first white man to live with the Aborigines in south west Australia. From the evidence it seems myth. There is no clear reason to reject Commander Milius's observation, who was an eye-witness and averred that Vasse died in the wild surf and undertow when he went overboard.

Departure from Geographe Bay

Baudin sent the small boats off to their own ship, but immediately regretted it. The wind again blew up, endangering them. At the same time the *Naturaliste* was seen to drag its anchor. Baudin consequently gave orders to sail immediately. He set sail on his ship and close hauled he made north west heading for the open sea with the wind blowing from the north north east, slightly forward of his beam, with the *Naturaliste* following.

It was already dark when the ships left. They got under way at 10.30 at night. That was the last the vessels saw of each other until they met at Timor on 21 September. In the meantime, from June 9 until September 21, both ships made separate surveys as they sailed north along the western coast of Australia.

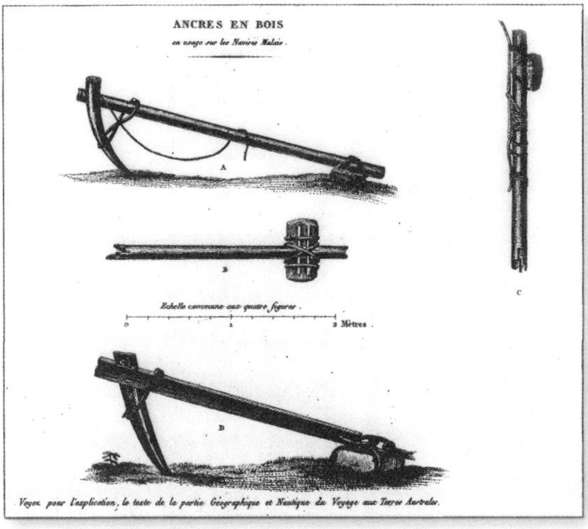

Sketch of wooden anchor used by Indonesian fishermen.
The Baudin expedition 1801.

Aborigines, King George Sound by Louis de Sainson 1826

CHAPTER 6

Baudin and Western Australia Section 2: Baudin and Hamelin's exploration of the Swan River and Shark Bay and Baudin and Freycinet's exploration of King George Sound, Bunbury and Shark Bay, 1801 and 1803

Separation of the *Géographe* and the *Naturaliste*

Baudin reached the open sea west of Cape Naturaliste by daybreak on 10 June, 1801. The early morning light revealed no trace of the *Naturaliste* which had been lost sight of during the night. Presuming that Hamelin had not managed to clear the cape and had stayed on in the bay, and with the weather still threatening, the still worried Baudin continued to drive out to find deep water. Once there, in safety, he set a comfortable course steering south with the wind on his quarter and the current astern, heading back the way he had come at the end of May from Cape Hamelin to Geographe Bay.

Baudin cannot be criticized for standing on and leaving Hamelin behind. This was a wise and seamanlike decision. The weather was foul and the expedition was in strange waters on a hostile coast[1] which had an unenviable reputation for shipwrecks. In the circumstances, Baudin's duty was to see his ship and the scientists and men in his charge reached safety. He did not make the mistake of seeking shelter, which has led many mariners into trouble. He sailed out to find sea room.

Baudin sails back to Cape Leeuwin

But it is difficult to justify Baudin's subsequent actions and commend his later decisions. When the storm dropped, he did not immediately return to see if Hamelin had successfully weathered the storm. Nor did he keep the rendezvous set earlier for the ships to meet at Rottnest Island in case of separation. Baudin instead went direct to the second rendezvous set for Shark Bay in case the ships missed each other at Rottnest. However, Baudin did not stay at the second rendezvous for long. He left there and went to Timor before Hamelin arrived. Baudin and Hamelin, therefore, did not meet again until 21 September when the *Naturaliste* reached Timor. Geographe Bay thus was the only place in western Australia where the expedition worked as one team. Hamelin and the *Naturaliste* did not return to western Australia on the second visit in 1803. He went back to France from Sydney.

The long separation of the two vessels on the western Australian coast did little to enhance Baudin's reputation and his popularity among members of the expedition. This was not because he had left Hamelin behind. He eventually looked into Geographe Bay on the way north and found the *Naturaliste* gone and therefore apparently safe. What caused ill feeling about Baudin, and envy of those who sailed with Hamelin, by the scientists couped up on the *Géographe*, was Hamelin's report of what he had done on the way to Timor. This revealed that while Baudin was beating about at sea, well off shore, Hamelin had made thorough surveys of the Swan River, Rottnest Island area and of Shark Bay. To the chagrin of the scientists this was done with few scientific resources. The bulk of the scientists were on the *Géographe* with Baudin. Hamelin in fact had so little in the way of scientific resources at his disposal that he used the ship's pharmacist to map the mouth of the Swan River. More significantly, Hamelin's reports revealed that he provided good opportunities for his scientists and men to make sustained and detailed surveys. He invariably anchored his ship for a considerable time in one place, shifting only when necessary and then not far. From this base he sent out small survey boats to make sustained surveys lasting for days. Baudin in contrast let his men go ashore seldom and for a short period only. He liked to have the men recalled and snugged down at night. He left his scientists ashore only in one place — Shark Bay — on the way to Timor although he anchored at other places in north western Australia.

Baudin's independent trip and survey therefore did not contribute much to the overall achievements made by the expedition up to August 1801 when he arrived at Timor, although the *Géographe* had the better resources.

Baudin's abortive attempt to solve the riddle of the location of Cape Leeuwin

When the wind dropped and the weather became fine after Baudin left Geographe Bay and headed south, Baudin found himself in the vicinity of his first landfall. He therefore made the inexplicable de-

BAUDIN EXPEDITION, 1801

ROUTE FOLLOWED BY THE GEOGRAPHE (BAUDIN) 1801

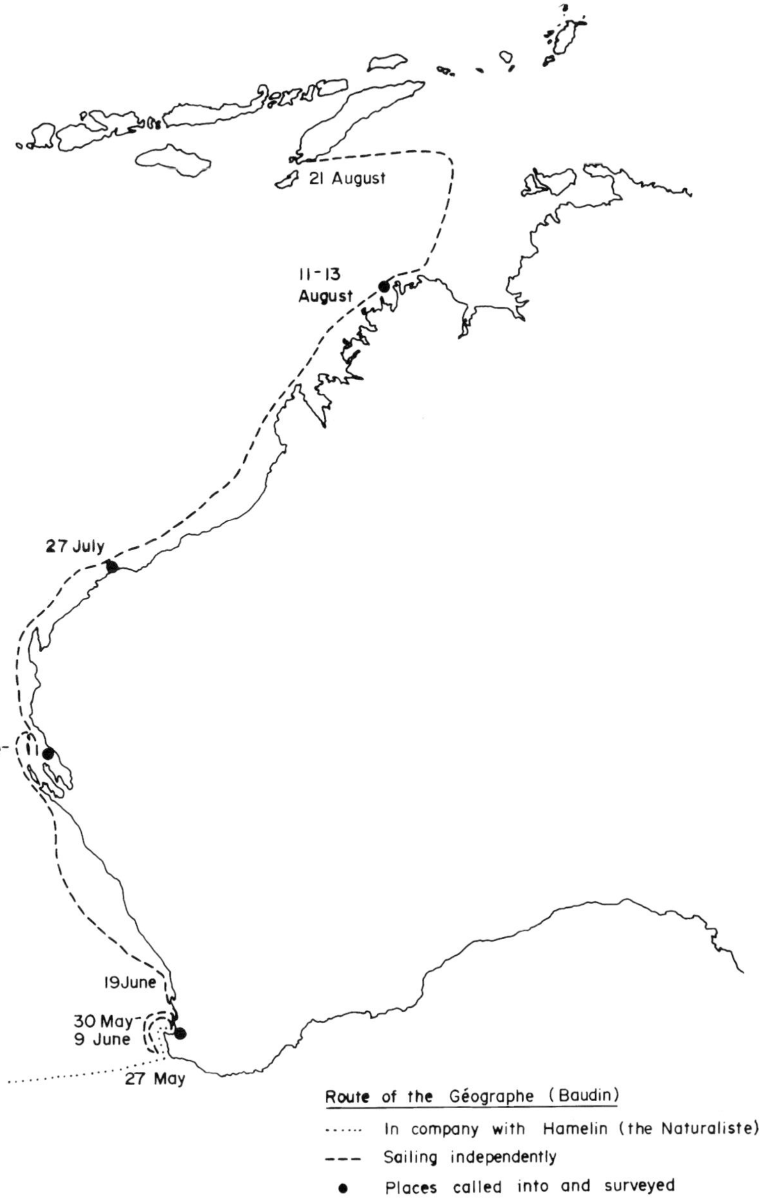

Location of anchorages and areas surveyed:

Baudin (Géographe):
Geographe Bay, 30 May–9 June 1801. Shark Bay, 26 June–11 July 1801. Dampier Archipelago 27 July 1801. Bonaparte Archipelago 11-13 August 1801.

cision to stay there for a while to try to solve the riddle of Cape Leeuwin. The exact location of this was still not determined. Baudin and Hamelin had had a difference of opinion about the location of it just after the first landfall was made.[2] Hamelin was of the opinion the more southerly cape which they saw when they made their landfall, which in fact is Cape Hamelin, was Cape Leeuwin. Baudin believed a cape more to the north, that is either the present named North Point or Freycinet Point was Cape Leeuwin. Both were wrong. But Baudin did little to resolve the problem at this time. He stood too far out westward from the coast and did not make far enough south to determine the correct form of the coast in that area. He consequently left the matter to be examined when he returned there after going to Tasmania as originally planned.[3]

Baudin returns to Geographe Bay

Baudin thus again turned north without achieving anything, and once more sailed along the coast between Cape Hamelin and Cape Nauraliste which had been examined the previous month. He re-entered Geographe Bay on 14 June, found no trace of the *Naturaliste*, and headed north, keeping well out from the coast so that only occasional glimpses were had and general impressions were made.[4]

On 15 June, the *Géographe* approached the latitude of the Swan River. From well out to sea the look-out saw a point which appears to have been Cape Peron, which Baudin concluded swept in to form the entrance to the Swan River to the north. But bad weather again set in and Baudin again turned his stern to the wind and steered south, back along the way he had come. When the wind eased he again turned north, keeping out from the coast. On 19 June, he sighted and sailed past Rottnest Island close enough to be seen by the men on the *Naturaliste* which was lying at anchor between Rottnest Island and Cottesloe Beach on the mainland, although he failed to see the *Naturaliste* at the rendezvous.

Baudin sails past the rendezvous point

Using Rottnest Island as a fixed departure point, Baudin set a seaward course to take his ship outside the infamous Abrolhos reefs and islets. He did not come near the land again until he was close to Shark Bay. In the meantime he endeavoured to keep his scientists occupied and happy by netting. This way he collected a lot of jelly fish and other specimens of sea life. But this did little to placate those aboard who were interested in the resources of the continent.

Arrival at Shark Bay

Land was sighted from the *Géographe* on 22 June not far north of the Murchison River approximately sixty nautical miles south of Dirk Hartog Island. The steep, dark, harsh looking cliffs, which Boulanger immediately called "the iron coast" because of its formidable appearance, offered no apparent landing place or shelter.[5] Baudin therefore continued north, coasting in sight of land. On 23 June, the south point of Dirk Hartog Island was passed. The land still appeared sterile and uninviting "like the lands of Sinai". Baudin did not take the Naturaliste Channel usually used by the north bound ships. He continued on past Dorre and Bernier Islands, which he called "the Barren Islands" and, no doubt frightened by the north winds he had experienced in Geographe Bay, he took an unusual course and entered Shark Bay from the north through what is now called Geographe Channel.

Survey of Bernier Island

He came in to anchor north of the islands on 26 June commenting that they seemed very sterile. This anchorage was not comfortable, but was good for the mind of a man who feared north winds. If one rose he could easily reach out to sea. However he did not stay at his first anchorage. He shifted the *Géographe* gradually south, seeing the weather was good, until he found a satisfactory anchorage near the north end of Bernier Island on 27 June. Two small boats were launched. One with Riedlé the head gardener, was sent to Bernier Island to see if a satisfactory landing could be made. The other took soundings around the ship. Riedlé returned at 2 a.m. the next day with a collection of plants and seeds not seen before. Baudin, however, was asleep and was not presented with a report until the following morning. Midshipman Bonnefoi who commanded the boat which took Riedlé ashore reported that there were two good sizeable coves suitable for landing. Baudin therefore immediately had the boats prepared for a further trip. Baudin commanded one and Sub Lieutenant Picquet the other. By then the differences between these two which occurred at Geographe Bay had been patched up, and Picquet was once more allowed to command a small boat.

The boats were loaded with equipment for hunting and fishing, but left so certain that they would catch supplies that no provisions except some biscuits and water were taken. Although the distance to be covered was short, the trip took two hours because, Baudin bitterly commented, not one of the rowers selected could row.

It was consequently nearly lunchtime before the boats were safely anchored in the cove at the north of Bernier Island. Unfortunately no fish were caught. The party therefore ate only biscuits and water and then dispersed to collect specimens and hunt until 5 o'clock which Baudin set as the deadline for return and dinner. The enthusiastic Péron alone did not keep to the schedule. He wandered until dark, making a collection of shells. He became lost, spent the night out under the stars, lost his collection of specimens and returned famished the following morning to find Baudin gone, but the small boat under the command of Picquet waiting for him.

Baudin and the others meanwhile had had a successful afternoon. They collected a large number of specimens and enough oysters, crabs, fish and crayfish to have "a delicious evening meal". When the party returned to the ship that night they found that the crew who remained on board had had even more success. Over 600 pounds of fish had been landed with handlines. Although this news was pleasing, Baudin was irritated. Péron appeared to be up to his old tricks. He was missing. Péron came back late in the morning on 29 June. Baudin in anger immediately recorded that never again would he be allowed to walk alone on shore. He determined in future to send someone with him to keep him constantly in sight. The main trouble was, as usual, Baudin had planned to weigh anchor and move off the night before. It was too late to do this when Péron came aboard, so he stayed at anchor.

Departure for Peron Peninsula

The next morning 30 June, anchor was weighed after Péron returned aboard and the *Géographe* headed eastward to try to find an anchorage near the mainland. A suitable place was found near the coast not far from the mouth of the Gascoyne River. A small boat was sent to make a brief survey inshore, but Midshipman Bougainville who was placed in command, was given strict orders not to land. He had recently incurred Baudin's wrath for his "lack of seriousness". According to Baudin he seemed to be interested only in hunting and fishing.

The mainland from the ship and from the survey boat looked unattractive, and so no more surveys were made after Bougainville's short sally. Baudin thus missed the opportunity of discovering the Gascoyne River. Instead he moved his ship south to examine "Dampier Inlet", that is Peron Peninsula , in the middle of the bay. The trip there was slow. This was mainly Baudin's fault. Baudin had read Dampier and believed there was too little water for him to go to the east of the Peron Peninsula. In fact it is quite deep once the banks to the north of the peninsular are rounded. Baudin consequently always sent a small boat ahead sounding. In the meantime a breeze from the north arose which frightened Baudin and he decided to stand off the island. He had had enough of northerly winds by that time. In fact he had no reason to fear. Shark Bay is in the tropics and does not get the northerly gales which dominate only in the temperate latitudes in the southern region of western Australia in the winter.

Baudin anchored well off Peron Peninsula on 2 July, but the boat which was sent ahead reported no good landing place existed. A more detailed survey made by the small boat the following day revealed shallow banks north of the peninsula which were impossible to clear. What Baudin should have done is to have gone further east and sailed down into Hopeless Reach towards Hamelin Pool. He kept too close to the north point and spent all his time tacking about the Cape Peron Flats seeing if he could find an anchorage there. On 4 July a swell came from the north. Fearing an impending gale, Baudin gave up the attempt to land on Peron Peninsula where he intended to set up an observatory, and made back towards Bernier Island. His scientists thus never landed on the mainland.

Return to Bernier Island

Baudin anchored off Bernier Island for the second time on 5 July and remained there until 11 July. This was an ideal place for setting up an observatory. Unlike the waters north of Peron Peninsula, those on the east coast of the north part of Bernier Island are deep and shoal free. This meant Baudin could get the *Géographe* close in to shore to have the equipment rowed ashore, and that communications between the ship and the camp ashore would be easy.

Observatory established

The observatory was established on the small beach between Wedge Point and Wedge Rock, which was calculated to be at 24° 47′30″ south and 109° 8′ 43″ east of Paris. The latitude Baudin assessed is correct. That parallel runs almost through the centre of the beach. But his longitude assessment is 1° 42′ 7″ in error, to the west.[6] The beach where the observatory was located is actually on 113° 10′ 50″east longitude from Greenwich (110° 50′ 50″ east of Paris).

The calculation and awareness of this degree of error is of great significance to all interested in Baudin's survey of western Australia. Bernier Island is the only place where Baudin's astronomers landed on the coast at that time, and took careful observations and recorded the determined latitude and longitude of their observatory. As the site of this observatory and its correct latitude and longitude is known it is possible to assess the degree of error in Baudin's calculations. By using this figure, allowing for a little degree of error to

BANDED KANGAROO, SHARK BAY

compensate for diurnal changes in Baudin's time pieces, it is possible to assess with some degree of accuracy the other places where Baudin anchored and surveyed such as at Geographe Bay where he failed to set up an observatory and left no precise records of anchorages and locations.

While the astronomers were busy making observations, the rest of the scientists and officers spent the time making a thorough survey of the island and examined Dorre Island to the south, although Baudin recorded that he was dissatisfied with the efforts of the officers who "appeared to be more interested in hunting for game".

Baudin departed from Shark Bay on 11 July after collecting specimens, and making a thorough survey of Bernier Island.[7]

Northwards survey to Timor

After he left Shark Bay Baudin did not let his scientists land again in western Australia. They remained on the ship until it got to Timor. Thus all Baudin achieved after leaving Geographe Bay is that he made his scientists the first world experts on the subject of Bernier Island, a small piece of land in the north part of Shark Bay. This

BONAPARTE ARCHIPELAGO NEAR CAPE VOLTAIRE:
BAUDIN EXPEDITION

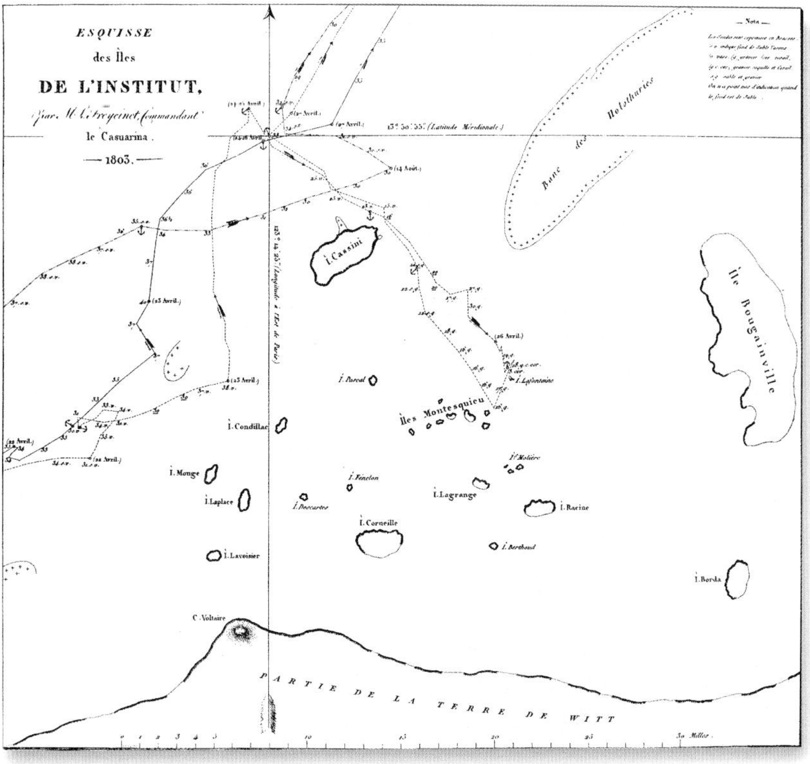

On the way to Timor he made close inshore surveys at selected places. He stood in to the North West Cape looking for the William River. This had been named by early Dutch navigators and its location remained a mystery since then. Baudin did not find the river, but he found an extensive opening and properly concluded that this was a deep bay and not a river mouth, although he did not investigate.[8]

He reserved his next more detailed survey for the islands of the Dampier Archipelago, near Roebourne. There on 27 July, he sent his naval engineer Ronsard off in a small boat to reconnoitre. But none of the scientists were allowed off the *Géographe*. However, Ronsard and his men collected specimens for them. Baudin then continued coasting north-east along the coast, dredging and making a cursory survey. The survey was ended on 19 August near Bonaparte Archipelago. Only four days wood then remained for the cook-house. Timor and fresh food and other supplies were near. Baudin consequently decided against going on towards Melville Island as planned

DAMPIER ARCHIPELAGO:
BAUDIN EXPEDITION

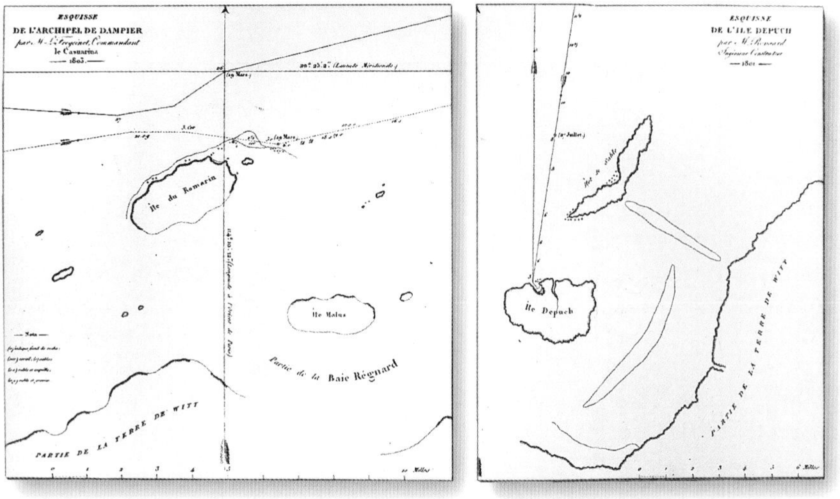

and headed north for Timor and refreshments. He anchored at Kupang on 21 August, 1801.

Hamelin departs from Geographe Bay for the rendezous at Rottnest Island.

On 9 June, 1801, Baudin had reason to feel apprehensive about the *Naturaliste* as he ran out to sea from Geographe Bay by himself. The trip from Europe had revealed that the bluff shaped *Naturaliste* sailed like the barge it was designed to be.[9] It could not be easily sailed into the wind and was slow and handled badly downwind. All the way to Australia it had consistently held up the progress of Baudin's better shaped, better performing, faster corvette.

Hamelin's success in holding the *Naturaliste* off the lee shore in storm bound Geographe Bay, and his eventually reaching the open sea is a tribute to his fine seamanship. It took three days for him to achieve this, tacking every two hours, under constant threat of destruction in the surf.

When he raised anchor, Hamelin took a calculated risk. He set a large sail area, beyond what the *Naturaliste* would normally carry in those weather conditions, in an endeavour to pull the ship's head off the coast and into the wind to wear the ship past Cape Naturaliste and Naturaliste reef. This placed an undue strain on the masts and rigging which was added to by the violent rolling of the ship which constantly threatened to snap the ropes or timbers. But Hamelin's gamble paid off. Taking advantage of lulls in the weather and wind shifts, Hamelin weathered the bay on 13 June and directly bore away for Rottnest Island to meet Baudin at the appointed rendezvous.[10]

Arrival at Rottnest Island

Hamelin arrived at Rottnest the following day, 14 June, but surprisingly for the crew of the *Naturaliste,* there was no sign of the faster corvette.[11] Hamelin determined to wait. He moved in towards the east of the island and dropped anchor in ten fathoms of water east of Rottnest Island. It is not possible to determine from his calculations the exact place he anchored. He estimated he was approximately six miles off the island and approximately nine miles from the mainland when the ship was snugged down. He took bearings, but these are not of much use. The ship was in an open road and rolling badly. His compasses were therefore unsteady and not well corrected. Where he was positioned made it difficult to take occiduous and ortive amplitudes (east and west risings and settings of the sun) which were then used to assess local magnetic variation. But if we take his angle of bearings on two points on Rottnest as being reasonably correct, that is that his ship was positioned at a place where there was an approximate angle of 32° between the northernmost part of Rottnest in sight and the southernmost part of Rottnest in sight, then it was lying nearly a third of the way to the mainland on a line from Bathurst Point on Rottnest to Cottesloe.[12] There is nearly 10 fathoms of water to be found there, eastward from Kingston Spit, which is as Hamelin recorded.

Where he chose is an uncomfortable and insecure place to anchor in winter weather conditions. It is on upward sloping ground. Nearby to the south is a shallow weed and sand bank covered by four to five fathoms of water. When the wind, waves and weather set in from the north it becomes a particularly uncomfortable place as the waves and swells steepen and shorten when they near the shallows. In these conditions an anchored ship is soon placed in a position resembling a bad lee shore.

Hamelin soon found this out. When the weather set in from the north not long after he anchored, which put his survey parties in danger in their little boats, he had to shift the *Naturaliste* to a better anchorage in deeper water further north and further east.

From the first anchorage, the land appeared attractive and inviting. The mainland appeared to be well wooded "with trees of medium height". The coast seemed to be made up of sandy beaches and white cliffs. Rottnest Island, closer by, appeared to be somewhat similar, with woods extending right down to the shoreline. Shore parties were therefore quickly despatched to make preliminary explorations.

ROUTE FOLLOWED BY THE NATURALISTE (HAMELIN) 1801

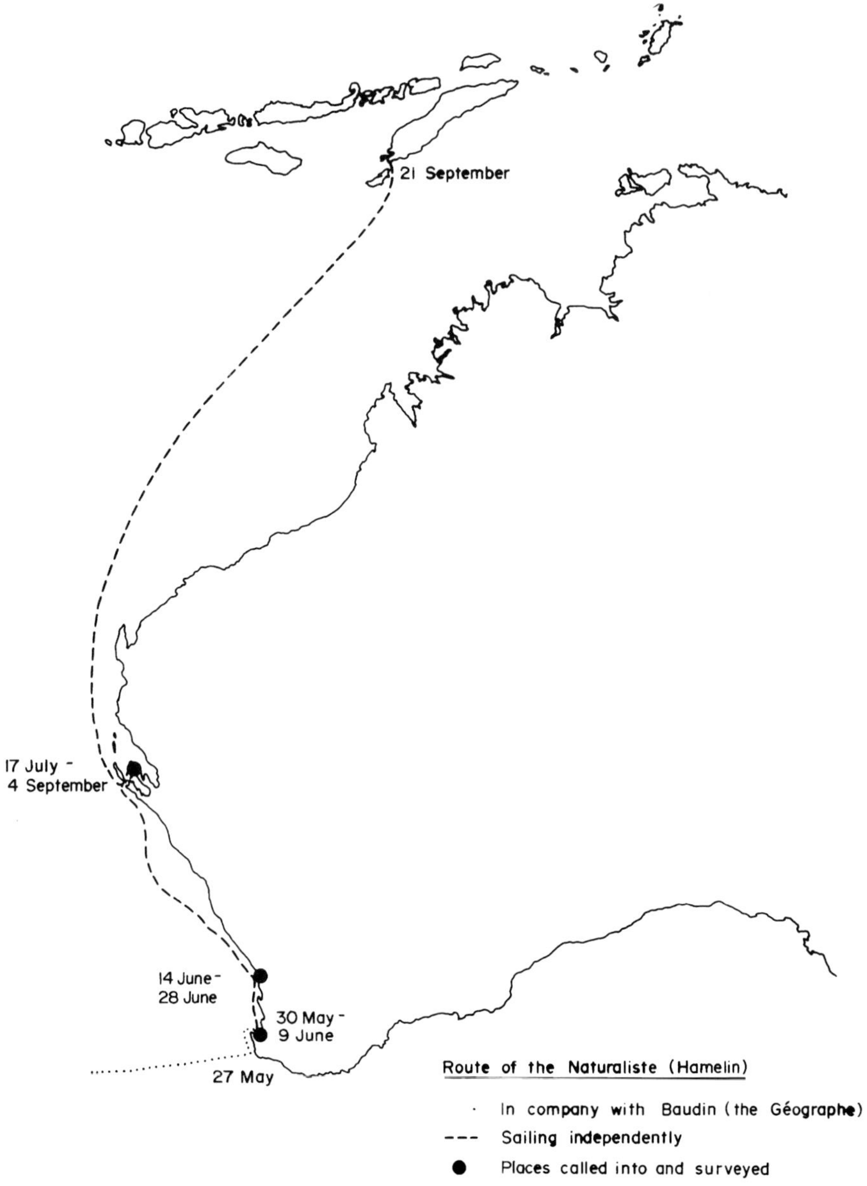

Location of anchorages and areas surveyed:
Hamelin (Naturaliste):
Geographe Bay 30 May–9 June 1801. Swan River 14-28 June 1801.
Shark Bay 17 July–4 September 1801.

Surveys of Rottnest Island and approaches to the Swan River

On 15 June, a small dinghy was despatched to Rottnest with a fish net and with the naturalists in order to take stock of the situation, and get fresh supplies. Not many fish were caught, but two large seals were easily killed and brought back and relished. Hamelin in particular was pleased at the finding, recording that "he ate much".[13]

One intriguing find was made on the island, besides the seals. Some burnt wood was found and this raised the question of whether or not the island was inhabited.

There was another intriguing mystery. The mouth of the Swan River had not been located. In fact it could not then be possibly seen from where the ship was anchored. The Swan River flows towards the south-west at its mouth. The north shore formed by a low headland runs in the same direction, and this shields the river mouth from the view of those looking for it from the north.

On 16 June, Sub-Lieutenant St Cricq was despatched to the south in a dinghy to clear up the mystery, but returned unsuccessful. At the same time, Sub-Lieutenant Heirisson was sent to make soundings around the anchored ship, as then was the normal custom, to see if there were any hidden shoals or dangers.

Freycinet's detailed survey of the east part of Rottnest Island

Preparations were then made to despatch more serious, far reaching exploratory missions. Four areas were selected for concentrated study, Rottnest Island and its resources was given special attention as requested by the French naval ministry. The course of the Swan River was also given close attention. The two other survey parties were sent to examine the islands south of Rottnest, and one to map the bar at the mouth of the Swan River.[14]

These surveys were all made with difficulty and danger. June is the season for fierce northerly winter gales in the Swan River region as well as in Geographe Bay, and the anchorage Hamelin selected is just as open to the north as the one near the Vasse River. The only relatively safe anchorage in winter before the present harbour at Fremantle was built, was in Cockburn Sound, behind Garden Island where the present Western Australian naval base is. But the way into there was narrow and tortuous, and dangerous in winter and was then not known, although it is shown as an anchorage on Dalrymple's map of 1797.

Hamelin consequently did not venture south of the latitude of Rottnest Island in the *Naturaliste*. Not wishing to be again caught in the confines of a bay, he stayed in the north approaches to the Swan River and even then ran into trouble. The same storm which kept

Baudin out to sea hit the *Naturaliste,* and not only caused further damage to the rigging and ground tackle of the *Naturaliste,* but also wrecked two survey dinghies — one in Thomson Bay at Rottnest — the other on Cottesloe Beach. Fortunately for the future work of the expedition, they were both salvaged after the gale.

Louis de Freycinet left on 18 June with a party including Faure the geographer to survey Rottnest[15], to help make a map of all of the Swan River region and to assess its resources. Freycinet originally intended to make this survey by boat. But by the time he got into Thomson Bay just after dawn, he was beset by strong winds. He tried to make south to get around Bickley Point into the lee of the island, but found the way barred by rocks and surf. He consequently made for the beach at the south end of Thomson Bay and was cast ashore in the surf approximately half way between the Natural Jetty and where the army jetty now is.

Wreck of the survey boat in Thomson Bay, Rottnest Island

This was a mistake on Freycinet's part. The way to the south of the island is barred by the Natural Jetty over which the surf pounds during a northerly gale. This danger and the tide can be easily seen and felt in the bay. He would have been better off if he had made north in Thomson Bay, where a promontory, reefs and banks offer good shelter from north winds and seas. However, it is difficult to do this in a small boat when it is far to the south in the bay after a gale has started. A very strong current sweeps south through the bay, sweeping across the Natural Jetty. Freycinet was fortunate he was not swept onto this. If he had been, he would have been in serious trouble. As it was he lost only his dinghy and provisions. Near the beach his boat filled with waves and disappeared. Only a few biscuits were saved.

With no hope of rescue while the wind blew, and not wishing to waste time, Freycinet and his party commenced a survey of the island on foot. They started to walk northwards, but by taking that direction they soon came to cliffs on the north coast of the island which they found to be too steep to climb and pass. They consequently gave up their attempt to walk along the coast and turned inland. But before they left the rocks on the north, they made a noteworthy discovery. They found evidence of a then recent shipwreck. Freycinet estimated that the vessel that had gone ashore was 300 to 350 tons. Unfortunately, he does not give the exact location of the place. From the evidence it seems to have been not far from Bathurst Point. He does not mark the fine little sheltered harbour at Geordie Bay on his sketch map and therefore does not appear to have gone that far west.

The identity and location of the wreck therefore remains a mystery. It could likely have been a sealer or a whaler. These were working the western Australian coast at the time seeking seal skins for the Asian trade. No doubt some part of its wreckage is still underwater in the reefs on the north of Rottnest.

After leaving the coast, Freycinet and his party headed south and discovered the salt lakes for which Rottnest is famous. These were examined and later sketched and named Duvaldailly's Ponds after the midshipman who discovered them. After the survey was ended, the party returned to their wrecked boat. Fortunately, while they were away it was washed on to the beach. Work was immediately commenced to repair it, but the party had to wait until 20 June to get back to the ship. In the meantime they built a shelter of the boat's sails in the hills near the stranded dinghy close to the site of the present army camp and killed seals to sustain the party while the survey was continued. A trigonometric base line was marked out on the shores of Thomson Bay extending from where the army jetty now is to half way from there to the present hotel site. From there and from high points, Thomson Bay was thoroughly mapped and compass observations made of the ship and other islands and prominent features on the mainland coast. A detailed map of the salt lakes was also drawn.

Rescue of the explorers marooned on Rottnest Island

On 20 June, Hamelin sent his remaining small boat, a punt (pousse pied), ashore with supplies and the carpenter to fix the dinghy. After this was done the party returned to the *Naturaliste* with a comprehensive study of the Thomson Bay area completed, and a not too good impression of Rottnest. Before leaving, Freycinet left a message on Rottnest for posterity. He nailed a piece of lead to a tree near where his dinghy was swamped reading:

Ensign Freycinet -Geographer Faure 28 Prairial - 1 Messidor
An 9 (17 June–20 June, 1801).

At the foot of the tree he buried a bottle bearing a message —

"28 Prairial year 9 the small dinghy of the Naturaliste sent under the command of Ensign Freycinet with Faure the geographer, to make a geographic survey of this island was wrecked on the coast. The crew was obliged to live for three days with the seals, the spray and the rain.

Duvaldailly, midshipman
David, gunner
Bourgeois, sailor
Monier, sailor
Ozane, butcher.[16].

This bottle has never been recorded as being found. Unfortunately, where Freycinet camped is now the site of an army camp and a lot of the area has been changed and disturbed.

Later on in the stay, on 27 June, Bailly the mineralogist landed on Rottnest to specifically study the mineral and other resources, after he returned from the Swan River expedition, to complete the island survey. While surveying the minerals he noted also the presence of the special species of wallaby that inhabits the island and the local snakes, and also a then unknown species of lizard which he collected for Péron, and large quantities of seals which he commented could be valuable commercially. The Baudin expedition did not make any further survey of the island or of the Swan River region after the survey by Hamelin. When he returned in 1803, Baudin did not stay or go ashore in the area. He considered the area to be not worth further examination.

Heirisson's exploration of the Swan River

On 17 June, the day Freycinet went to Rottnest, a more extensive and important survey mission under the command of Sub-Lieutenant Heirisson, left for the Swan River.[17] The aim of this party was to move as far as possible inland along the river, and report on the resources found there to see if the Swan River could be recommended as a satisfactory port of call for vessels seeking refreshments. To do this, Heirisson was provided with six days' supplies and the services of the mineralogist, Bailly, the only natural history scientist on the *Naturaliste,* who was delegated to make a description of the region for the benefit of science.

Hamelin considered Heirisson's to be the most important and sensitive of the missions sent. He consequently issued special orders and advice to the party. Their main task was to chart the course of the Swan River as far inland as they could go.

No doubt with his unfortunate personal experience on the Vasse still fresh in his mind, at the beginning of his written orders Hamelin warned the party to be careful if they ran aground for that, he indicated, is the time natives would choose to attack.

He then went on to request the party to find out all they could about the Swan River, to see how large it was, and whether it contained fresh water. Aware of the problems in Geographe Bay, he informed Heirisson specifically to let the naturalists ashore to do their work. Heirisson was also advised to keep a guard always on the boat, and for him not to leave it at the same time as midshipman Moreau. There should always be an officer aboard to be in charge.

In the case of meeting natives, Heirisson was advised, he was to attempt to communicate with them but was not to compromise his position. He was to ask where water and food could be found; he was to see what the natives ate, and should offer presents. He was further advised not to approach large numbers of natives, and was to see what they were doing. He was ordered not to use his guns except in an extreme emergency. "It is the intention of the French government, and a desire close to my own heart, not to shed the blood of the natives at the places we visit. Our object is natural history research. You should brood on that", he told Heirisson.

Heirisson was advised in case of delay or accident, to light fires at night and fire rifle shots in the day so he could be found. It was added, that he was to put the party on two-third rations if he was away more than four days.

Heirisson's party made a quick passage southwards before the north wind which had arisen, and arrived at the Swan River mouth at 8 a.m. There they encountered their first obstacle and disappointment which helped them to conclude that the Swan River was not a suitable refreshment port for France. The river mouth was obstructed by a shallow rocky bar which denied access even to the shallow draft long boat. If water and other resources existed in the interior, these could not be easily ferried to waiting ships because of the barrier.

Discovery of the Canning River mouth

With difficulty Heirisson got his boat and party across the bar, into deep water, and moved inland through the limestone hills and cliffs which form the banks of the river in the lower reaches, and then sailed into the extensive water basins which make up the river inland near Perth. He sailed around Point Walter Spit and reached the foot of Mount Eliza, which now stands above the city of Perth, and camped there for the first night. From Melville Water, the party saw another river entrance southward on the other side of the river basin. This was not closely examined, but was named Moreau Entrance after a midshipman with the party. This discovery caused speculation about whether or not the Swan River had several mouths, and whether the Moreau Entrance was another branch that led to the sea. It appeared to sweep towards the west in the direction of the coast. And it left the Swan at a large basin of water. The discovery of this made the French feel that the narrow, barred river entrance which they had used to get inland was possibly not the only river outlet. It seemed too small to be the entrance to the big body of water behind it. In fact they were wrong. The Swan River has extensive water areas along the lower reaches primarily because it is an old sunken valley. It is tidal and salt water, and more of an inlet than a river. And the so called Moreau Entrance which they saw, is in fact the Canning River, which after a short run towards the west from its mouth, turns eastward to become a fresh water river flowing from the same range of hills that form the head waters of the Swan.

Arrival at Heirisson Islands

The following morning, 18 June, the party climbed the heights above present Perth, which is now King's Park, and saw the river meandering inland towards the Darling Ranges which appeared quite close. They set out immediately to reach them. Unfortunately after crossing the last big water basin, Perth Water, they came to a second barrier in the river formed by a shallow mud bar with islands, which they named Heirisson Isles, now Heirisson Island. With difficulty the boat was hauled across this and the party continued inland. It was near Heirisson Island that the first fresh water stream, Claise Brook, was found and used but not named. The party also saw their first black swans in this region. Up to then they saw only pelicans and other water birds. Inland from Heirisson Island the party also observed that the river lost some of its saltiness.

Discovery of the Helena River

There is no clear indication where the expedition reached and camped on the night of 18 June. From the map and other evidence it seems they stayed on the bank of the Swan River near the entrance to the Helena River, which Heirisson and Moreau followed on foot for half a league (nearly two miles) where it appeared to end. There they discovered "a man's footprint of extraordinary size". But they saw no sign of aborigines, which added to the impression gleaned at Geographe Bay that western Australia was very sparsely populated.

By this time the expedition was in flat plain lands, which were composed of clay and sand and frequently covered with water. These areas later became good farm lands after the British settlement was made.

Arrival at furthest point of the survey near Henley Brook

The next day the expedition pushed on up the river in an attempt to reach "the mountains" which seemed invitingly near. But the distance of the low, smooth, time worn hills is deceptive. Consequently that night the party seemed to be no nearer to them than before, although they had travelled a considerable distance. By this time the river was narrow and had shallowed to seven or eight feet. They camped the night at a place that again cannot be accurately determined, but judging by the map and reports it was in the vicinity of Henley Brook, the last fresh water stream which the party discovered and marked on the chart. This area is now in the centre of the rich wine growing and producing industry of western Australia.

Return to the *Naturaliste*

The party by then was running short of supplies, although members shot water birds to supplement the rations. Heirisson consequently stopped his survey at that point and headed back down river on 20 June.

On the morning of the following day they again ran into trouble at Heirisson Isles. This time they took a different course, keeping close to the right or north bank. There they met shallow bank after shallow bank. They struggled continually in the mud, unloading the boat to lighten it, and pushing it through the shallows for some thirteen hours. The situation seemed critical. The party was wet, dirty and fatigued. There were provisions only sufficient for one more meal, and the *Naturaliste* was still a whole day's trip away. To add to the problem, they were confronted by a sudden terror. Night came on the party suddenly. "We were just preparing to land and dry ourselves, and fresh our exhausted strength by a little rest, when all at once we heard a terrible noise which filled us with terror; it was something like the roaring of a bull, but much louder, and seemed to proceed from reeds which were very near to us", the party reported. The mystery of this noise has never been cleared up. There were no heroes to be found in the crew by this time. No one in the party went to investigate. According to Dr. Dom Serventy, an ornothologist with a specialist knowledge of wild life in western Australia, the "terrifying sound" heard by Heirisson's crew was no doubt the cry of a brown bittern. This bird is a member of the heron family and inhabits swamp and marshland areas. It is a wader and has the ability to stay still and blend with its environment when in danger. Heirisson's crew consequently would have had little chance of seeing it. The noise they heard, if it was a brown bittern, is eerie, sounding more like a boom than a cry. It was, incidentally, the frightening sound of the brown bittern which gave rise to the myth held by early settlers that an unknown animal, the bunyip, stalked the Australian bush.

The famished travellers on hearing the noise lost all interest in going ashore to rest and eat. They consequently stayed in the boat, uncomfortable in the wind and rain.

Early the next morning, 22 June, they again found themselves aground. They managed to push the boat off with difficulty, and went ashore, lit a large fire, dried out, ate a meal and then headed back for the *Naturaliste,* which they reached that night.

Bailly's later report on the minerals of the Swan River region is most interesting and useful. His survey, together with Heirisson's, provide descriptions, made in the case of Bailly by a trained scientist, of an unspoilt river area which was soon to be settled and changed by the British, which is therefore of great use to historians and natural scientists.

Collas' survey in the Swan River mouth

Two other parties were sent out to make surveys in coastal waters at the same time as Heirisson. Sub-Lieutenant Cricq had not discovered the river mouth on 16 June, so the mystery of the position of the river mouth remained a mystery for those on the ship. To clear this up, the ship's pharmacist, Collas, was sent on 18 June making the third expedition. He was sent to discover the position of the river mouth and chart it. He arrived there at 9 a.m., after a quick passage helped by the north-west gale. He completed his examination, reporting that the north bank was sandy and sterile but that the south bank was bushy and had colourful flowers and trees which were large and impressive. He reported that the land there appeared fertile. He collected some wild celery. He drew a map of the bar at the river mouth and returned to the *Naturaliste* with the specimens he collected for the scientists.

Milius' survey of Carnac Island and Garden Island

The other, fourth party met with more serious challenges and adventures. This group, under the command of Lieutenant Commander Milius[18] accompanied by Levillain the zoologist, went to survey the islands south of Rottnest. They left the *Naturaliste* on 18 June and headed south south-east, passing the Straggler Reefs examining the small rocky island of Carnac which they called Bald Island (Ile Pelé) and later named Berthelot. They then discovered a third large island, Garden Island, which they called Buache Island, and judged it to be almost as large as Rottnest.

By the time the party reached Garden Island the north-east wind which got Freycinet into trouble and difficulty, had turned to become a north-west gale. Milius consequently took his boat around the north-east corner of Garden Island, to find shelter in a secure little bay there, now called the Pig Trough. Here he set up his instruments and camp and commenced surveys. He constructed a base line on a flat piece of beach in the centre of the Pig Trough, stretching from the little point of land a third of the way from Beacon Head to Second Head, to a second little point in the bay approximately two-thirds of the way south to Second Head, that is under the present cottages on the beach front there, near the little jetty. Using this base he made close surveys of the off-lying rocks and nearby coast. He also climbed the high hills on the north end of the island and took compass bearings of the ships, Rottnest and the other islands and rocks, which were very accurate. This information was to be used with that gleaned by Freycinet and others to complete the planned comprehensive map of the area.

Milius left a most favourable report on the environment and resources of Garden Island, which place was later used by the British as the base for their first colonists at the Swan River Settlement. He recorded that there were plenty of seals for meat. These were so numerous they even seemed to live in the woods. The woods themselves were thick, pleasant and sweet smelling, Milius reported. He collected plant specimens for the scientists to better describe the resources, and recorded finding a species of plum. The soil appeared to be rich and black, about 4 to 5 inches in depth and was considered to be suited for agriculture. He reported that there were not many birds, but the crows were reported to be most eatable.

Departure of Milius from Garden Island

Milius left Garden Island at sunset the same evening. With his geometric survey completed, he placed a bottle with a note about his visit in the vicinity of his survey camp at the Pig Trough (Bay).

To get back to the *Naturaliste,* Milius had to plough straight into the north gale. Progress was not too bad up to near the Swan River. Before that is reached there are extensive, shallow banks stretching across the sea from the Straggler Reefs to the mainland just south of the Swan River mouth. These and the outer shallows break up the waves and swells so that sailing there is relatively comfortable. But once north of these banks, the sea becomes unpleasant and dangerous for small boats in a gale. By midnight Milius was in rough water and in trouble. He had passed the Swan River mouth not long before, and tacked out to sea. By midnight he had made as far north as Cottesloe Beach. There he anchored approximately three-quarters of a mile offshore, in sight of a large sand patch.

The wind then shifted west and sail was again hoisted to reach out to the *Naturaliste* which was then bearing north-west. But trouble struck. The mast broke after the anchor was raised, and the boat was swept towards the beach. Oars were put out, but these and the men who used them were found to be useless. Milius reported somewhat bitterly about the crew, "when we took to the oars only one took orders and acted correctly. The rest lost their heads".

Milius' long boat washed ashore at Cottesloe Beach and his party marooned

The long boat was soon caught in the surf. The boat was lightened as much as possible. Food and equipment were thrown overboard. But the efforts to hold the ship off the beach were futile. By 8 a.m. on 19 June the party was cast ashore without food, "and not even the means to light a fire". Fortunately they managed to get the boat through the waves onto the beach without it being broken up. But it was damaged and needed caulking and repairing if it was to get them back to the *Naturaliste.*

It is possible to assess from bearings taken by Freycinet and others at Rottnest Island and on the *Naturaliste* that the place where Milius was cast ashore was the main beach at Cottesloe near the present bathing pavillion. He came ashore at a prominent sand patch which lay due east of Bickley Point on the south end of Thomson Bay, and was about three miles from the Swan River Mouth.

A comparison of Hamelin's charts with later ones drawn by the British Admiralty and with later written sailing guides for the region indicates that the conspicious sand patch mentioned by Milius, Hamelin and others was the extensive patch of sand north of Mudurup Rocks at Cottesloe. The white sand patch was used as a marker for ships entering Gage Roads from the Indian Ocean after the British settlement was established at Fremantle.

That Milius landed at Cottesloe Beach is also evidenced by the fact that that beach is the only place in the vicinity where a boat could be washed ashore onto sand, and could be refloated. When a north-west gale blows the surf runs high at Cottesloe Beach. But once the wind shifts to the south-west, as it does there after a gale, the beach is relatively protected by the reef ledge which runs into the sea out from Mudurup Rocks.

South of there, for some miles, the beaches are rocky with offshore dangers, with no conspicious sand patches then existing. North of Cottesloe, the situation is somewhat the same.

Fortunately for the party marooned on the beach, cold and wet, the sun came out at 4 o'clock that afternoon and a fire was lit by means of a magnifying glass.

Milius explores the coast and inland areas from Cottesloe

Milius used his six days of enforced idleness, which lasted until he returned on board on 23 June, to advantage. He and his party made local surveys along the coast and inland. A walk was made along the coast southwards to the Swan River mouth where a tricolour was hoisted on the north headland of the river, and a message left in a bottle for Heirisson, telling him of their plight.

Inland from Cottesloe was also examined. The party walked to the river, which is not far distant from the beach at Cottesloe, and ventured "fifteen miles into the bush".

Good drinking water was found in rock holes near the Swan River, and good supplies of fresh water were found in quantities at two feet under the surface.

It is not clear where on the river Milius went. He gives no indication of the direction he took from the beach. He estimated that he was on the river bank about 12 miles from the river mouth. This would put him near the present city of Perth. It is doubtful if he got that far. He was more likely in the vicinity of the rocky areas around Peppermint Grove and Claremont.

It would be of interest to know the exact route he took. Going along the river bank he reported finding two aboriginal cabins, two spears and some fish. The trees there he found most impressive, being much bigger than those near the coast. He also recorded finding nuts like chestnuts which the naturalist and sailors roasted and ate with relish, commenting that these tasted like European chestnuts. In fact these were the nuts from zamia palm which are quite poisonous and had caused trouble to the previous visitors to western Australia. Consequently Milius' party became severely ill, "vomiting blood".

Later the French found evidences of the aborigines eating these nuts (*Cycas* seeds) and concluded that they must be a special type of human being. In fact the aborigines were not immune to the poison in the nuts. They carefully prepared them for cooking, somewhat like the Japanese do in the case of poisonous blowfish. The aborigines place the nuts in running water which leaches out the poisonous substances.

Milius' marooned party rescued

Milius and his party were joyed to see a boat coming for them early on 21 June. Hamelin was aware of their plight, but had to wait for the gale to subside before he could help them. The rescue party landed, the longboat was put into a seaworthy condition and was launched that evening. The party immediately set out for the *Naturaliste,* and reached it the next day after what Milius called "six days of fatigue, worry, pain and privation".

FREYCINET AND HEIRISSON'S FINAL MAP OF THE SWAN RIVER AND APPROACHES TO FREMANTLE WITH THE LOUIS NAPOLEON ISLANDS (ROTTNEST, CARNAC AND GARDEN ISLAND)

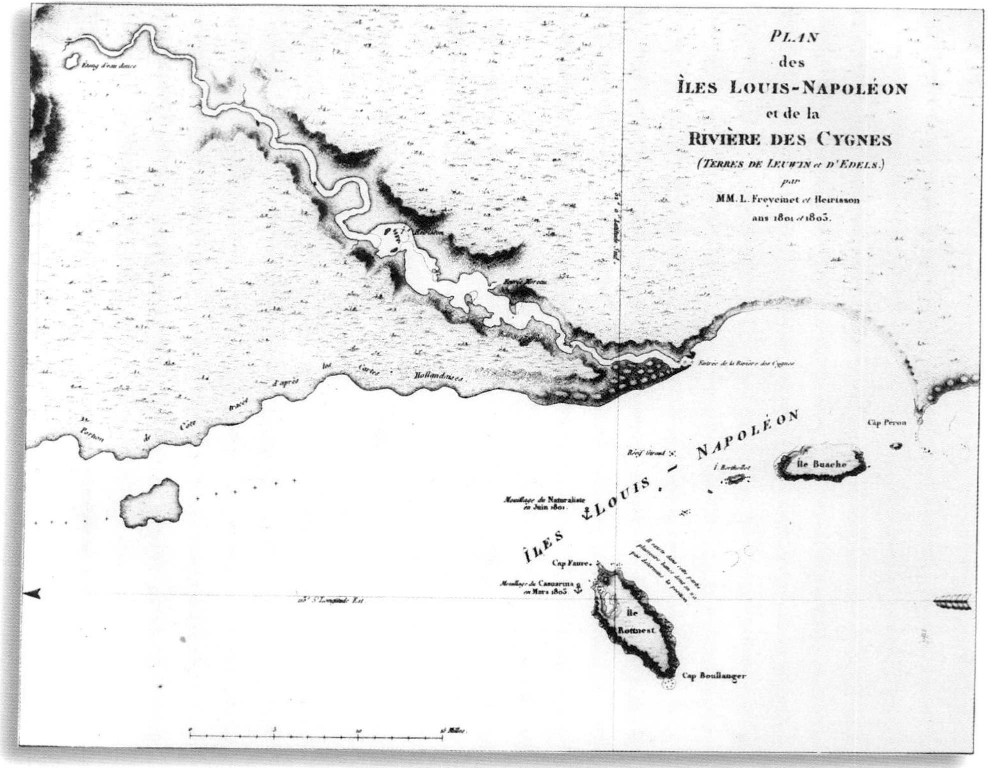

Favourable reports of Swan River region presented

But Milius and his party brought back valuable information about both Garden Island and the mainland. They alone of all the parties had found evidence of natives. Milius gave a detailed description of the cabins he found, and the life style of the aborigines he observed near the Swan River. He also gave a report on the environment, indicating it was fertile and productive, adding to the impression being created that the Swan River region was a suitable base for supplies and refreshment for French ships in the Indian Ocean.

In consequence of what Heirisson, Bailly, Collas and the others reported, which was added to Milius' account, Hamelin concluded that the Swan River would be a satisfactory place of refreshment for ships in that part of the Indian Ocean. There were plenty of fish and seals and the flesh of the latter was judged good. There were good birds, including swans inland, and fresh celery and herbs available. And the wood was easy to cut. Besides this there was no sign of hostile natives. They seemed to have disappeared. Consequently it was easy to land in the area, which was an important consideration.

The main difficulty was fresh water. It did exist, but was hard to get. It was a long way inland and the bar at the mouth of the river was dangerous to cross in rough weather. But to compensate for this, the game ashore seemed plentiful.

Hamelin concerned on the *Naturaliste*

Meanwhile on the *Naturaliste,* as preparations were being made to send out the different parties already on 17 June, the barometer dropped and the wind set in from the north. This is a reliable sign in those waters at that time of year that a gale is on the way, usually hitting within twenty-four hours. By the next day it had set in, with strong gusts coming from the north-west and west. Nevertheless the different parties were despatched to complete their surveys.

By the 19 June, the situation for Hamelin and the others left on the *Naturaliste,* anchored on the slope near the shallows was uncomfortable if not serious. The clouds had closed in and the barometer continued to fall to reach 27.1 which is low and meant a big blow. Hamelin consequently raised anchor and shifted to a second better anchorage in the fairway, off Swanbourne Beach, near the present optional pilot ground off Fairway Buoy. He was fortunate in his choice. There is some reef and coral in the vicinity of where he anchored. As a result he was not only more comfortable, but also better fed because of the move. He recorded catching large fish. The spot is in fact a good ground for snapper, dhufish, groper and a variety of rock and bank fish.

From the ship, that night, on 19 June, Hamelin saw a fire on the beach to the east, and wondered if it had been lit by natives or by his men. In fact it was Milius and his boat crew who were wrecked on Cottesloe Beach, which he did not then know.

It was also at this time those on the anchored *Naturaliste* saw the *Géographe* in the distance, but they could not move to give chase. Their boats were all away, and their crew were depleted. In any case at that stage, those on the *Naturaliste* did not realize Baudin had not seen them. They believed that because of the bad weather, Baudin had made out to sea to ride out the gale, and would return to the rendezvous.

When the ten days period of grace Hamelin had given to Baudin to make the rendezvous at Rottnest ended on 28 June without any further sign of the *Géographe,* Hamelin weighed anchor and set a course for Shark Bay hoping to have better luck at the second rendezvous.

Although Hamelin was worried about the fate of his commander, and disinclined to proceed alone, it is fortunate for posterity that Hamelin was left on his own after Geographe Bay. He and his crew made a distinctive contribution to the scientific knowledge of the Swan River region and the nearby offshore islands in the most difficult circumstances. With only three scientists, and a limited number of small boats and a relatively small crew at his disposal, and plagued by wild winds and foul weather, Hamelin and his men carefully surveyed and charted the area from fifty miles inland along the river, to the limits of the offshore reefs and islands, making the previous reported anchorage known and usable for other navigators.

Besides this the survey parties made a sizeable collection of plants, flowers, seeds, fruits, birds, shells and other specimens designed to portray the natural history of the mainland and offshore islands at the Swan River region. Collas and Bailly helped notably with this work, making forays to collect specimens from the reefs and islands and the mainland, while waiting for Baudin to appear. These specimens were sorted out and boxed up ready to be sent to the research scientists waiting in France, as the *Naturaliste* sailed northwards.

Hamelin leaves a letter for Baudin on Rottnest Island

Before Hamelin left the Swan River region he wrote a rather disconsolate letter to Baudin, put it in a bottle, and left it on Rottnest Island in case Baudin turned up.

In this he told Baudin:

"It was with regret that on the night of 8 June, I lost sight of your

light in Geographe Bay. I fear you had lost your corvette, especially the following day. I had trouble with my anchor etc. I therefore decided to make for our first rendezvous. I arrived safely and Heirisson went eighteen leagues up the Swan River. At 2 p.m. on 18 June we saw the Géographe. We waited ten days for you. We must be in Timor by 2 October, 1801. I have a fear we'll not meet you. We suffered a lot of wind damage."[19]

Hamelin raised anchor at 8 p.m. on 28 June and sailed north towards Shark Bay.

Hamelin sails north to Shark Bay on the *Naturaliste*

After leaving his anchorage, Hamelin, unlike Baudin, hugged the coast as closely as possible on the way to Shark Bay, letting his natural historians and mappers view the form of the land along the coastline, providing them with the opportunity to acquire knowledge and make contributions which was denied to those sailing with the more cautious commander of the *Géographe.*

No harbours or safe anchorages were found between the Swan River and Shark Bay. From the ship the coast appeared to be well guarded by off shore reefs and shoals, and dangerous to navigation as Dampier and the Dutch had warned it was.

The land beyond the lines of reefs continued to appear uninviting as it had from Geographe Bay, having "the same melancholy appearance" as the coast further south. There were few outstanding features to be noted in the log. A short way north of the Swan River, near Whitford's Beach, it was noted that the coastal hills rose to a height not previously seen. And amidst these, at Mullaloo, the large, conspicuous bare sand dune patch was observed and noted as a landmark.

The *Naturaliste* was then sailed close to the coast past Breton Bay, Lancelin Island and Jurien Bay. At this point Hamelin made out to sea to check the position of and sail outside the infamous Abrolhos Group. On 8 July Hamelin reached and sailed close to these islands, making a brief survey.

As a result he corrected the position of these on the navigational charts of the western coast of Australia, noting for the benefit of mariners that they were closer to the mainland than had been previously thought, and that there was deep, apparently reef free water, and not a shallow broken bottom on the western or seaward approach as the Dutch charts indicated, although Hamelin cautiously commented "we saw the islands in calm weather", implying breakers could run in bad conditions.

On 11 July Hamelin turned towards the mainland and reached it near Red Point (Pointe Rouge) on 14 July, sailing past Gautheaume

Bay along the coast to Shark Bay.

Arrival at Shark Bay

On 16 July, Hamelin reached South Passage — then called Thorny Passage by the French — at the southern end of Dirk Hartog Island. This tortuous passage was not then used for navigation. Hamelin consequently bore north along the island coast to enter the Naturaliste Channel between Dirk Hartog and Dorre Islands. He reached the entrance on 17 July and anchored approximately a mile out to sea off Cape Levillain on the north-east corner of Dirk Hartog Island, which was then known from the charts to be a safe anchorage.

As soon as he rounded the north end of Dirk Hartog Island to make his approach to the anchorage, Hamelin became apprehensive. There was no sign of the *Géographe,* nor was there a flag to be seen on the island to show that Baudin had passed through the entrance to anchor elsewhere in the Bay, as was usually done by ships making a rendezvous at that time. Hamelin considered he had come to the right place. Shark Bay covered a large area. However, he followed the passage into Shark Bay taken by St. Allouarn, Dampier and the Dutch and headed for the only place near the entrance which was renowned as an anchorage. Unknown to Hamelin, as has been described above, Baudin had sailed out to sea past the Naturaliste Entrance to take the broader north facing entrance, Geographe Channel, in Shark Bay. The bad experience in Geographe Bay had been enough for Baudin. He had no liking for the north gales he had met and had no desire to be caught weatherbound in an entrance which could be difficult in north and west winds. He consequently made north to keep close to that entrance, as Hamelin had himself done at the Swan River. But at Shark Bay this was an unnecessary precaution which lost Baudin the chance of making the set rendezvous, and cost him dear in regard to his reputation as an expedition commander. He left a message for Hamelin as required, but this was done on an obscure island far away from the normally used entrance. There was, therefore, little chance of Hamelin ever finding this, and he did not. Baudin, by rights, should have left a mark at both entrances, and in particular the normally used one at Naturaliste Channel. He had the time and the opportunity to do this.

Hamelin decides to wait for Baudin and to survey

The apprehensive Hamelin immediately called a conference of officers to discuss the matter. The situation was worrying. Baudin was known to be a stickler for orders and regulations. He demanded that his subordinates keep strictly to their schedules and that they should be punctual. Hamelin therefore expected Baudin to do the same and set an example. It was consequently concluded that the *Géographe* had been delayed. Therefore, Hamelin again decided to wait at the set rendezvous point.

Initially Hamelin intended to stay for ten days. Fortunately for posterity and science he ended up remaining in Shark Bay for more than six weeks, from 17 July until 4 September, surveying, mapping and collecting specimens of the natural history of the region.

Value of Hamelin's surveys

Besides making a permanently useful collection of specimens and reports on the natural history of parts not visited by Baudin's scientists who concentrated only on Bernier Island, Hamelin and his crew made three distinct contributions to knowledge about Shark Bay. Firstly he found that the so called Middle Island seen and referred to by Dampier and St. Allouarn, in fact was a peninsula which was renamed by him Peron Peninsula in honour of the eminent hard working scientist on the *Géographe*. This discovery was missed by Baudin although he was near Peron Peninsula.

Secondly, at this place on the mainland, Hamelin and his men again came into close contact with aborigines as they had at Geographe Bay. This enabled his men to leave a detailed and useful description of the aborigines and their life style in that region, which is of value to posterity.

Thirdly, the comprehensive survey made of the region by Hamelin's men, revealed that Shark Bay in fact was not a bay, but "a heap of gulfs, harbours and coves"

The comprehensive surveys made to discover this were despatched from two separate anchorage areas. On 2 August, Hamelin moved the *Naturaliste* from near the entrance to Shark Bay at the Naturaliste Channel, to Dampier Road which is on the north-west side of Peron Peninsula. This made survey work in the eastern section of the bay easier, as well as providing a more comfortable, secure anchorage. This is where Baudin had made for, but he anchored near the shoal waters on the other eastern side of Cape Peron Flats, a month before, and had to shift.

The extensive and thorough work Hamelin and his men did from these anchorages is commemorated not only in the first accurate chart they made of the Shark Bay region, but also by the present nomenclature of the region. Almost every outstanding feature in the area still bears a French name which form lasting memorials to the individual scientists and other members of the Baudin expedition. The two large inlets which form the greater part of Shark Bay are named after Freycinet and Hamelin. The one noticeable absence is a place named after Baudin. His name is not noticeably commemorated there, indicating the regard with which he was held by his colleagues.

Hamelin commenced surveying immediately the *Naturaliste* was anchored. The first trips made were mostly short ones. Hamelin had no desire to be caught short handed again, with his crew mostly away on long trips, as had happened at Rottnest when the *Géographe* had sailed by.

Once the *Naturaliste* was securely anchored, Heirisson was sent to sound around the ship and to the east to explore the anchor ground, as was then the custom for ships in strange waters. At the same time Freycinet was sent to Dirk Hartog Island to find a suitable landing place and a good camping ground for a small group Hamelin decided to put ashore to keep a look out seaward for the *Géographe*.

Freycinet returned on 19 July with the news that he had found a suitable place on the north-east corner of the island. The next day a small party of three, made up of the chief coxswain Jean Marette, who was in charge of the group, Stanislas Levillain, a zoologist and Bourgeois, a sailor, went ashore and set up camp. The aim of this party was to leave a note in a conspicuous place for Baudin to find if he missed seeing the *Naturaliste* when he came.

Discovery and repair of the Vlaming Plate Memorial

This party made a significant discovery which brought subsequent fame to Hamelin and the French. At the north-west end of the island they discovered an engraved pewter plate on the ground at the foot of a rotted post. This was the sign left by the Dutch navigator Vlaming after he sailed from the Swan River to Shark Bay in 1697. The engraved plate also recorded that Dirk Hartog had landed at the same place previously in 1616.

Aware of the fact that their finding was of interest, Marette's party returned to the *Naturaliste* with the plate which had not been seen for 104 years. Hamelin who was most impressed by the find, took immediate action. He approved the party's action in bringing the plate back to the ship, but indicated that if they had taken it off the post it would have been vandalism. He consequently had another suitable post made by the ship's carpenter and went ashore himself to erect the memorial on the exact spot where it was found on Inscription Point.

Heirisson's survey and difference with Hamelin

The same day that the camping party landed, Sub-Lieutenant Heirisson was detailed to make an extensive survey of the bay to the south of the ship. He was supplied with food and water for eight days and ordered in particular to search for a river reported to run into Shark Bay somewhere to the south. This was a pointless task. Heirisson had no chance of emulating the work he had done on the Swan River. There are no rivers running into Shark Bay. The story about a river was based on the belief that an opening in the coast meant a river coursing into it. There are certainly many openings in Shark Bay, but these are all sea inlets and not river caused.

This mission resulted in a bitter difference emerging between Hamelin and Heirisson[20] which was similar to those which affected the harmony of life on the *Géographe.* The incident which resulted in the arrest of Heirisson and his confinement to his cabin for 22 days, where he took ill, indicated that neither the *Géographe* nor the *Naturaliste* were what sailors call happy ships. Both Baudin and Hamelin were good seamen, but both were very strict disciplinarians, and somewhat autocratic in attitude. They took personal control of

SKETCH OF ABORIGINAL CAMP, CAPE PERON, SHARK BAY

Muséum d'Histoire Naturelle, Le Havre

their ships and crews, and enforced discipline on scientists, officers and men alike, to weld the men into an amenable team. This gained them respect for their positions, but often cost them the support of their colleagues who were all volunteers.

At Kupang, for example, at about the same time Hamelin and Heirisson fell out, Sub-Lieutenant Picquet again fell foul of Baudin while the *Géographe* was anchored there.[21] As a result he was goaled in the Dutch fort there. Trouble had been brewing between these two since near Shark Bay. On the way north from Rottnest Island, Picquet had clashed with Baudin and was relieved of his post. Tempers flared in particular because of Baudin's interference with Picquet's work. Baudin did not hesitate to come to deck and give direct orders

ABORIGINAL CAMP, CAPE PERON, SHARK BAY

to the crew during Picquet's watch. When this happened, Picquet usually walked off to let Baudin do the watch which meant Baudin could not leave the deck, which did not please the commander.

After he arrived at Kupang, Baudin moved cautiously in regard to the man he viewed as a trouble maker. Picquet was popular among the crew. He had had a prominent landmark in Geographe Bay named after him by the crew who acted spontaneously after Baudin publicly reprimanded Picquet for failing to come back on time from a survey mission there. Baudin therefore waited until the bulk of the officers and scientists moved to live ashore at Kupang before he had Picquet arrested, confined in goal, and then sent home in disgrace.

Heirisson did not suffer this fate, he completed the voyage, but the clash at Shark Bay was serious.

If what Heirisson wrote in his journal is correct, trouble between the two was brewing for some time, and the way Hamelin handled the matter reflects badly on him, and shows a further difference between him and Baudin. Baudin's problem was he spoke out against his men openly and quickly, recording these in his log for officials at home to note and act on. Hamelin, according to Heirisson, was not open in his criticism but indirect and criticized by innuendo and listened to and spread tales.

He recorded that he discovered Hamelin was casting hurtful remarks about his abilities. The trouble was mainly that Heirisson did not submit what Hamelin considered to be a satisfactory report on the survey of the Swan River. Certainly the report is not of good quality and scarcely deserves the name of a report. Heirisson also upset Hamelin by returning quickly from his Shark Bay survey with no results. Heirisson in fact does not appear to have had the ability to write good reports. If this was a fault then Hamelin also was at fault in not recognizing this and in not acting wisely to get his junior officers to produce the results he wanted. If he was displeased with Heirisson he should have told him this and immediately denied reports by others that he was making innuendos and so correct a situation which can cause feelings on a small ship to become volatile. Men in any situation find little comfort or reason to give respect to superiors who appear to engage in tale telling.

Despite this clash, the rest of the survey work was carried out at Dirk Hartog Island in particular by the mineralogist Bailly and other scientists and specimens were collected.

Before moving eastward to Dampier Road on 2 August, Hamelin had his own inscription made as a lasting memorial to the expedition. The plate, which read:

> *The French Republic. Expedition of discovery under the orders of Captain Baudin, the corvette Naturaliste, Captain Hamelin 27 Messidor An 9 (16 July 1801)*[22]

was erected on a post similar to Vlaming's memorial.

Freycinet discovers and explores Freycinet Inlet

As the *Naturaliste* was being prepared to shift, Sub-Lieutenant Freycinet was sent off in a long boat to make a long and detailed study of the bay to the south, with orders to rejoin Hamelin at Dampier Road. In the fifteen days he spent away from the ship, Freycinet made a thorough survey of Denham Sound and of the estuary which bears his brother's name. The main outcome of his voyage was that the so-called Middle Island was found to be a peninsula. The only other noteworthy event was the sighting of some Aborigines in the vicinity of Useless Inlet. The islands and seas in the bay also were observed to be productive. There were oysters, birds and fish in quantities to name but a few of the products.

Meanwhile from the deck of the *Naturaliste* at its new anchorage to the north-west of Peron Peninsula, smoke was seen rising from the hills near the beach. Sub-Lieutenant Cricq and the mineralogist Bailly were consequently despatched to the beach on 5 August to examine the situation. On landing they were attacked by a group of approximately thirty Aborigines armed with spears. Not wanting to cause bloodshed and have trouble Cricq fired shots over the heads of the Aborigines and the group made off. A brief survey of the locality was then made and on the following day, 6 August, a further party landed and set up an observatory on the beach. The long boat damaged at Cottesloe Beach was also taken ashore and repaired.

Observatory set up on Peron Peninsula and Aboriginal camp discovered

Hamelin himself landed at the observatory site south of Cape Peron, "having a great desire to talk to the natives". He did not see any, but found their tracks and those of a dog which he noted with interest. He also saw open oysters on the rock indicating a source of aboriginal food. At the far north of the peninsula an aboriginal camp was found. Hamelin has left a description of these cabins which he describes as "beehive style", equipped with a stove and a rubbing stone.

The *Naturaliste* lay off Peron Peninsula for over one month, until 4 September. The ship's alembic (salt water purifier) was set up ashore. The party camped near this. Frequent surveys were made from here with the men, locating native camps and observing the natural features of the north end of the peninsular, recording this for posterity.

Moreau surveys Hamelin pool to complete the survey of Shark Bay

One further extensive survey mission was sent from the ship in Dampier Road. On 22 August, Midshipman Moreau and the geographer Faure were sent to complete the survey of Shark Bay, commencing where Freycinet had left off at the north end of Peron Peninsula, sailing to the east of Peron Peninsula, south into Hamelin Pool and then along the mainland coast to the east of the bay to a spot opposite the anchorage. This area was mapped and charted by Moreau in the same thorough manner as Freycinet did in the western section of the bay. The large extensive pool that was discovered was named after the ship's commander, Hamelin. Faure Island, which was found to be rich in turtles, was named after the geographer. Laridon (Lharidon) Bight was named after the ship's doctor on the *Géographe,* and the narrow stretch of land at the base of this bight, which was found to join Peron Peninsula to the mainland was named Taillefer Isthmus, after the second ship's doctor on the *Géographe.*

THE NATURALISTE ARRIVING AT KUPANG TO REJOIN THE GEOGRAPHE

After the *Geographe* and *Naturaliste* were separated in a gale in Geographe Bay on 9 June 1801, Baudin failed to meet Hamelin (the *Naturaliste*) at Rottnest Island or Shark Bay as arranged. Instead he went to Kupang where Hamelin was surprised to find him on 21 September 1801.

By the time Moreau and his party returned, supplies on the *Naturaliste* were running low, and there was still no sign of Baudin and the *Géographe*. Hamelin consequently decided to weigh anchor and sail direct to Timor to keep the next rendezvous set for there in October, when both vessels were meant to be there refreshing.

Departure from Shark Bay

Hamelin departed from Shark Bay on 4 September. He headed for Timor and arrived at Kupang on 21 September where he at last saw the *Géographe*. Six shots were fired on the cannon. The crew on the *Naturaliste* gave three cheers, shouting "Long Live the Republic", and anchor was dropped. Hamelin then learned that Baudin had actually been anchored there for a month, and that far from being refreshed, the crew of the *Géographe* including Baudin, were ill with dysentery. Hamelin's crew were to suffer the same fate as a result of replenishing their water barrels from the dirty river water.

The two vessels stayed in this uncomfortable position at Kupang until 13 November. They then left in company, making a course towards the north coast of western Australia, then sailing into the Indian Ocean heading southwards to pick up the westerlies to go to Tasmania.

Arrival in Tasmania

The expedition arrived at Tasmania on 13 January, 1802. There, Baudin carried out his orders in a most meticulous manner. The anchorage he selected for the expedition on the east coast of Tasmania, near D'Entrecasteaux Island, was safe, the weather good and the surroundings attractive. There was plenty of fresh water, food and wood for the stoves available. There was an abundance of wild life to be studied. There were natives, in the main friendly, whose culture seemed of great interest. And the girls were attractive and not bashful. Among the many who reported being coyed into the woods, was Midshipman Heirisson who reported graphically that he was led off into the trees by a curious girl to see if he looked and functioned like native men.

There was thus little cause for complaint by any on the expedition. The scientists who were invariably grumbling and discontented were given ample opportunity to go ashore to collect specimens and conduct researches. The work done in Tasmania in fact is a highlight of the Baudin expedition, justifying the sending of it, and revealing the competence of the men selected.

Survey of Bass Strait

After the survey of Tasmania was completed, Bass Strait between that island and Australia was surveyed. Here Hamelin and Baudin were again unexpectedly separated and took independent courses to Port Jackson. Hamelin sailed no further west, but made from Bass Strait towards Port Jackson to refit his vessel. Baudin continued with a survey westward, reaching as far as the gulfs in South Australia, past Kangaroo Island, near where he met Flinders who at that time was making his survey of the south coast of Australia from Cape Leeuwin eastward.[23]

Running short of water and with his men on short rations and ill, Baudin made back from the gulfs without surveying them, heading for Port Jackson where he arrived on 20 June. He and Hamelin stayed there for nearly five months refitting the ships and nursing their scorbutic and other diseased crews back to health.

Hamelin departs on the *Naturaliste* with scientific specimens for France

When the expedition finally left Port Jackson on 17 November, 1802, to make a summer passage along the south coast back to the western coast of Australia to complete the work there, a change had been made to the expedition. In Sydney Baudin bought a small Sydney built schooner, the *Casuarina,* to replace the *Naturaliste* which was ordered home to take back the specimens collected so far, so that the waiting scientists in Paris could use them. The *Casuarina* was placed under the command of Sub-Lieutenant Louis de Freycinet who had accompanied Hamelin in the *Naturaliste* on the trip out.

Freycinet given command of the schooner, *Casuarina*

This was a great compliment to young Freycinet. He was a relatively junior officer. There were many officers above him on the expedition who could rightly have expected a separate command. Freycinet made this point, considering declining the appointment when he was made the offer, to avoid hostility. But Baudin prevailed, and the fact that Freycinet remained popular with his colleagues says a lot for his personality and abilities. This command marked the commencement of a distinguished career which was to see him among other things, lead the next French naval and scientific mission sent in 1817 to further survey western Australia.

Although he was pleased to have a command, Freycinet was not happy with the ship he was given. He left a brief and critical pen picture of it in his journal.

"The Casuarina is a 30 ton schooner, very badly constructed and worse fitted out. It is too short for its masts. It takes in five inches of water a day. We have to continually pump, it is a very poor sea boat," he wrote.[24]

There were frequent comments on the unseaworthiness of the *Casuarina* in the log of the voyage thereafter. Nevertheless Freycinet sailed the vessel as far as Mauritius, making surveys on the way.

The three vessels left on the first part of their journey in company, heading southwards from Port Jackson for King Island. From there Hamelin sailed to France and the *Géographe* and *Casuarina* then made along the coast to complete the survey from there to St Francis Island near Ceduna in South Australia, close to the head of the Great Australian Bight, where D'Entrecasteaux had reached sailing from the west.

Freycinet surveys the gulfs in South Australia

Baudin had to complete a lot of this part of the voyage alone. When he was back near the gulfs of South Australia, he transferred Boulanger, the engineer-geographer to the *Casuarina* and ordered Freycinet to chart the area, allowing him twenty days to complete this. The task was a gigantic one for a thorough surveyor like Freycinet, and he did not get back on time. Baudin consequently sailed for St Peter Island without him "where he had plenty to do", expecting Freycinet to follow. Baudin reached the Nuyts Archipelago where St Peter Island is, on 7 February, 1803, and remained there until 11 February, 1803, surveying and collecting specimens on the mainland near where Ceduna now stands, and on the offshore islands there. Freycinet had not arrived by the time this survey was completed. Consequently Baudin again weighed anchor and made direct for King George Sound believing it to be not worthwhile to survey the coast examined previously by D'Entrecasteaux.

The *Géographe* and *Casuarina* separated

Freycinet in the meantime made for St Peter Island after he surveyed the gulfs, but claims he could not find the island because of its incorrect location on the chart. He consequently took the leaking *Casuarina* straight to King George Sound to make repairs, arriving there at 1 p.m. on 13 February, five days before Baudin.

Both men independently made an important discovery on the way. They had expected to make a long and uncomfortable trip westward, pounding into headwinds and seas. Instead they picked up the east winds and land breezes which blow off the continent near the coast, and as a result made a quick and easy trip. This was a major find which upset the then existing British belief that a westward route south of Australia was not practical, and resulted in the opening of a new sea lane westward.

Apart from this, the contribution made to knowledge on the second visit to western Australia was neither great nor distinctive. The two places Baudin selected for the expedition to stay at, had both been surveyed several times previously. The harbours and good supplies of wood, water and food at King George Sound which had been missed by D'Entrecasteaux in December 1792, had been discovered and made known by Vancouver who was so impressed with the region that he took possession of the land there for Britain on 29 September, 1791. The region was later more thoroughly surveyed by Flinders in December 1801.

The other place of call, Shark Bay, had been explored by Baudin himself, and by Hamelin. However, Baudin claimed he wanted to go there primarily to catch turtles to eat.

Achievements of Baudin's second visit to western Australia

The two main contributions to knowledge made as a result of the revisit to western Australia in 1803, were the discovery of Koombana Bay and Leschenault Inlet the site of present Bunbury Harbour which formed an excellent, safe small boat harbour on the east side of Geographe Bay; and the making known of the correct shape of the coast of western Australia from Cape Hamelin northwards. The earlier chart of the area made in 1801 by Baudin's men was mis-shapen, primarily because of the difficulty to correctly determine longitudes and compass directions. By the time Baudin returned his men were experienced observers. Consequently the charts they redrew on the last visit presented a shape which looks more like western Australia.

Of course French scientists gained further as a result of the second visit. Large numbers of specimens were collected from the areas visited to help make up a comprehensive collection of the natural history of the continent, even though they came only from a small area.

Freycinet arrives at King George Sound and his difference with Baudin

At King George Sound, Freycinet and his crew paid for their folly

BAUDIN EXPEDITION, 1803

ROUTES FOLLOWED BY THE SECOND BAUDIN EXPEDITION TO WESTERN AUSTRALIA 1803

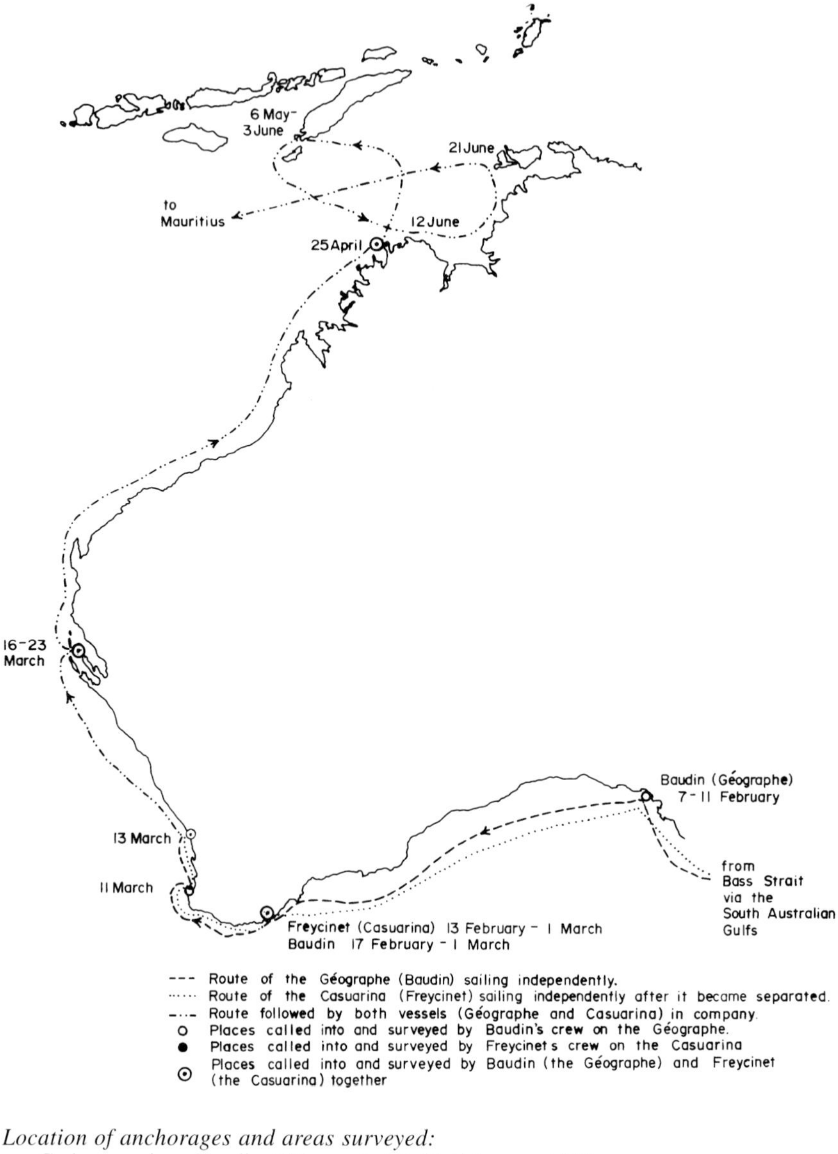

Location of anchorages and areas surveyed:

Ceduna region, Baudin (Géographe) 7-11 February 1803.
Albany region, Freycient (Casuarina) 13 February–1 March 1803.
Baudin (Géographe) 17 February–1 March 1803.
Geographe Bay and Bunbury region, Baudin (Géographe) 11 March 1803.
Rottnest Island, Freycinet (Casuarina) 13 March 1803.
Shark Bay, Baudin (Géographe) and Freycinet (Casuarina) 16-23 March 1803.
Bonaparte Archipelago 25 April 1803.

in losing sight of the *Géographe.* They made only a limited contribution to the scientific work of the expedition there, which even then was not officially recognized and accepted by Baudin, because Baudin gave Freycinet no chance to participate in the research programme carried out there.

This was caused in part by Freycinet having to take time off to repair his ship. But after the *Casuarina* was repaired and ready to be worked, Freycinet who had asked for nails and other supplies, was refused this, and was viewed by Baudin to be an incompetent waster, and consequently was set to work counting and listing his supplies while others made surveys.

The whole stay at King George Sound in fact was marred by a bitter division between the two men. The cause of this was very much the same as the cause of previous differences Baudin had with his

VANCOUVER'S MAP OF KING GEORGE SOUND

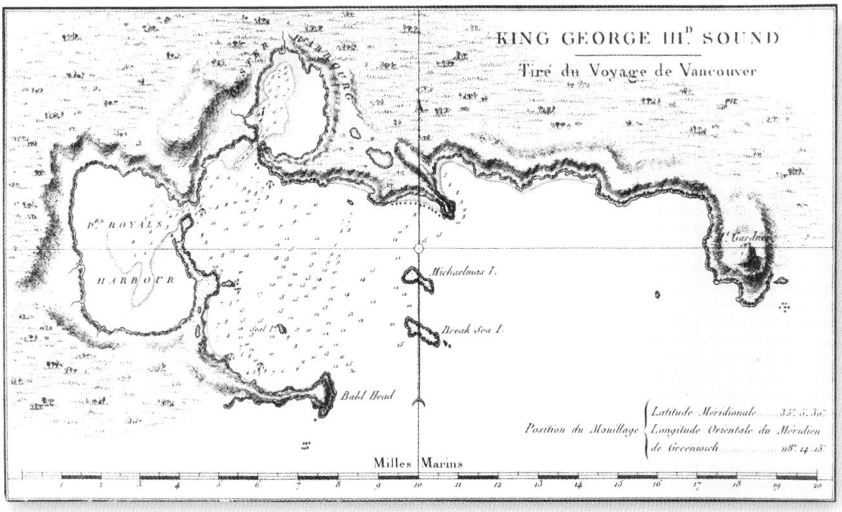

officers and the scientists. Baudin maintained that Freycinet could not keep appointments and was therefore insubordinate. Freycinet, on his part, maintained Baudin would not listen to reason, nor accept explanations.[25]

What irked Baudin at King George Sound was not only Freycinet's separation at sea, for which Freycinet was soundly reprimanded, but also the fact that Freycinet took the *Casuarina* into the inner harbour of Princess Royal, where Baudin did not want to go. This denied Baudin the opportunity to directly supervise all of the work of the expedition, which he did not like.

This situation was in part Freycinet's fault. He first went to the usual watering place on a bay south-west of Seal Island in King George Sound, where there were two fresh water streams used for supplies for ships. However, he then shifted the *Casuarina* into Princess Royal Harbour before the *Géographe* arrived, and left no sign or message that he had been at the outside watering place. Baudin found him by firing his cannon.

Certainly the situation in the inner harbour was better for careening and repairing the *Casuarina.* The waters were shallow and still which made beaching the craft safe. The wood and food supplies were also better in Princess Royal Harbour, and the fresh water was found to be cleaner and more pure. Freycinet had scorbutic men in his crew, and had a duty to find the best place to nurse his men back to health, but he also was moved to anchor in the inner harbour by the desire to be free and independent of the *Géographe,* which made Baudin's assessment of him correct. He was not prepared to be kept at heel. In acting as he did, finding a lone anchorage where Baudin could not follow, he raised the suspicions of his commander. To make matters worse Baudin's scientists and officers used Baudin's boats to visit Freycinet, where they had jovial parties which were not available on the *Géographe,* thus making the *Casuarina* seem like a refuge which further upset Baudin.

Freycinet claims in the history of the voyage he wrote with Péron, that he was one of three parties detailed to survey and chart the harbours. This was not so. Baudin sent out only two official survey parties.[26] Freycinet certainly made a survey of Princess Royal Harbour, but he did this on his own initiative without it being recorded in the log of surveys ordered by the commander. The commander's log merely records that Freycinet received two sound dressings down — one for coming alone to King George Sound — and the other, "delivered for the good of Freycinet and for the good of the naval service" — for failing to give a satisfactory report of the stores on his ship which he was made to do while others were surveying.

Freycinet made one important find in Princess Royal Harbour, apart from the plentiful supplies of geese, ducks, oysters, seals and fish supplies and good fresh water in the vicinity of his anchorage off Pagoda Point on the west of Princess Royal Harbour. Before he left the bay near the former whaling station in the outer harbour, he found on 15 February a bottle with a message left by Flinders on the *Investigator.* This recorded that the *Investigator* had called and had found water and wood. It also advised that the entrance to Oyster Harbour was guarded by a shallow bar. It then stated that the *Investigator* was bound for Port Jackson via the south coast, and requested the finder to advise the British Admiralty of this.[27] Freycinet later gave this to Baudin, on 28 February, just before departing from King George Sound, for him to report the matter.

Baudin's arrival in King George Sound

In the other, outside anchorage, Baudin settled down to hard work straight after anchor was dropped and the *Casuarina* had been signalled for. Two springs were found in the cove east of Waterbay Point. The water there was somewhat discoloured but tasteful, so tents were sent ashore and erected, one for the sick, the other for the naturalists.

Men were then set to work to dam one stream to make a pool to wash the ship's laundry. Others were set to work filling the ship's water casks from the second spring. Baudin had a viaduct built from this spring to the beach to help on the latter task, but this did not work so the job had to be done by hand.

On 20 February two expeditions were despatched and Bernier the astronomer was sent to Mistaken Island — called Observatory Island by the French — to set up an observatory. This was subsequently moved to the point on the mainland opposite, after a bushfire caused by the camp fire almost destroyed the observatory on the island.

Ransonnet surveys the coast eastward to Two People Bay

One expedition left in a long boat with ten day's supplies, under the command of Midshipman Ransonnet, to survey the coast eastwards between Mount Gardner and Bald Island. Baudin had passed in sight of these on the way in, and had noticed some openings. Ransonnet's task was to see if these were usable harbours.

This expedition returned at 6 p.m. on 27 February with good news. Two People's Bay and other good anchorages had been discovered and charted, proving the trip worthwhile. But more exciting for those who stayed in King George Sound was the news that the party had made close contact with friendly Aborigines. This was the first time this had happened to the Baudin expedition in the south-west part of Australia. Previously the Aborigines had been either timid or hostile. In this case in a bay near Bald Island, eight Aborigines were seen. These came and shared the food prepared by the party, and were given buttons, old coats and other presents. Ransonnet presented a pen portrait of them when he returned. He reported that they were tall and very agile, they had long dark hair, black eyebrows, a short flattened nose, deep set eyes, a big mouth, jutting lips and very sound clean white teeth. The inside of the mouth was observed to be black like the outside skin. Unlike the Aborigines in Tasmania, the men in the vicinity of Bald Island would not let their women near the French sailors, Ransonnet added.

Midshipman Baudin surveys Flinders Peninsula

The second survey party left in a longboat with six days' supplies, under the command of Midshipman Baudin, who was not related to the commander. He was accompanied by Faure the geographer and was ordered to map the coast from Flinders Peninsula, near the anchorage, northwards and then east to Mt Gardner to link up with the plan being made by Ransonnet. This work was done without incident.

Captain Baudin explores Oyster Harbour and the Kalgan River

The day after these survey parties left, on 21 February, Baudin himself took a party in his small boat to examine the inside harbours.

PRINCESS ROYAL ARBOUR AND OYSTER HARBOUR WITH THE KALGAN RIVER (RIVIÈRE DES FRANÇAISES) 1803

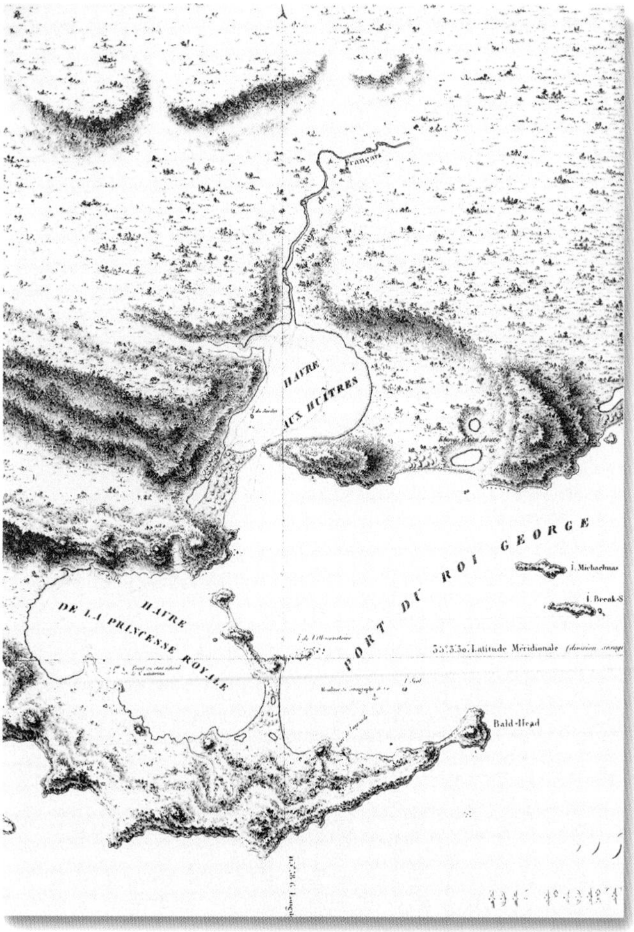

FINAL MAP OF KING GEORGE SOUND AFTER THE COMPLETION OF SURVEYS BY THE BAUDIN EXPEDITION, 1803

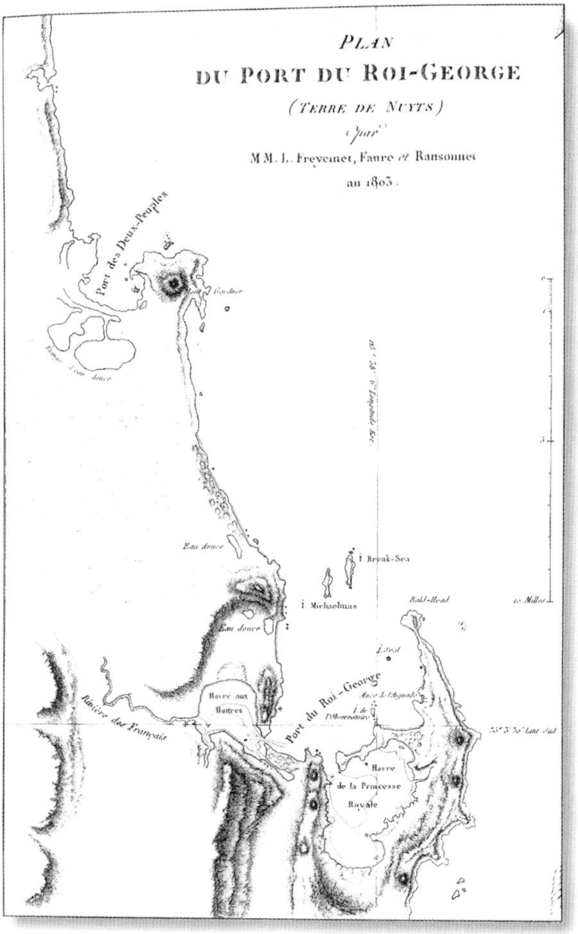

He first visited Princess Royal Harbour where he was not impressed, neither by the bay, nor by the anchorage Freycinet had selected, nor by the *Casuarina*. He found the schooner dirty and disordered, the bay shallow and shoaly and the anchorage too far from the beach. From the *Casuarina* almost a mile of shallow flats had to be crossed to reach the beach near Pagoda Point. This, he complained, meant men wasting time. He consequently ordered Freycinet to hasten repairs and shift the *Casuarina* to the *Géographe's* anchorage within two days.

Baudin then headed for Oyster Harbour which had been favourably commented on, but not thoroughly examined by Vancouver, which Baudin knew about. The party reached the mouth of the harbour that evening and set up camp on the left bank, under Mt Martin. From there some Aborigines were seen that evening but disappeared quickly, much to the disappointment of the party.

Young Baudin was in the same vicinity at the time. Faure the geographer was consequently invited to join Baudin's party to help explore inside the harbour. He had already been inside as far as Green Island with Midshipman Baudin to see if the seeds planted there by Vancouver had taken root, but found only local native growth.

The following morning Baudin set out to go to the northern limits of the bay. The highlight of this trip was his discovery of the Kalgan River, called Frenchman's River by the French, which he followed for nearly six miles, until his progress was stopped by Aboriginal fish traps across the stream.

Aboriginal "monuments" discovered

On the way up, about four and a half miles from the mouth, Baudin's party reached a freshwater stream and found traces of Aborigines which provided one of the most interesting points of speculation and controversies in the whole voyage. On each side of the stream they discovered "monuments". These were nothing like the "temple" found by Péron near the Vasse in 1801. Each of these was about seven or eight feet from the bank of the stream and set in a bare patch several feet wide. Each "monument" was surrounded by eleven spears with red tips. These were tested to see if it was blood, but found it to be resin.

A variety of different interpretations based on different opinions appeared to explain the "monuments", which in the end added to earn Baudin further disrespect. He entered into the speculation and controversy, although he was a seaman and navigator and not a trained scientist, claiming that in his view the monuments were the graves of two great warriors who had fallen either in battle or in personal combat with each other, and that the spears across the stream indicated their continued defiance of each other. Baudin consequently put a medal and some beads on each "grave", and nearby he planted maize and other garden seeds.

The other more general theory put forward about the "monuments" was no less imaginative but less open to ridicule. This was that the "monuments" marked the limits of tribal territory. Unfortunately the ship's artists did not visit the site to draw exactly what was seen so that no clear picture of what the "monuments" were is available.

Baudin returned to the *Géographe* on 23 February, leaving Faure to continue his survey towards Mt. Gardner. On the way back he called in to Bernier's observatory to see the situation there. One further incident served to highlight the rest of Baudin's stay. An American sealer, the *Union,* under Captain Pendleton came to the anchorage the day Baudin returned. A close rapport developed between the two, reflecting the continuing friendship between the two revolutionary, anti-British nations. Dinner was arranged and the American captain was given charts and information to help him hunt seals eastwards, although he was asked not to kill the pigs and poultry Baudin had put on Kangaroo Island to produce supplies there for ships to take as stores.

Baudin departs from King George Sound

By 1 March all the surveys were completed and the scientists were back on board, and anchor was weighed. The general impression gained was that King George Sound was a most satisfactory refreshment point, by far the best so far located in western Australia. It seemed in fact the only oasis in what Boulanger called "a silent desert" which he felt western Australia to be.

The scientists were most satisfied about the stay. The botanists alone collected nearly two hundred new species of plants, and the zoologists were no less satisfied, particularly about the collection of shells.

Before the expedition sailed, Freycinet was given strict orders to stay in Baudin's company, and was told that Baudin would take the *Géographe* to St Allouarn Island to stay for a day, after which he would go to Rottnest Island for three or four days and then on to Shark Bay. Rottnest Island was set as a rendezvous point in the case of accidental separation.

The *Géographe* and the *Casuarina* again separated

It is not easy to say who was responsible for Freycinet getting lost a second time on that trip not far out of King George Sound. The only thing clear is that it did not improve Baudin's temper, nor his faith in Freycinet.

After passing West Cape Howe on the way to Cape Leeuwin, Baudin saw some large lagoons inland and despatched the *Casuarina* to investigate and map the region. Freycinet set out to do this on 5 March, and did not see the *Géographe* again until it arrived at Rottnest on 13 March where, fortunately for him, Freycinet's excuse that mist had obscured his view was accepted by Baudin and he was not relieved of his command as Baudin promised himself. Freycinet in

PROFILES OF THE COAST

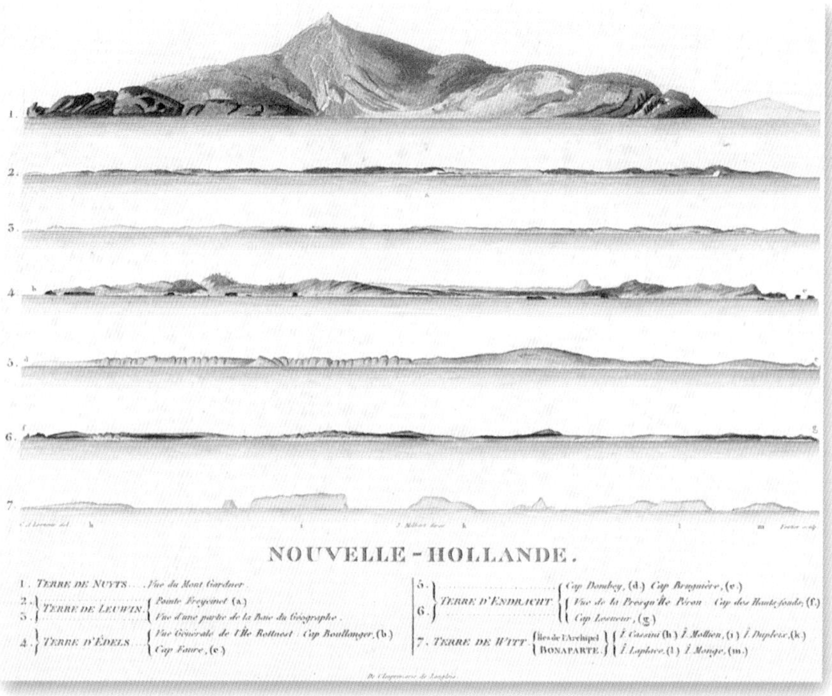

the meantime made a detailed survey of the coast from near William Bay westward, and then after sailing past Cape Leeuwin, moved into Geographe Bay where on 9 March, 1803 he discovered Casuarina Point which forms Koombana Bay on the east side of Geographe Bay, but which he did not closely examine. He then sailed close along the coast, exploring and mapping Warnboro Sound near Cape Peron on 10 March, 1803 and then moved on to Rottnest Island where he anchored to the north of Duck Rock off Thomson Bay to await the *Géographe.*

Baudin consequently went to St Allouarn Island alone where he did a brief and unimpressive survey.[28] He did not clear up the mystery of the location of Cape Leeuwin. He seems to have confused this with Cape Beaufort or some point further east. He consequently quickly fixed the position of St Allouarn Island, placing this far to the west of his supposed Cape Leeuwin, leaving the mystery of the actual form of that part of the coast for British navigators to solve.

He ran into one danger near here. He nearly ran onto rocks south of Cape Leeuwin, which frightened him and the crew, who feared to be cast away on "the sterile coast".

Baudin then followed the same course as the *Casuarina.* He

PROFILES OF THE ISLANDS AND COAST,
NORTH WESTERN AUSTRALIA

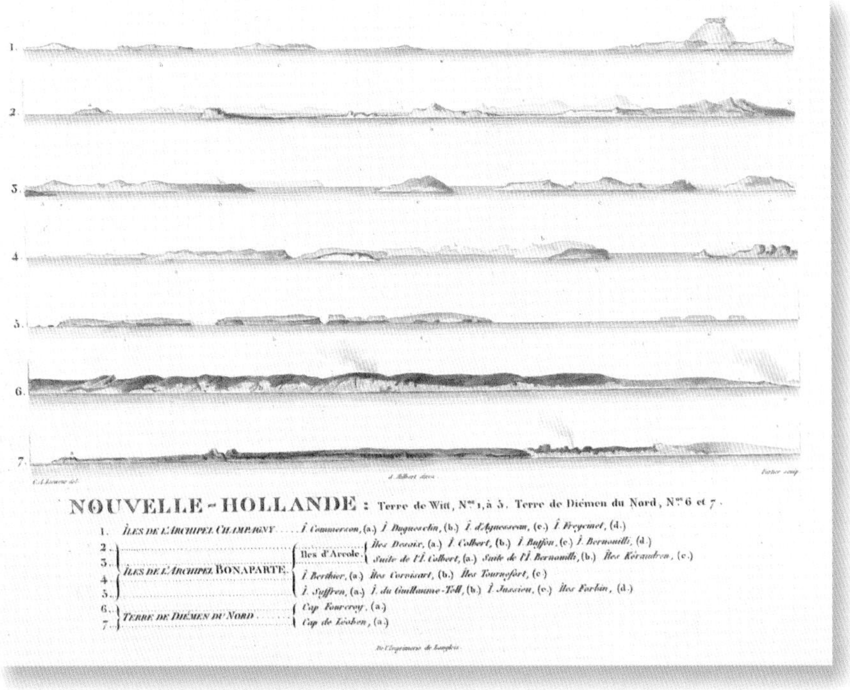

entered Geographe Bay two days after Freycinet, on 11 March. He endeavoured to land to pick up the salvage equipment he lost when trying to free the wrecked longboat near the Vasse, but was prevented from seeing this by the mist. He therefore sailed on and independently discovered Koombana Bay, noting that he saw many more fires on the way on this voyage than on the previous one.

Baudin surveys Koombana Bay and Leschenault Inlet

The *Géographe* was anchored at midday on 11 March, west of north from Casuarina Point and Midshipman Bonnefoi (called in the printed history of the expedition Montbazin) was sent in a boat to survey and chart the area. Bonnefoi not only charted Koombana Bay which he found to be a shallow but satisfactory shelter for small boats, but also discovered and partly mapped Leschenault Inlet to just beyond Middle Island which he named. The inlet itself was named Port Leschenault after the ship's botanist. Bonnefoi had little time to collect specimens, but wrote an enthusiastic report about the wild life seen in the area.[29]

FREYCINET'S MAP OF THE COAST FROM NEAR MANDURAH NORTH TO NEAR MULLALOO

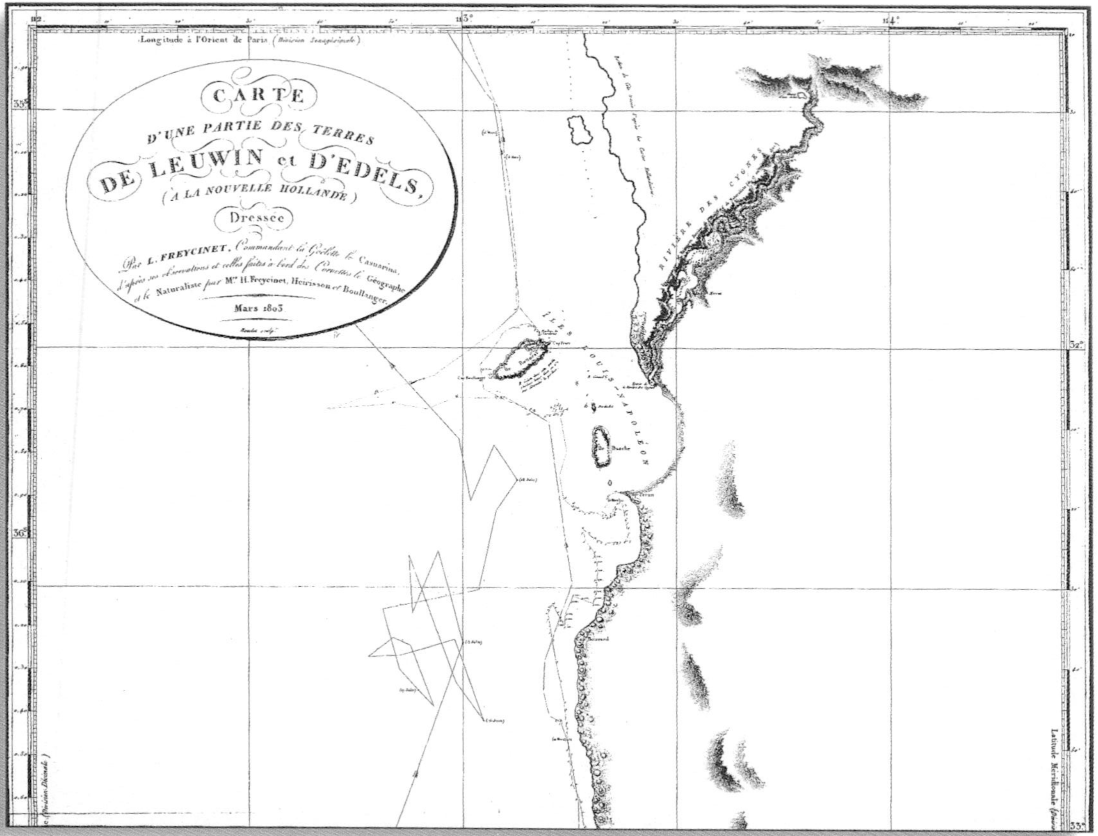

The *Géographe* and the *Casuarina* meet at Rottnest

After this brief stay Baudin sailed to Rottnest, keeping well out from the coast, meeting Freycinet there on 13 March, where Freycinet received his expected reprimand.

Baudin did not stay at Rottnest as planned. He considered it to have been sufficiently well surveyed by Hamelin, and not an important nor a useful enough place to spend more time on. He consequently immediately left with the *Casuarina* at heel, and sailed out to sea from the coast towards Shark Bay with Freycinet complaining about the distance out and the lack of opportunity this gave to observe the form of the coast.

Arrival at Shark Bay

The expedition arrived at Shark Bay on 16 March. This time Baudin took the Naturaliste passage in. His main aim there was to get turtle meat, as has been noted. Midshipman Ransonnet was sent off in the longboat to get these in Hamelin Pool on 17 March, after the ships were anchored in Dampier Road. Other boats were sent to catch fish. One of these on its return, reported a hostile demonstration by a large group of Aborigines on the beach. Baudin was consequently moved to investigate.

The *Casuarina* in the meantime was sent off to survey the bay to the north to see if it was possible to sail from Dampier Road to the northern passage. Why Baudin viewed this as a mystery is unknown. St Allouarn had sailed that way in 1772 and had marked his course on his chart, although he did not go all the way and take the northern exit. Nevertheless he showed it was possible to sail between Dorre Island and Peron Peninsula which is what Baudin wanted to find out.

Aboriginal settlement discovered

While waiting for Freycinet's report, Baudin sent the scientists ashore to see if the Aborigines seen could be contacted. One party was sent under naval engineer Ronsard. His party discovered a considerable settlement on the north end of Peron Peninsula consisting of twelve to fifteen huts. But there were no Aborigines. They had departed, apparently hurriedly, for fires were still burning and utensils remained.

The following day, 19 March, Midshipman Bonnefoi was despatched with a party of scientists and the artist Petit to draw the camp and the Aborigines, if they could be found, and to boil water to get salt to lay down the fish being caught. This expedition resulted in

THESE THREE MAPS SUMMARIZED FOR BAUDIN THE EARLIER SURVEY WORK
WHICH HAD BEEN CARRIED OUT AT SHARK BAY

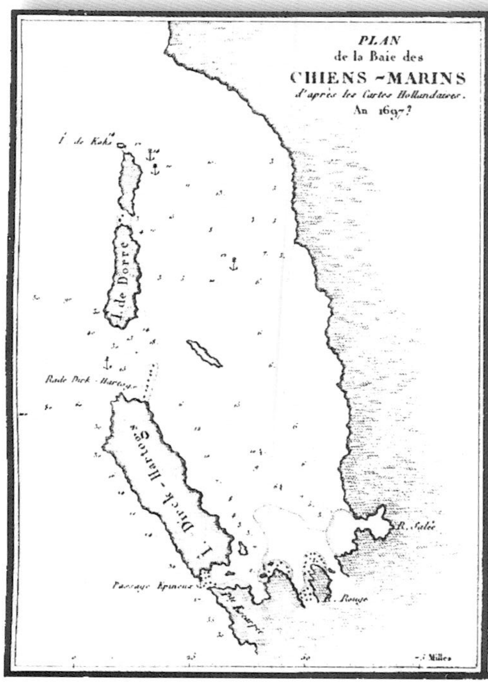

DUTCH MAP OF
SHARK BAY, c1697

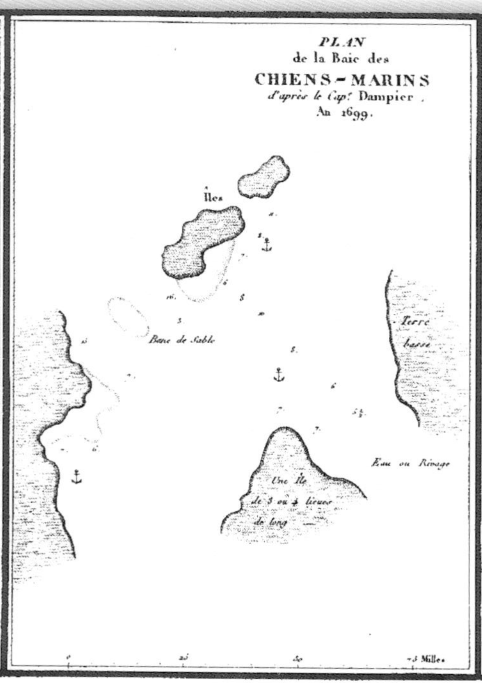

DAMPIER'S MAP OF
SHARK BAY, 1699

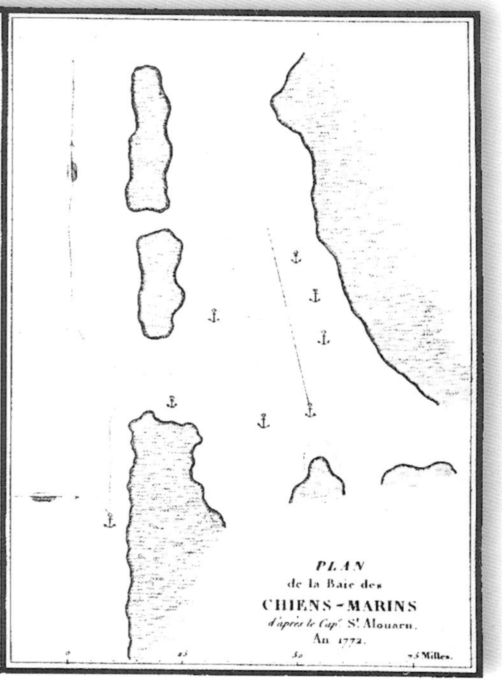

ST. ALLOUARN'S MAP OF
SHARK BAY, 1772

FINAL MAP OF SHARK BAY AFTER THE COMPLETION OF SURVEYS BY THE BAUDIN EXPEDITION, 1803

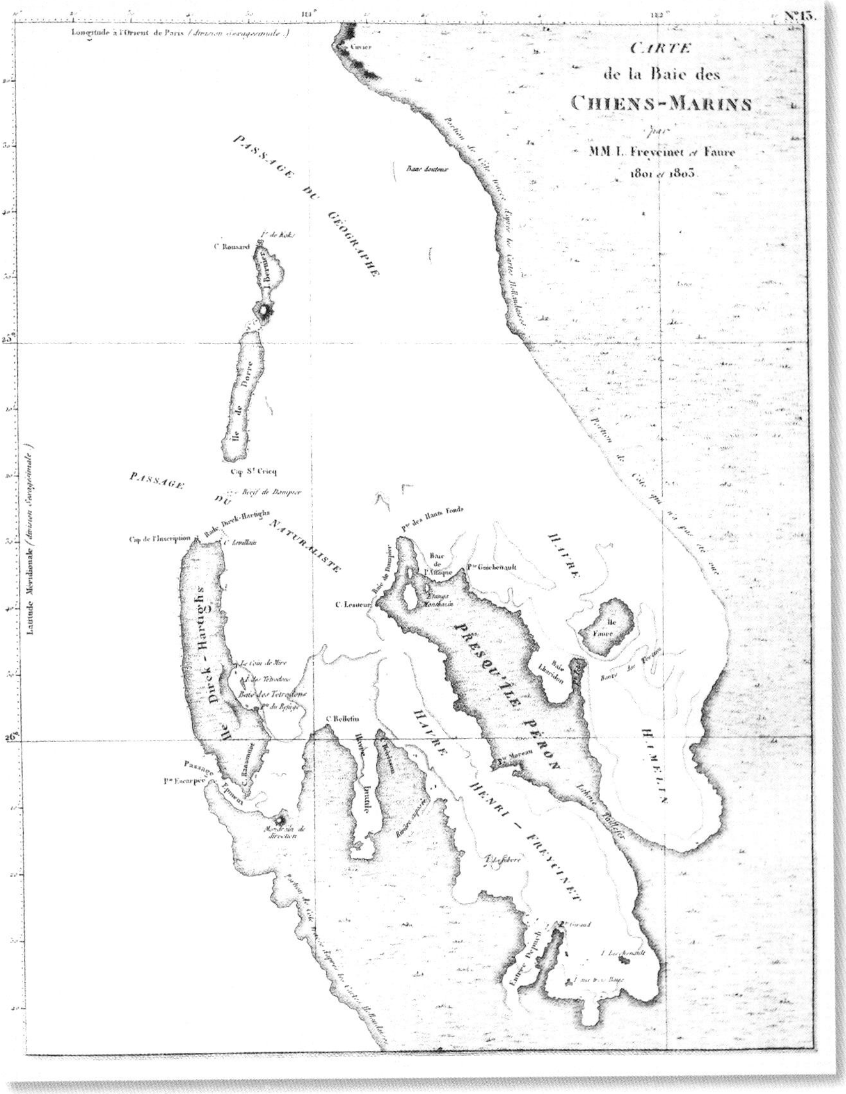

further upsets for Baudin which did not improve his temper or his failing health. Bonnefoi had orders to come back that night. He did not return till the next day to receive a dressing down and to be informed that the cost of the gunpowder used in the signal cannon which Baudin had fired, was added to his expense account.

The real cause of the trouble was Péron had again got lost. Once ashore he talked Petit the artist and Guichenot the gardener's assistant into walking across Peron Peninsula. This proved to be a longer walk than Péron expected. They consequently returned late and exhausted. To make matters worse Petit had sketched only an empty village and nothing else, making his trip ashore seem futile to Baudin. He was expected to make detailed sketches of all he found in the camp, but did not do this. He did see some Aborigines during his walk with Péron, but these had adopted a threatening attitude, and he did not stop to sketch them.

Bonnefoi thus had been placed in a difficult position. He did not want to leave the lost scientists alone and unarmed in what appeared to be hostile territory, so he stayed and incurred Baudin's wrath for not coming back on time, which was not soothed by his excuse, nor by the news that a large supply of salt had been gathered from a dried rock pool to be used for the fish.

Midshipman Ransonnet returned from turtle hunting on 22 March with a disappointing catch of twelve turtles among which were only three large ones. Plenty of turtles had been found but these proved hard to catch. The crew had to go into the water to catch them, which was not easy and was made more difficult on one occasion by the presence of a shark which battled with them for the prey.

Departure from Shark Bay

Freycinet also returned with bad news. He had not had time to survey the passage. Baudin consequently called a halt to the operations at Shark Bay and weighed anchor on 23 March. The expedition then headed north along the coast for Timor making a cursory survey on the way. Kupang was reached on 6 May, 1803. There Baudin took ill, vomiting blood. In a poor state of health he set off on 3 June to complete his work by making a survey eastward along the north coast of Australia from where he had left off in 1801 at Joseph Bonaparte Gulf, intending to go to Cape York and examine the straits between Australia and New Guinea. However, his illness prevented him from carrying out his task. He merely sailed as far as the western side of Melville Island. Then seriously ill he turned back on 28 June 1803, and headed for Mauritius, where he arrived on 7 August 1803. Baudin died there several weeks later. The *Casuarina*

WESTERN AUSTRALIAN BLACK SWANS,
EMUS AND KANGAROOS IN THE GROUNDS OF
NAPOLEON'S PALACE AT MALMAISON

was left at the island, the expedition sailed home under the command of Milius, reaching Lorient on 25 March, 1804.

As far as science is concerned, the Baudin expedition made a great contribution, bringing back to Europe far more specimens than Cook had done. The charting done was not precise. But the shape of the Australian coast was made known, although Baudin had to share the honour of doing this with Flinders.

However, the Baudin expedition did not really solve the question of whether or not the western part of Australia was of possible use or interest to Europeans. Opinions remained divided. Some Frenchmen like Rosily, who visited the area with St Allouarn were impressed. Others, including Baudin, viewed it as a large silent desert with little potential. His surveys revealed only one safe, sheltered anchorage on the west coast at Shark Bay. But this had no water. The only other anchorage he found to be of use was at King George Sound.

There were whales, seals and fish which the expedition believed could be exploited commercially. Some graphite which could be of use had been discovered among other minerals at the Swan River.[30] But in all the resources did not excite interest. In any case after Baudin's return Napoleon's France faced troubles. The real French interest in western Australia as a place of settlement was consequently left until the Restoration period, when new outlooks emerged in France.

D'Entrecasteaux Expedition, Esperance 1792

PART III
Restoration France and western Australia

Dessiné par Marchais d'après A. Pellion. Gravé par Forget.

NOUVELLE-HOLLANDE, Baie des Chiens marins. CAMP DE L'URANIE, SUR LA PRESQU'ÎLE PÉRON.

Freycinet's survey camp on Peron Peninsula with a fresh water distillery, managed by his wife Rose.

CHAPTER 7

Freycinet's Further Exploration of Shark Bay, 1817

The end of the Napoleonic link with Western Australia

Baudin's unenthusiastic report about the lack of facilities and the scarcity of water and other resources in western Australia, which were not substantiated by his research workers, brought an end to the Napoleonic link with the region. The whole of Australia, in the view of the French, was not worth a priority rating among the places viewed as possible strategic bases and support areas sought and developed as Napoleon's empire rose to its height at home and abroad in the years after the Baudin mission visited Australia. The records of the French voyagers showed that both the eastern and western parts of the continent in the temperate zone where France preferred to establish self sustaining support depots, were inhospitable. The western part appeared unsuited for European man to colonize, and the greener looking more attractive eastern area appeared unsuited for the French because of the touchiness and hostility of the British who claimed it was all theirs. A Napoleonic settlement in close proximity to the British was out of the question. Such a move would only serve to recreate the tensions and troubles and costs experienced in Canada which led to bitter colonial rivalries and wars there. Napoleon therefore maintained Mauritius, unsatisfactory as it was, as the main French strategic and supply base in the Indian Ocean.

It was to there that French strategists were appointed,[1] and it was from there that French power was extended as Napoleon expanded France and French influence in the Indian Ocean region as part of his imperial expansion policy after the Treaty of Amiens of 1802 brought a peace of sorts to Europe.

The scientists who had earlier been to the forefront, pressing for a French interest and activity in Australia, faded into the background where they busily engaged themselves in research, using the rich collection of Australian specimens brought back by Baudin's men. In this sense the Baudin mission was a complete success outdoing that of La Pérouse, and far outshining the scientific work of Cook. Baudin, with all his faults, succeeded in bringing back the largest collection of new specimens gleaned from foreign parts ever to come to Europe for study.[2]

Even if Napoleon had ambitions to make a section of Australia part of his world empire, he would not have succeeded. Although he was yet to rise and bring France to its height when the Baudin mission was ended in 1803, the seeds of decay and failure for France were already evident.

Primary among the factors which brought about the downfall of Napoleon was Britain's continued command of the seas, and

FREYCINET EXPEDITION, 1818

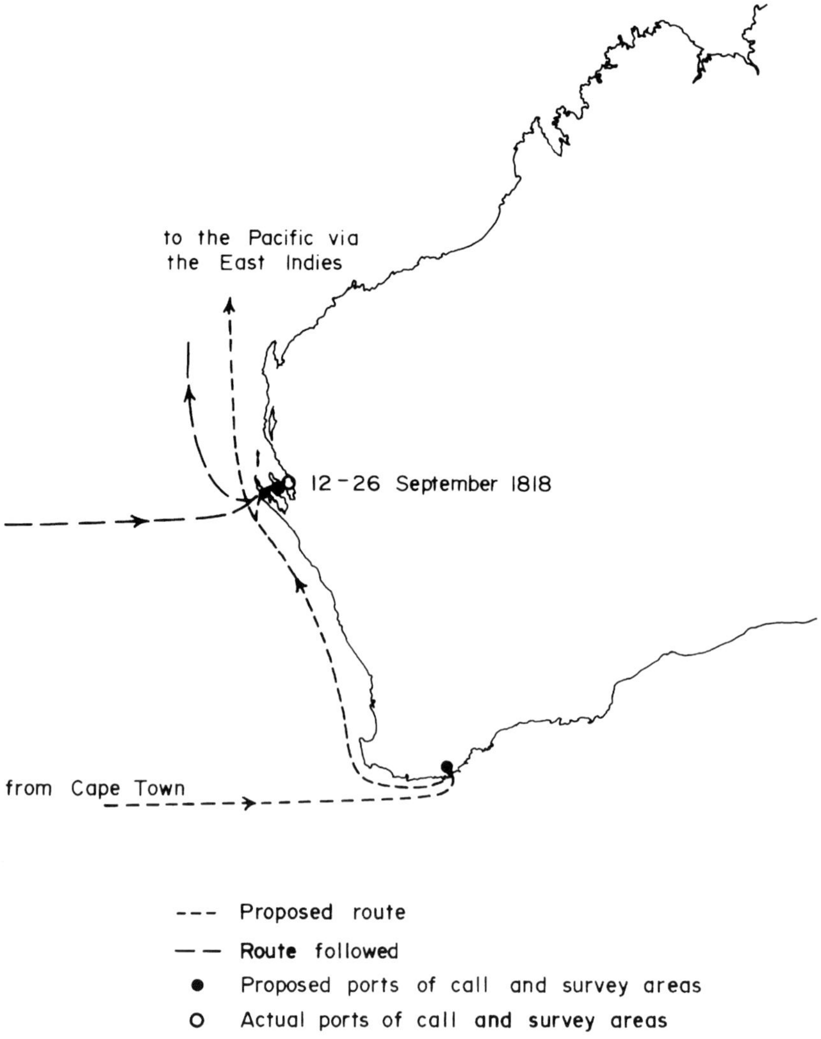

Location of anchorages and area surveyed:
Shark Bay, 12-26 September 1818.

Westminter's continued reluctance to let France hold power there either in war or in peace. The uneasy peace of Amiens made between the two powers in 1802, once more gave French naval squadrons and merchant ships free access to the world. Napoleon immediately miscalculated and made a disastrous move for France in using this new found freedom for his large naval fleet. Immediately after the peace treaty was signed he despatched a large force to Haiti to crush the black revolt led by Toussaint Louverture[3] against the French there. This war was not only unwinnable. It had severe repercussions for Napoleon. His attempt to re-impose the colonial rule of France and condone slavery by force once more in the West Indies, lost him the respect of intellectuals and others who believed in the rights of man. By contrast, Britain with its anti-slavery crusade and message of democracy appeared to be the power genuinely applying the principle of liberty, equality and fraternity which had been the rallying cry of the French revolutionaries. The despatch of a large French force to the sensitive area of the West Indies, from where Britain got its sugar supplies immediately served to alert Britain to the continued threat of France. The peace established at Amiens consequently did not last long, and France as a result, lost its power at sea once and for all at Trafalgar in 1805. After this Napoleon did not have the ability or resources to support military or colonial ventures overseas. His plans to move into the Middle East, hatched by Sebastiani, and his schemes for India which were the main centres of interest in Napoleon's imperial schemes for the Indian Ocean area after the Baudin mission, were doomed to failure. The fault was primarily that of the French navy. It was a formidable force and had the added power of Spain and the forces of other French allies attached to it. But under the dour indecisive Decrès who built up the force, but who did not seem to know how to use it, the navy was ineffective and seemed to lose confidence that they could achieve victory. The French navy often won battles, but these were short lived gains and seldom of strategic importance. In the Indian Ocean in particular France under Napoleon lacked a Suffren. The active commander there, Linois[4] certainly kept the British busy in India and about the trade routes there, but the real hero and popular figure was the rich merchant privateer Robert Surcouf[5] who worked out of Mauritius, raiding British convoys. In 1809 he returned to St Malo with prize cargoes reported to be worth five million francs, to help the ailing French economy. In consequence of this, Britain reacted by occupying the French bases and colonies, including Mauritius which the French navy did not and could not prevent. Thus by 1810 Napoleon had no overseas empire to rely on.

A second factor contributing to Napoleon's failure to succeed in expanding eastwards towards Australia was his continued preoccupation with Europe, and his need to give a priority to that area. The result of Trafalgar made little difference to his continental progress. He continued to rise and expand until he reached the height of his power at the Treaty of Tilsit in 1807, when Russia was humbled and joined his continental system. But after this Napoleon declined, and his Europe system was soon discarded. Peoples who had been offered help by French revolutionaries in 1792, to achieve "liberty", by the end of the first decade of the nineteenth century had turned to demand "liberation" from Napoleon's continental system dominated by France. A new Europe consequently was struggling to emerge as Napoleon's empire faced troubles and collapsed.

France suffered economic difficulties primarily as a result of the British blockade, and the policies adopted by Napoleon in consequence cost it the support of its allies and lost it the sympathy of its colonies.

To solve the economic problems Napoleon at first, at the time of Amiens, turned to neo-Colbertism to save the situation. He attempted colonial reorganization designed to weld his empire into a whole, and sought to build economic security for all behind protective trade barriers. This effort was destroyed by the British at Trafalgar. From the end of 1805, France and its allies were blockaded. Not even the French coast was secure from the British navy, and French traders could no longer use even the Mediterranean where Britain secured Malta and other bases.

Napoleon consequently resorted to protectionism, creating a continental co-prosperity sphere. French scientists helped materially to make this common market into a success by using their laboratories to find new European commodities to replace those lost when the colonies were cut off. Sugar beet, for instance, was developed by French scientists to replace the lost West Indian sugar.[6] But the scheme for substitute commodities was neither popular nor a success. It eventually cost France its satellites and allies. Spain and Portugal were the weak points in the French continental scheme. France had no chance to police the large coastline of the Iberian peninsular after Trafalgar, and was no match for the Spanish guerillas who rose to fight for independence and liberation from Napoleon's rule. Britain took the opportunity and moved into the area, opening a war front on the continent. Napoleon consequently found his frontiers receding and his occupied territories revolting. By 1813 the die was cast. Britain and Europe joined in a common cause to end the French dominance which was eventually achieved at Waterloo.

Renewed interest in scientific exploration after the restoration.

The restoration of the Bourbon monarchy in France after the fall of Napoleon was followed soon after by renewed efforts to complete the exploration of the southern oceans. This new quest for scientific knowledge, which brought the French back to examine then still little known western Australia, was made primarily for scientific reasons, and was a logical move. French scientists for decades before had held a dominating position in Europe, sharing prominence there with Napoleon and his empire. When the emperor fell, French scientists had no desire to see their research institutes collapse and their disciplines and studies disappear. They consequently exerted themselves to promote French science which soon became one of the accepted ways to restore French prestige and respect. Scientific missions designed to emulate the circumnavigation of Bougainville and La Pérouse which had brought glory to earlier Bourbon monarchs, were despatched abroad not long after the new government settled down in Paris. Four fact finding missions were specially ordered to survey south western Australia between the restoration of the Bourbons and the final acquisition of western Australia by Britain which marked the end of French interest. Freycinet was sent on a first mission in 1817. Following this Duperrey was sent in 1822, Bougainville in 1824 and D'Urville in 1826, although only two of these succeeded in reaching their objective.

Most significantly, in the course of these explorations there was a sudden shift of interest among French administrators from science and knowledge for its own sake to politics. In the period after Freycinet's mission, the French government for the first time made specific plans to colonize western Australia, in order to realize the long held Bourbon dream of having a temperate base in the Indian Ocean to match British controlled South Africa, which makes the restoration period of particular interest.

Difficulties in sending out missions of exploration

The sending out of the first scientific mission which served as the background effort to the later political drive for colonization was not easy. A series of barriers existed that made the despatch of missions overseas difficult. Firstly, many of the emigrés who returned to serve in the restored government had little interest in science and little desire to serve scientists who in many cases had supported the revolution and the empire which made them into exiles. Consequently when it came to the point of allocating funds, scientific research and exploration was not given a high priority by the restoration government. In the case of the Freycinet mission, it was the approval and personal support of the monarch Louis XVIII which helped get the ships allocated and despatched.[7]

Secondly and more significantly, the French navy which previously had supplied scientists with the means to go abroad to make researches, was no longer available for this purpose. As a result of the bad experiences on the D'Entrecasteaux and Baudin missions, civilian scientists were no longer accepted as crew members by the French navy. Scientific studies and the collection of specimens were consequently left for naval officers to make.

More significantly when peace was established in 1815, France did not have the naval and other facilities needed to support long range maritime expeditions. The restoration government was not only unwilling to allocate scarce resources to promote science at the expense of other projects aimed at reconstruction. It was also disinclined to spend on the navy. Many of the emigrés who served in the restoration government had landed interests and had no cause to help the navy re-establish itself. There was little reason for them to do this. The Treaty of Paris had stripped France of its overseas empire. In the Indian Ocean, for instance, France kept only the small island of Reunion which is volcanic and has no natural harbour and so therefore is not suitable for a base. In India, France kept only the right to trade with the five ports of Chandernagore, Yanaon, Pondichéry, Karikal and Mahé. Besides the loss of colonies, France also had little in the way of overseas commerce. This had been virtually ended by the British blockade and by Napoleon's continental system which aimed at European self-sufficiency. In these circumstances it was difficult to justify large grants for the French navy, and to justify using the scarce resources it had for voyages of circumnavigation.

The *Meduse* Affair

To make matters worse the morale in the naval service was low. After the Bourbon monarchy was restored, emigré naval officers were brought back into the service and given commands because of their loyalty, rather than because of their abilities. The bad effects of this and the tensions this caused was publicly and widely evidenced by the *Meduse* affair. On 17 June 1816 the ship *Meduse* was despatched to Senegal, the one remaining French colony in Africa, with a detachment of guards. The vessel was under the command of a former emigré, Captain Duroy de Chaumareix, who had little recent experience at sea. On 1st July he was nearing the African coast and standing into danger. He was warned of this by an experienced naval officer serving under his command, but the expert advice was ignored and the *Meduse* was wrecked on shoals. As there was insufficient life saving equipment a raft was built, and over 150 survivors boarded it.

This was not found by rescue vessels until 17 July when only 15 of the survivors remained. Feelings about this tragedy were naturally intense.[8]

But despite these reverses and bad features, there were avenues opened which soon led to hope for France as a maritime nation and for successful naval reconstruction. French traders and shippers, in particular those in Bordeaux, who had long suffered the limiting effects of the blockade and the Continental Decrees, restored French commercial links with the outside world. Trading ventures were sent from Bordeaux and other ports to Cochin China, China and the Pacific by venturers such as the Balguerie family. To consolidate and improve trade in the east, which came to be of special interest to France, Achille de Kergariou was despatched on the *Cybèle* in 1817 to restore links with Cochin China and to show the French Bourbon flag in eastern seas once more, and to report on trade opportunities.[9]

Portal and the growth of French naval power

At home while the Duc de Richelieu[10] worked at the Foreign Ministry to have restoration France accepted as a power in the new Europe after Napoleon, the Baron Malouet[11] worked at the Ministry of Colonies to reorganize the small empire France was allowed to keep, to make it into a workable empire. His early work was subsequently built upon by the notable administrator Pierre Barthélemy Baron Portal.[12] Under his administration the colonies were given a new change of prosperity. Crops such as cotton, which found a ready market in France, were encouraged in overseas areas to serve the interests of colonists and Frenchmen alike. Commerce was encouraged. Expansionists such as Sylvan Roux who worked to re-establish France in Madagascar were patronized and supported. Thus within a few years after the restoration France was once more a world wide commercial power with colonial interests.

Portal's efforts did not stop at colonial and commercial reorganization and reconstruction. In 1818 he took over the Ministry of the Navy and by forceful persuasion had the naval vote increased so that by 1823 France was once more a naval power to be reckoned with. This was not done without exciting bad feeling and opposition among former enemy nations. But Portal handled the matter skilfully, adding to the navy in order to help in the international effort to suppress the slave trade which was then an approved crusade. By 1823, as a result of Portal's efforts in particular, France was sufficiently powerful at sea to be delegated by the Congress of Powers at Verona to intervene in Spain on behalf of the Bourbon monarchs there who were threatened by revolutionaries. And by the end of that decade, the French navy was strong enough to play a key role at the Battle of Navarino (1827), to help resolve the Middle East problem. Subsequently the navy helped France expand to Algeria and other places which were acquired to form the next French empire.

Side by side with this development, a new interest in the sea was stimulated and kept alive by the poet Chateaubriand who rose to play a prominent part in restoration life and society.

Freycinet's voyage of circumnavigation on the *Uranie*

The navigator selected to show the Bourbon French flag around the world on the first naval scientific mission conducted in the restoration period was Louis Claude Desaules de Freycinet who had fallen foul of Baudin when on his mission, but had made a distinctive contribution in the field of chart work in particular.

After the end of Baudin's mission, Freycinet returned to France where he was employed in hydrographic work. On the death of his friend and colleague, François Péron in 1810, Freycinet was invited to complete the unfinished history of the Baudin expedition.[13] This was a monumental task, but was completed by 1816, after some troubles in the previous year about the dedication to Napoleon which caused a delay.[14]

When the work was completed Freycinet proposed another voyage of exploration, which was given the support of the Institute of France which recommended the matter to the French government.

The proposal soon found supporters in the naval and colonial ministry which was expanding its interests, and in the monarch who personally approved the scheme.[15]

Proposed Itinerary

The plan submitted was primarily for scientific research.[16] Freycinet was to sail by way of Rio de Janiero and Cape Town to south western Australia so as to arrive there in January 1818. There he was to examine King George Sound, sail west around Cape Leeuwin and make north along the coast to Shark Bay where he was to establish an observatory.

After this he was to sail north through the East Indies to the Pacific, then down to Port Jackson, and from there sail home by way of Cape Horn.

Freycinet was ordered to concentrate on six areas of research on the voyage. He was to study the geography of the Earth — the shape of the Earth — physics — natural history, in particular comparative anatomy — botany and mineralogy, in particular layers of rocks.[17]

To help in determining the shape of the earth which was then still a mystery, the vessel was ordered to make along a general route in the vicinity of the Tropic of Capricorn and establish observatories on a parallel close to that of Rio de Janiero which is primarily what led Freycinet to Shark Bay.

Proposal to make a scientific study of man

There after his observatory was set up, as in other places, he was to make a most comprehensive study of the native people and their environment for which a detailed plan of operation had been worked out in consultation with scientists in Paris. Observations were to be made of the resources and the fertility of the country. This study was to be added to by a study of the physical conditions, the physical qualities, the ages, the illness, the domestic life styles, the clothing, habitation, furniture, utensils and methods of warfare and social and moral outlooks of the native people.[18]

This information was to be collated and presented under seven separate headings with 596 sub-classes listed under these, which makes his expedition one of the most significant anthropological expeditions conducted by the French.

The seven headings were designed to provide a most descriptive analysis of each human society visited. These consist of:

1. The history of the people visited.
2. A geographical description of the country.
3. Their production.
4. Observations of the human species.
5. Their industries and arts.
6. Their commerce.
7. Their government.

Departure from France

Equipped with these orders and methods for research, Freycinet sailed from Toulon on 17 September 1817, on the small bluff built *Uranie,* a vessel of 350 tons. He had with him a complement of competent scientific naval officers. The ship's surgeons Quoy and Gasmard were naturalists as was Duperrey, one of the officers who was later selected to lead the next expedition. The ship's pharmacist, Gaudichaud was a competent botanist and Jacques Arago, a competent draughtsman, went in that position. He subsequently wrote voluminously about the voyage.

Among the rest of the competent crew there was one noted addition. Freycinet smuggled his wife Rose aboard with him to make the trip. She subsequently wrote her own delightful account of the voyage which helped add to the frame of the expedition.

The *Uranie* was well equipped with stores and materials. By then the French had learned to place great emphasis on health on ships.[19] Strict regulations for cleanliness were given to Freycinet to observe, and anti-scorbutics and good water containers made of iron were supplied to avoid the health problems which had adversely affected earlier missions.

Freycinet sailed the *Uranie* to Rio de Janiero by way of Gibraltar and Teneriffe, arriving there on 6 December.[20] The observatory was set up as planned, and scientific investigations on the subject of the shape of the earth was commenced in earnest. On 29 January 1818, anchor was weighed and the vessel made for Cape Town which was reached on 7 March. From there the *Uranie* was sailed to Mauritius which had been taken over by the British. There the vessel was overhauled and made ready for its long Pacific journey.

The journey Freycinet made to western Australia after he left Mauritius and Bourbon which he briefly visited, is disappointing for the historian and research worker. Worried about the state of his ship and his supplies Freycinet decided against going to King George Sound and made direct for Shark Bay which is the only place he visited in the region.

Arrival at Shark Bay

The *Uranie* approached the coast near the Naturaliste Channel at Shark Bay on 11 September 1818, when seaweed was observed floating. The coastline was seen by one of the look-outs at midday and the vessel made shorewards in high seas and a strong wind from the south-west.

The following day, 12 September, anchor was dropped approximately six miles off Dirk Hartog Island, the sight of which did not impress the crew who recorded that it was arid and lacked rivers and streams.

Removal of Vlaming's plate

On 13 September, a party was despatched under Fabré, accompanied by Quoy to get the historic plate left by the early Dutch navigator Vlaming, which had been restored by Hamelin. Freycinet, when he arrived, decided that it was unwise to leave the plate in an exposed position where it could be damaged or taken by irresponsible people, and therefore determined to take it to France for safe keeping.[21]

The search party found the plate with difficulty and brought it back to the *Uranie* where it was kept by Freycinet and then handed over to the authorities in France. There it was lost for over a century. It was eventually found in 1940 and was subsequently presented by France to the Australian government in 1947.

While Fabré and Quoy were ashore on Dirk Hartog Island, making for Inscription Point, the *Uranie* was moved across the inlet to a second anchorage off Peron Peninsula. There the vessel stayed from 13 to 26 September.

By that time water was short, so the alembic, designed to distill water, was sent ashore and lit. An observatory was set up close by on the north end of Peron Peninsula, and parties went out to search nearby to find local natives.

Contact with the Aborigines

These were seen not long after camp was established on the beach. A group of natives carrying spears was seen in the hills above the beach. These called out and demonstrated. In reply the French visitors danced and sang to show their good intentions. In order to con-

MEMBERS OF THE FREYCINET EXPEDITION MEETING ABORIGINES, SHARK BAY, 1818

solidate the establishment of friendly relations the resourceful Arago brought out his castanets and played a tune, which scene he has recorded in one of his sketches of the encounter with aborigines at Shark Bay.

LESUEUR'S SKETCH OF WONNERUP ESTUARY AND THE VASSE RIVER, GEOGRAPHE BAY

Unfortunately little information about the aborigines was recorded. The parties who met suffered from a communication barrier. Both sides called to each other but could not comprehend what the other meant. However, a description of what was seen was left by the ship's secretary, Gabert and others.[22] These descriptions are not flattering to the aborigines. "They are perhaps the saddest savages in creation" it was recorded. They were observed to be without clothes and poorly armed and of "poorly appearance". They appeared to live on shellfish and appeared to have nothing but the most simple tools. What intrigued the French was their apparent ability to live without fresh water, which was nowhere to be found by them in the vicinity. The general conclusion reached, therefore, was that they had discovered a race of man which could exist on salt water.

In order to prove the point Arago reported conducting an experiment. In full view of the aborigines, a cup was dipped into the sea and filled with salt water. The Frenchman then pretended to drink it. As the aborigines showed no surprise at all, he recorded:

> *"It is to be presumed, therefore, that these poor people drink only salt water, and live wholly on fish, shellfish and a kind of pulse resembling our French beans which is met with here and there in the interior of the country."*[23]

There is no record available to indicate if the aborigines jumped to the same conclusions about the Frenchmen they observed drinking salt water, indicating that they had also come across a unique specimen of human being.

ABORIGINES, SHARK BAY, 1818

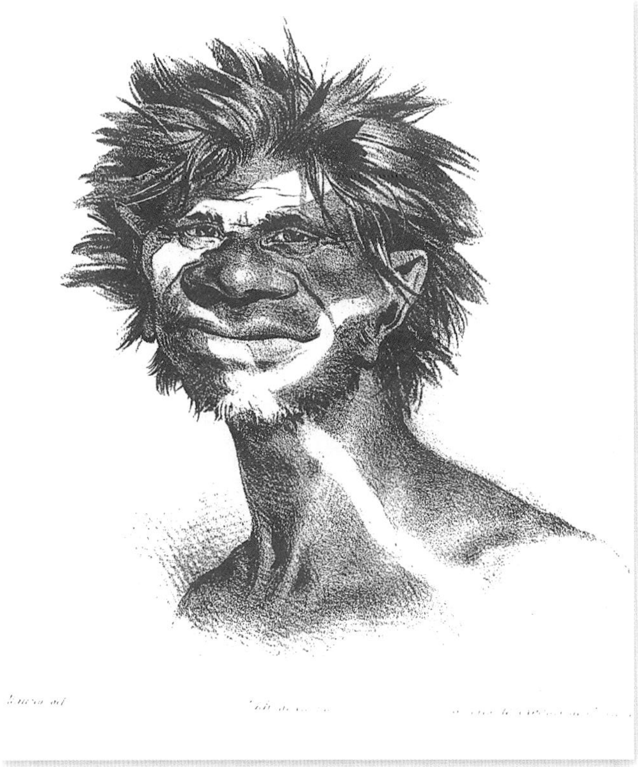

An effort was made to locate the camp used by the natives. A small party moved inland and found this near the tip of Peron Peninsula and left a description of it.

"The huts are formed of a few branches crossing each other, covered with brushwood and clay," Arago records.

"They are six feet in depth, four or five in breadth and three and a half feet above the ground.

The entrance is always on the side facing the wind that most commonly blows. The natives make their fires in the centre, and sometimes around them. The best are very rudely constructed, and insufficient to shelter them from the heat of the sun or violence of the wind.

On some high points they erect also a kind of observatory, formed of a few trunks of trees on which they post themselves to observe the distant country. It is build as rudely as their huts and cannot be of more use to them."[24]

While these observations were being made, a crisis situation had developed on board the *Uranie*. Fabré had been left on Dirk Hartog Island with two days' supply of food and a small boat to get back to the *Uranie*. He was due back on 14 September but did not arrive. A sloop was therefore sent to look for the lost party on the 16th. Fortunately they had merely been windbound and all arrived safely later that night with the Vlaming plate.

Only one further survey expedition was sent out. Duperrey was sent to Hamelin Pool to complete the mapping of the area, which Baudin had not completed. When he and the other parties returned, and the scientific observations were completed, the men packed up the camp and returned on board and the *Uranie* sailed for Timor on 26 September.

Departure for Timor

Before they left the beach, the French left looking glasses, scissors and a knife with half opened oysters to show the natives how to benefit from tools, with Arago commenting:

When I consider the resources of these unfortunate beings, their mode of existence appears to me to be quite a problem."[25]

Freycinet and his mission arrived at Timor on 1 October, where he reported home on the reason for his failure to carry out his orders and survey from King George Sound northwards to Shark Bay. He indicated, like Arago, that western Australia had neither resources nor good harbours suitable for careening his vessel and for refreshing

THE WRECK OF THE *URANIE* AND THE STRANDING OF FREYCINET'S EXPEDITION IN THE FALKLAND ISLANDS

The sinking of the *Uranie* resulted in the loss of the collection of specimens collected at Shark bay for scientific investigation in Paris.

his crew. He consequently informed his superiors that he had decided to make for the Spice Islands instead of King George Sound.[26]

After refreshing in the East Indies, Freycinet sailed north of New Guinea to Guam and Hawaii and then to Port Jackson where a most comprehensive study was made of the aborigines and colonial society. From there he sailed on 25 December, 1819, heading along a south passage to Cape Horn.

From Cape Horn, Freycinet headed for the Falkland Islands where the mission ended in disaster. Making into Frenchman's Bay where Bougainville had tried to establish a French colony fifty years before, the *Uranie* struck a submerged rock and had to be beached on 15 February, 1820. Freycinet and his company were marooned on the Falklands for seventy-three days. Unfortunately the *Uranie* was a total loss, and a great number of the scientific specimens collected, including the live animals being brought back, were lost.

Freycinet with difficulty managed to purchase the vessel which rescued his party, and renaming this the *Physicienne,* he sailed for home with the remnants of his expedition, reaching Le Havre on 13 November, 1820, to achieve fame.

Scene on the Kalgan river showing a native fish trap, by Lois de Sainson, 1826

CHAPTER 8

The Transportation Committee of 1819 and Blosseville's plan for a penal colony in south western Australia, 1819-26

The desire for a French Botany Bay

One of the direct results of the end of the allied occupation of France in 1818, which came not long after Freycinet sailed from Toulon on the *Uranie,* was the Bourbon government's move to establish a penal colony in south western Australia.

Once they were free from the fear of restraint by the resident forces of the victorious allies, the newly independent French restoration government lost little time preparing to create a new French empire. This time, having lost several empires previously, which restoration economists were loathe to commend as beneficial to the *ancien régime,*[1] the restored Bourbons discarded the idea of recreating the old French colonial system as France expanded, and turned instead to copy the British example which was held by all to be the model for success. As soon as the allied forces were withdrawn, a committee of investigation into transportation and colonization which was viewed as the basic reason for Britain's success was therefore immediately set up to report and make definite proposals.[2]

Interest in transportation and its use by Britain in the process of colonization was not new in France. The British penal settlement established at Port Jackson not long before the Paris mob tore down the *ancien régime's* prison at the Bastille, soon caught the attention of French revolutionaries.

Early criticism of the British transportation of criminals to Australia.

At first, in spite of their claim to be social reformers with a scientific outlook, French revolutionaries were not impressed by the new British experiment in punishment and colonization in Australia and lost little time looking closely at what was going on there. This was brought about in the main by a series of critical reports in the influential *Moniteur* which dwelt on the hardships and shortages in New South Wales and created the impression that the British policy of sending criminals to vacant lands to make both of these useful and productive, was a failure.[3] This view was neither fair nor carefully conceived, being more the reflection of anti-British sentiment than the result of scientific analysis and understanding, and served to inhibit further interest and study by the French in the British project.

French acceptance of transportation as a punishment

An interest in the idea of transportation as a punishment for crime, however, was revived by the French revolutionaries and kept to the fore as a result of domestic happenings. The occasion for this was the proposal for a new revolutionary penal code[4] which was made after the question of the "rights of man" was settled and after Montesquieu and Rousseau's political beliefs had been used to fashion the new instruments of French government by means of a constitution. By then, in the early 1790s, the French revolutionaries who had released the prisoners immured by the *ancien régime,* found they had criminals and prisoners of their own, who with the outbreak of the revolutionary wars and opposition, were soon added to by increasing numbers of people charged with being subversive, treasonable, royalist in attitude, ecclesiastical or some other types of anti-revolutionary belief.

On Monday 6 June, 1791, when the penal code and its provisions was being discussed by the new French government, Adrien Duport, a deputy in the Constituent Assembly who interested himself in criminal affairs and the treatment of prisoners, proposed that the transportation of criminals be added to the list of punishments to be included in the new code. This was approved.[5]

This was no new departure for France. Deportation or transportation was one of the punishments used, although not often, by the government of the *ancien régime* before the fall of the Bastille. The old practices, however, were not simply revived and copied. The new penal code systematized the punishment for the first time in France, making quite clear who would suffer or benefit from transportation. The new rules laid down that transportation was for criminals and enemies of the state who did not deserve to pay the extreme penalty for their crimes and who could correct their ways and be rehabilitated by being forced to live in isolation away from the pressures and temptations found in metropolitan France and Europe.

This proposal was made at that early stage of planning French revolutionary society primarily because of humanitarian and quasi-scientific reasons. At the time, in mid-1791, there was no pressing need to arrange for the transportation of criminals. These then did not exist in great numbers. The pressure of numbers came later with the terror and with political extremism and oppression. The basic idea behind the early proposal was that crime was the product of a hostile and evil environment, and that perfect people could be produced in a well ordered agricultural society where man could live "naturally" away from temptations. Consequently transportation was viewed as a suitable punishment for crimes committed against the new revolutionary society, as it aimed at social reform.

Change in the French attitude towards Botany Bay

In this sense the British experiment at Botany Bay obviously stood out as a practical model when the French penal code was being considered. However, in the discussions that took place in Paris, the British experiment was not referred to.[6] But it is significant to note that just over two months before the debate on transportation took place, a translation of Captain Phillip's book about New South Wales was being distributed by Buisson's bookshop in Paris,[7] and this work offered a more attractive picture of the penal settlement at Port Jackson than the one provided by the *Moniteur.* The publication of this book marked a change in attitude by the French to the British experiment and to Australia.

This became clear in the subsequent discussions held in Paris about punishment by transportation. The new penal code had merely opened the topic and not finalized it. The code merely stated that transportation was one of the punishments for serious crimes. It left to be decreed later, the place to which criminals were to be sent.[8] This led to lengthy discussions about the most suitable places. In the years that followed a variety of different proposals were made,[9] but none was acted upon for two reasons. Firstly, opinion in France was divided on the matter when the problem became acute in the purges and the later terror. Those who believed that environment could mould character continued to support the idea of transportation to less civilized and more natural places, while "humanists" of a different type, such as Danton, while agreeing that France should rid itself of its "pests", believed that ideal societies in the new world should not be polluted with them.[10]

Also at the time France had little to offer in the way of a choice. The Bourbons had previously lost the greater part of their old colonial empire. When the new penal code was being discussed, revolutionary France possessed only a few tropical and populous settlements which were not ideally suited for penal colonies, and the outbreak of war in Europe denied France the opportunity to occupy other more suitable lands abroad as Britain had done in New South Wales.

The search for a site for a French penal colony

Definite proposals were made to use Guiana, Madagascar, Sierre Leone and Saint Dominique as convict settlements, but little was done about this. France did not have the maritime forces needed to develop and defend territories abroad. In any case, as the revolutionary wars progressed and the need for defensive and other works grew and the need for labour increased, there was an increased desire to keep prisoners at home to serve as *forçats*, working on state projects.

However, discussions on the theory of transportation proceeded and the search was continued for a suitable site for a convict colony. This evoked a new interest and changed outlook to the British experiment in New South Wales and eventually led to French proposals to establish its own penal colony in south western Australia.

The change of attitude in France is evidenced by the memoir on transportation written by Rollet on 14 October, 1792. In this, after a general discussion of the topic, Rollet described the techniques of colonization used by the Netherlands and Britain, stressing the value of the new British experiment which he urged France to copy.[11]

Growing interest in Australia

This appeal made no noticeable impact, but two other events served to turn French eyes to Australia. Firstly the reports about the disappearance of the popular explorer, La Pérouse, and the departure of D'Entrecasteaux to search for him revived an interest in the South Seas and Australia which again commenced to feature in French literature and reports. Secondly the *Moniteur* had a change of heart and commenced to present favourable reports about New South Wales. Of great significance in this regard was their report of two British success stories. French readers were informed late in 1793 that two men, Sidway and Barrington, who had been convicted of theft in England, had been transported to New South Wales. There, in their new natural environment, isolated from the evils and temptations of urban civilized society, both men were reformed, and became productive and responsible citizens, leading a healthy life contributing to the new society.[12] This information provided real evidence to support the claims made by those in France who supported transportation, that they were right in their belief that the weak could be made strong if they were made to live in a natural environment away from temptations and evil influences.[13]

The intensification and spread of war in Europe prevented France from following this line of thought with action. Further colonial expansion while France was engaged in continental wars was not

de Sainson pinx.
Jacottet lith. Fig.
VUE DU PORT DU ROI GEORGES .
(Nouvelle Hollande.)

PLATE I

PLATE II

de Sainson pinx
Tastu Éditeur
Lith. A. Bès
Tirpenne et Monthelier, Figures par V. Adam.
OBSERVATOIRE DE L'ASTROLABE
au port du Roi Georges, vû de la presqu'île Possession.
(Nelle Hollande.)

PLATE III

VUE D'UN ÉTANG
près la Baie du Roi Georges.
(N^elle Hollande)

PLATE IV

LE HÂVRE AUX HUÎTRES,
dans le Port du Roi George.
(Nelle Hollande.)

PLATE V

HABITANS DU PORT DU ROI GEORGES.
de Sainson pinx
Tastu, l'éditeur
Lith. de Langlumé
Boulanger lith.

PLATE VI

PLATE VII

PLATE VIII

possible. The matter was consequently shelved until after the fall of Napoleon and his empire.

The restoration government continues with the transportation of criminals

In the years immediately following the restoration of the Bourbons in Paris, the revolutionary system of sentencing selected criminals to transportation remained in force. French courts and administrators of justice accepted the punishment in principle, and no controversy arose about it.

However, as it stood in the penal code, the law in regard to transportation was still not easy to apply. There was still no place clearly designated to receive French convicts. The sentence given down to the prisoner was consequently a general one, with no guarantee that he would be sent abroad.

Increasing pressure on French gaols

This was not to last. Increasing pressure in French gaols and a revived interest in colonization and a desire for a new empire abroad brought a change. The French desire from this time was to emulate Britain and build a new strong empire based on the colonization of Frenchmen abroad.

In January 1816, therefore, proposals were made to use the few remaining colonies France possessed as recepticles for convicts. However none of these proved suitable. They were tropical, unhealthy and costly to maintain. What France really required was a new extensive colonial territory in the temperate zone, like the British New South Wales, where Europeans could live and prosper. But nothing could be done to acquire such a place for France in a vacant land, despite the growing desire and support for colonial expansion, while France remained under allied occupation.

The end of this occupation in 1818 was also marked by the appointment of the competent, expansionist minded Baron Pierre Barthélemy Portal as Minister for the Navy and Colonies in France. The impact Portal had on the restoration navy and French colonial empire, as has been noted, was great. A devoted, hard working, patriotic man with a rich experience in commerce and administration, Portal applied himself in his Ministry in the same vigorous and commanding manner he used when he built Bordeaux into a successful commercial centre in the Napoleonic period, which captured the attention of the restoration deputies.[14]

In 1819, as a result of his efforts, the vote made to the navy and colonies department was raised from 45 to 65 million francs, despite the economic hardships France faced. Portal's ambition was to reconstruct the French navy to make it a considerable force, put the colonies on a sound footing and equal Britain as a maritime power, which Portal believed was the true measure of national greatness at the time.

Maintaining prisons a problem for the navy

A naval construction program was consequently commenced, expeditions abroad planned and expenditure controlled to make spending effective.

In this latter regard there was one problem. *Forçats,* the criminals sentenced to transportation, came under the care of the Ministry of Navy and Colonies. By 1819, these had become a costly burden due to the increase in numbers.[15] Before the restoration *forçats* had numbered 5,400. In 1815, the number had risen to 8,578. In 1817, it rose to 9,381 and in 1818 to 10,815. There was consequently a fear in 1819 that the all time high of 15,000 *forçats,* reached in the peak year of 1812, would be reached and would place a strain on naval and colonial funds and resources.

The galleys at this time had been abandoned so the *forçats* were kept gaoled in naval controlled prisons at Brest, Rochefort, L'Orient and Toulon where they were guarded by marines which resulted in a further drain on funds.

Rising costs of supporting the convicts disturbed Portal and his departmental heads who gave a high priority to naval expansion. The best way out, in the circumstances, was to look for and establish a self-supporting penal colony abroad which would offer the two fold benefit of saving money at home and providing a strategic naval base abroad.

Australia was by then frequently before the eyes of those interested in such a type of overseas expansion. The volume on the history of the Baudin mission appeared in 1816. Freycinet and his wife sailed to the same region in 1817, raising wide interest in the region. A series of reports about New South Wales made at the same time also came to the notice of the French government and public. Consequently in 1817, as a result of the rising interest in the subject, the French Ambassador in London, the Marquis D'Osmond, was asked to get the rules and regulations of the British penal colony in Australia and send translations of the reports on transportation.[16]

Report called for on transportation

As a result of this and other stimulations, the French Minister of the Interior, Lainé, on 25 November, 1818, called for a report on transportation, indicating that France now needed to establish a convict colony abroad. Following this, the French government decided on 25 January, 1819, to establish a special committee to investigate and propose on the matter. It was this committee which decided on south western Australia as the ideal site for a French colony.[17]

On 1 February, 1819, the Council of Ministers wrote to the Comte de Simeon, a Councillor of State, that he had been appointed chairman of the committee of investigation. Nine others were to serve on it with him. These were Forestier, Jurien, Capelle and Gerando, all of whom were Councillors of State; de la Borde, de Rigny and Pelet, all of whom were Masters of Requests, and Willaumez and de Rossel of the navy.

The committee, with Pelet acting as secretary, met for the first time at 10 a.m. on Saturday 6 February, 1819, at the Hôtel de la Marine in Paris. Three further meetings were subsequently held — the second on Monday 8 February, the third on Wednesday 10 February and the fourth and final meeting on Wednesday 17 February.[18]

The committee went into the matter deeply and carefuly using varieties of reference materials and calling on expert advise to help them reach decisions. In all six questions were discussed — the question of substituting transportation for forced labour — the place to which convicts should be sent — the period of sentence — the question of replacing forced labour at home — the administrative arrangements — and social questions. In all these discussions, the model of New South Wales was never far away from notice.

The first topic was dealt with in the first session. It was agreed that transportation should be used as a form of punishment. The rest of that session and the next was spent discussing the best place to use. Some difference of opinion emerged on this. Willaumez supported the idea of using existing tropical colonies. This was opposed by Rossel and de Rigny who urged the need for France to use a place in the temperate zone, similar to the latitudes of France.

Rossel, however, did not press the point too far. He admitted France had no such place. Jurien then followed by suggesting France should annex a suitable territory. This gave Forestier a lead. He must have previously gone deeply into the matter. He spoke freely and at length on the topic of a suitable site, indicating that the best place was western Australia, which he indicated was possible as Britain had taken only the eastern part. Forestier was immediately supported by de Rossel.

Forestier suggests western Australia as a site for a French Botany Bay

At first, Forestier's suggestion made little impact. One committee member, not designated in the committee records, objected on three grounds, revealing the then current view of western Australia in France. He claimed that there was no water there, that the land was not fertile and that it was an enormous distance away. The committee then looked at other places and adjourned.

Forestier must have spent a busy week-end. When the committee met again on the following Monday, 8 February, the discussion which commenced along the same lines was again turned by Forestier towards considering the case of western Australia. This time, Forestier so impressed his colleagues with the value and virtues of western Australia that he was called upon to make a special report on the subject for the following meeting.

Forestier had only one day to prepare his memoir, and did it well. His report consists of twenty-two pages packed with factual information about western Australia and the value of transportation to such a place.

Rejecting all the existing French colonial territories as being not suitable places for penal settlements, Forestier gave three reasons for the committee to recommend western Australia as the most suitable site. Firstly he indicated that western Australia was fertile and productive and was not a waterless desert as Dampier implied. Secondly, he claimed that its isolation made it a favourable place as it would be easy to guard and defend. Thirdly, he claimed that the area in fact was not British. Britain, he indicated, claimed only to the 139° longitude east. He therefore recommended that the south western portion between 29° and 35° south be considered for French colonization.

He recommended three places as possible sites for the penal establishment — King George Sound, Flinders Bay behind Cape Leeuwin, and the Swan River.

Of these three Forestier favoured the Swan River site. The anchorage there, he indicated, was safe, being protected by Cape Peron and the offshore islands. To evidence this and the value of the area, Forestier tabled Bailly's report on the locality made when he accompanied the Baudin mission as mineralogist, which was glowing. Bailly at that time in 1819 was close at hand. He was serving as hydrographer in the navy department in Paris.

Forestier had only one reservation about the matter. France, he indicated, had no half way base to use to get to Australia. Her vessels were still not allowed to call at St. Helena for supplies because Napoleon was there, and Britain controlled the Cape and Mauritius, which therefore could not be relied upon. All France had on the way to Australia was Reunion which had no port. Forestier therefore recommended that a half way base be also acquired to make the total scheme practicable.

Lainé attended the final meeting held on 17 February, and presented a paper on the advantages of the Swan River as a French penal colony. However he was not as confident as Forestier. He agreed that the territory was not British, but in fact French by right of annexation made in 1772, but saw Britain as a continued stumbling block. Occupation of western Australia by France, he indicated, could give rise to tensions and feelings of fear by the British.

Western Australia favoured by the committee

After Simeon's committee ended its deliberations and completed its report, the matter of acquiring a suitable colony was left for the navy to decide and act upon. The person and office who played the most prominent part in this regard was Rosily who was in charge of the influential hydrographic office which helped determine overseas strategy and advice on strategic bases. Fortunately for those who pressed for France to colonize western Australia, Rosily knew the area and personally approved the scheme. He had sailed along the western Australian coast from Flinders Bay to Shark Bay in 1772 with St. Allouarn, and had been present when the territory was annexed for the Bourbons. He was also responsible for drawing the first excellent, accurate maps of Flinders Bay and Shark Bay which were the results of his close inshore surveys there.

Survey expeditions despatched.

In the years that followed, a number of expeditions were despatched from France,[19] equipped with specific orders drawn up by Rosily, to thoroughly investigate south western Australia and report on its suitability as a penal colony. These voyages are described in detail in the next chapter which deals with the practical aspects of the matter.

In the meantime, at home in France, one further major course of action was taken in regard to planning for the proposed colony in theory.

Blosseville's plans for a penal colony in south western Australia

At the beginning of 1826, Jules Blosseville,[20] a navigator and influential intellectual, was asked by the Director General of Police, to prepare a plan for establishing a penal colony at the Swan River.

This plan and the background to it is well described in the book about the history of Australian colonization written by Jules' brother, the Marquis de Blosseville,[21] after the early death of his navigator brother.

Jules Blosseville at that time was one of the most suitable and knowledgeable men who could be requested to draw up such a plan for France to take action. He was born at Rouen, Normandy, on 27 July, 1802, to a family with a background in politics, the navy and the colonies. His mother was born in Saint Dominique at a time when that colony was prospering under French role. In France, the whole family was well connected with the restoration court and with political society in Normandy. This placed Jules in a good position to have his schemes listened to.

After leaving school, Jules Blosseville attempted to enter the navy, which because of its size, was taking few recruits. He subsequently went as a volunteer to Cayenne on a seven month voyage in 1818, which introduced him to restoration colonial life and its problems. He then visited Brazil and sailed the Atlantic. On his return to France, he could not make up his mind whether to serve in the Mediterranean or go to the South Seas. However, as a result of his interest in science he was appointed in 1822 to accompany Duperrey on the *Coquille*, who was charged with the task of reporting on British, Dutch and Spanish methods of colonization in the region and with the task of surveying south western Australia to see if it was suitable for the establishment of a French colony.

Because of Duperrey's failure to carry out his orders, Blosseville missed the opportunity of seeing western Australia. However, he spent a considerable time in the British penal colony of New South Wales, which deeply impressed him with its "forest of masts" and its commercial activity.[22]

On the return of the *Coquille* to France in 1825, Jules Blosseville was appointed to do survey work at the mouth of the Seine River. His enthusiastic reports about the British convict experiment in New South Wales which he wrote while working there soon attracted attention and he was consequently called upon to make his proposal for a French penal colony in western Australia.

He completed two reports; the first in January 1826,[23] the second in March 1826,[24] both of which recommended that France immediately establish a penal colony either at King George Sound or some other suitable place in south western Australia.

Neither report is long. Both are approximately twenty pages, but each is informative and well constructed.

The first report commences with a general description of transportation and the nations which use that form of punishment in the process of colonization. He follows this with a criticism of French techniques of colonization by criminals, which he indicated had failed because the French had tried to use unsuitable tropical regions as receptacles. He therefore recommended that France establish itself in temperate south western Australia, which would fulfil the long held Bourbon drcam of having a reliable strategic and supply base in the Indian Ocean on the route to Asia.

Jules Blosseville suggested that only a relatively small area of western Australia be taken over by France. The proposed colony was to extend from the sea in the south to the Tropic of Capricorn in the north, and was to extend eastwards from the coast to 122^0 east longitude, which is near Esperance. He thus envisaged France acquiring only the now productive south-west corner of the state.

The capital of this territory, he recommended, would be best placed at King George Sound. This would mean that the hub of the colony would be well situated for Indian Ocean travel, and on the route from France east to the Pacific. King George Sound, which is on the west wind system, he indicated, is some 4,100 leagues from Brest, 1,600 leagues from the Cape of Good Hope, and a short distance from Pondicherry, Batavia and Timor.

Blosseville urged immediate action be taken, and that a suitable ship then in the harbour at Toulon be sent with an advance party, animals and supplies, to found the colony.

In the second report on the matter, Jules Blosseville spends considerable time reproducing details from early reports on King George Sound, the Swan River and Shark Bay, made by early British and French explorers such as Vancouver and Baudin. These reports are presented in a most detailed manner, putting forward the resources and virtues of each individual place including Port Leschenault (Bunbury), Rottnest, Garden Island as well as the better known major regions and bays.

The report ends with a detailed plan for the exploration of the south western part of Australia as a prerequisite to the establishment of a colony.

It was at this time that Britain got wind of the idea that France was interested in establishing a settlement in western Australia. Already relations between the two powers had deteriorated. The expansion of the French navy and its spread about the oceans abroad gave rise to considerable fears by Britain. On 1 March, 1826, the Earl of Bathurst consequently alerted Governor Brisbane about the threat to Australia posed by expansionist France. Major Lockyer was subsequently sent with a contingent to occupy King George Sound, the capital proposed by Blosseville for the projected colony.

The shift to New Zealand

France took no further action or interest in the matter after March, 1826. Jules Blosseville then turned to use his efforts to help found a French colony in vacant New Zealand where the British appeared to be not so much of a problem. This proposal eventually became a reality in 1840 when the French established themselves, albeit for a short time, at Akaroa just south of Christchurch in the south island of New Zealand. When that effort also proved abortive, the French moved to establish themselves at Tahiti and New Caledonia which became permanent parts of the French empire in the South Seas which the Bourbons and other French administrators had long dreamed of achieving for their nation.

The *Coquille* later *L'Astrolabe*

The 380 ton corvette *Coquille* was supplied to Duperrey for his 1822-1825 expedition sent to Australia to get facts for the Colonization Committee in Paris. It was renamed *L'Astrolabe* for Dumont D'Urville's 1826-1829 expedition during which Britain settled and annexed Western New Holland because of fear of the French. Some of the series of paintings of King George Sound (Albany) made by the expedition's artist Louis de Sainson, are reproduced after page 224.

CHAPTER 9

The French failure to settle western Australia and the shift of interest to New Zealand: the expeditions of Duperrey, Bougainville and D'Urville, 1822-1826

Duperrey's and Bougainville's failings

By rights, the names Louis Isidore Duperrey and Hyacinthe Bougainville should feature prominently in histories of Western Australia, in French colonial history and in this book. Both of these explorers were given prominent parts to play by the restoration government in France, in its move to establish a French colony in western Australia as part of the new Bourbon empire that was created after the fall of Napoleon. Both were ordered to make comprehensive surveys of south western Australia and report on its suitability for a settlement. But neither made the visit as ordered. Both by-passed south western Australia and sailed instead into the Pacific and into relative obscurity in history.

A short while ago, John Dunmore, a New Zealand scholar, took upon himself the somewhat hopeless task of trying to rescue these two navigators from relative obscurity, and provide them with a more prominent place in history by describing their achievements in the Pacific, which is Dunmore's area of interest.[1] Certainly Duperrey, who Dunmore tries harder to rescue from obscurity than Bougainville, made a valuable contribution to knowledge about the Carolines, which was his ambition.[2] For this he deserves and already received in his lifetime, recognition for his achievements. Because of his work he was awarded the Legion of Honour and admitted to the close circle of scientists in the French Academy of Sciences, which Duperrey as a man of science valued. Dunmore, incidentally, could have strengthened his case to have Duperrey revalued by also referring to the as yet undescribed but valuable political work Duperrey did for his country in Latin America when he should have been on the way to Australia.[3] The reports he wrote about that region were made at a crucial time when the United States was formulating its Monroe Doctrine and when the British Foreign Minister, Canning, was extending recognition to revolutionary regimes in Latin America in order to call "the New World into existence to redress the balance of the Old", as part of British policy against French expansionism and the Holy Alliance powers in Europe. Duperrey's informative reports home are valuable additions to those sent by special French emissaries at the crucial time when the restoration government in France was formulating its policy towards Latin America.

But not even these strong points provide historians with enough material to give Duperrey a greater place in history than he has received. And it is the same with Bougainville. He can be no further elevated with ease. The fact is, and what Dunmore has failed to see, both Duperrey and Bougainville the younger failed to carry out vital specific orders given to them by their superiors who needed the information they were sent to get. The Naval Ministry in Paris and a powerful inter-departmental committee there had decided by the time they were sent out to give a priority to establishing a French colony in south western Australia, as has been indicated in the previous chapter. By failing to go there as ordered, both men failed their country and their superiors in a time of need. Neither was consequently well rewarded for his services. Lasting glory was achieved only by Dumont D'Urville who was hastened out with somewhat the same orders for New Zealand in 1826, and carried these out efficiently. By then the British were aware of French interest in acquiring a colony and so they established a settlement at King George Sound, securing all of Australia for Britain.

Because Duperrey and the younger Bougainville unwittingly helped Britain secure territory at the expense of France, they can never be given a prominent place in French naval and colonial histories, although Bougainville achieved some fame and prominence in his lifetime.

The naval plan for co-ordinated surveys

It was originally planned by the French naval authorities, that Duperrey and Bougainville would both leave on voyages of circumnavigation, visiting different parts of the world, at about the same time.

The plans emerged after the return of the survivors of the ill-fated Freycinet expedition. Shortly after Freycinet's return, Duperrey, who had an ambition to survey the Carolines and other then still little known parts of the Pacific, worked together with Dumont D'Urville on a plan for a voyage of circumnavigation by a French naval expedition.

The opportune time for French action

By that time both the French navy and the government were prepared to accept and support such ideas. France then was rising strong under the Bourbons, acquiring power and prestige in Europe in particular when it came under the protection of the Holy Alliance. The measure of Bourbon success in this regard was clearly demonstrated at the Congress of Verona (1822) where, much to the chagrin of Britain, France was given the task, by the Holy Alliance, of intervening in Spain in order to save the Spanish Bourbons from the rebels there, as has been noted.

Abroad the story was much the same. The colonies France had been permitted to retain were administratively reorganized and made newly productive, adding to France's riches. The application of science to agriculture in Senegal was resulting in good crops of cotton and other goods as was the case in other areas which demonstrated the continued benefits of physiocratic policies both at home and abroad.[4] Beyond the colonies, in foreign waters, the French flag was again appearing frequently and in numbers.[5] Franco-China trade which had been cut off in the revolutionary war was re-established. French vessels once more sailed and traded in the Indian Ocean, the South Seas, the North Pacific and among the Spice Islands. There, France commenced to take a special interest in and about Annam, sending special missions such as that under Kergariou, who showed the French flag from India eastward to help restore French prestige and pave the way for a commercial revival after 1817.

Seeing it lacked a half way house to the east, France once again was faced with a two fold choice. To get a refreshment point for French ships, it could either link with the Iberian powers, which was a tenuous policy fraught with difficulties, and use their bases in Latin America and Asia. Or, alternatively or concurrently it could establish its own bases.

It was the latter course in particular the French restoration government determined to take, although the Spanish intervention by France led Canning and others to assume that the French Bourbons were up to their old tricks of using fraternal Bourbons to achieve French national aims.

While plans for new significant circumnavigations with distinct political aims were being formulated in France, a memorandum which was the result of inter-departmental consultations about the problem of prisons and prisoners in France, indicated that the authorities gave a high priority to having a colony established in western Australia. The naval Department of Plans which was responsible for drawing up the programmes for naval expeditions, and which was under the control of Rosily who was personally interested in the

Australian venture, stated in the memorandum, that the "French government has for a considerable time been searching for a place to establish a colony and had its eyes fixed on south-west Australia".[6] This, they claimed, was a place with a mild climate which was reported to be fertile and suited for European settlement. Among its other advantages was the fact that it was a place where Europeans could produce the consumer goods they required without having to rely on a native population, which was held to be the disadvantage of other potential base areas in Asia and the Pacific. What France wanted by then was a colony like Botany Bay which was a self reliant European settlement producing goods suited to the European palate.

The trouble was that more knowledge was needed about the suggested area before action could be taken. The French authorities, unlike the British, were most reluctant to send out colonists before a thorough assessment of the area they planned to settle was made. They never engaged in the practice of sending out families to areas which were not well-known, to suffer unforeseen hardships which they believed happened in the case of the early settlement by the British at Botany Bay.

In the case of south western Australia, the French authorities in particular wanted information about the anchorages, the ports, the resources, the soil types, and whether or not there was on the spot material that could be used for building construction.[7] The area of special interest in this regard was Rottnest Island and Garden Island, which appealed because of the ease with which they could be fortified and defended.

Duperrey and his voyages on the *Coquille*

Duperrey was finally selected to lead the first of the major circumnavigations, which were ostensibly for reasons of science. But before his plan was submitted to Naval Minister, Clermont-Tonnerre, it was amended by Rosily and others to incorporate the government's political-colonial plans. Duperrey was ordered, on his way around the world, to "stop at the Land of Leeuwin on the south-west coast of New Holland with the object of examining the nature of the soil, in particular at the Swan River and at King George Sound, to see if that part of New Holland was suited to receiving a settlement".[8]

Duperrey was assured this was quite in order to do because "the discovery and the act of possession of the land there was already made by St. Allouarn in the name of the King of France".

This latter point was the main trump card the French held. The victors in the Napoleonic wars had stripped France of most of the empire that had been created by the Bourbons before the revol-

EXPLANATORY MAP DUPERREY EXPEDITION 1823

THE DUPERREY EXPEDITION (THE COQUILLE) 1823

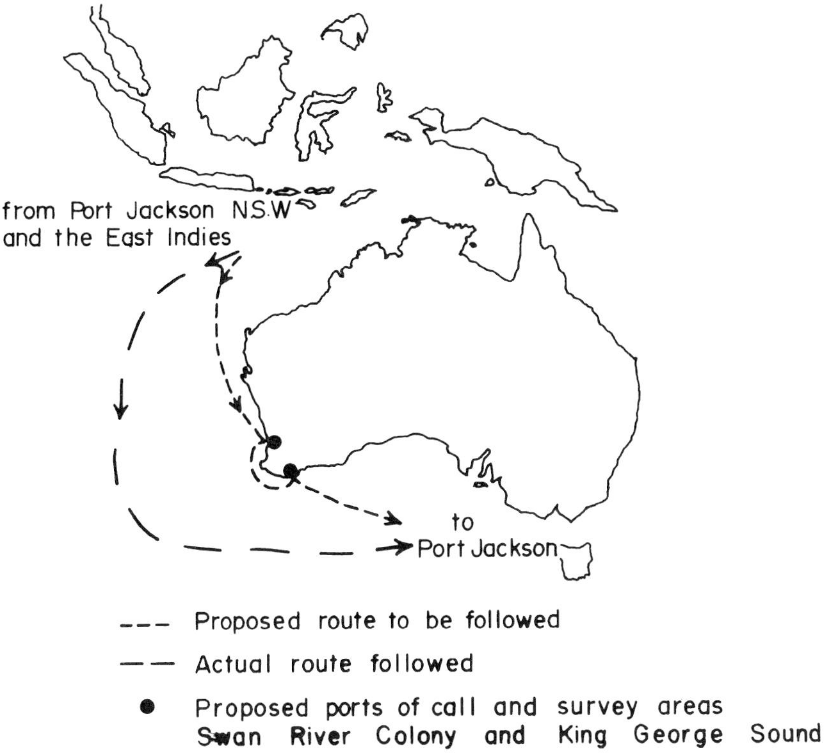

ution. But, being unimpressed by the concept of prescriptive rights to territory, which the French valued in international law, the victorious allies had not taken away France's claims to territory. Therefore in 1822, when Duperrey sailed, the French authorities considered they had a considerable empire at least on paper, which included western Australia.[9]

In the case of south-west Australia there was one complicating problem. Vancouver took possession of the territory at King George Sound, for the British monarch in 1792. But this was not followed up by a British occupation. Nor did Britain make a fuss about the matter. Publicly, Britain claimed only the eastern part of the continent. In any case when Vancouver had taken possession of the land, it had already been proclaimed to be French twenty years previously, which made the British claim doubtful.

As the specific plans for Duperrey to visit south-west Australia were being made up, there was some confusion about when he should make his survey there. Initially, it was planned for him to go via the Cape of Good Hope direct to "Leeuwin Land" and report, and then sail on. In the final plan, however, it was planned for him to make his study of the proposed site for a colony after he had seen British and other colonial establishments in the area.[10]

The final plan therefore given to him was for him to sail from France in the northern summer of 1822, on the *Coquille,* and head south in the Atlantic to the Cape of Good Hope, to reach there by the southern summer of 1822. From there, he was to sail into the Indian Ocean and look for the missing islands of Marseveen and Denia which were still shown on the maps as being south of Cape Town, but had not been seen for centuries. Duperrey was then to sail to Port Jackson. There, he was to contact Governor Brisbane who was regarded by the French as being a scientist and a Francophile who could be of service and help. After surveying the British colony and making contacts, Duperrey was to sail into the Pacific to complete his scientific work and then make for the Spanish, Dutch and Portuguese colonies to refresh his men and replenish his stores and to examine the colonial system and situation there. From the East Indies, he was to sail south into the Indian Ocean to "Leeuwin Land", arriving there in July 1824. After making a thorough survey of the designated parts there, he was to go back to Port Jackson and then sail home either by way of Cape Horn or the Cape of Good Hope, whichever he preferred.

Judging by the information on his personal dossier in the Naval Ministry,[11] the authorities made a satisfactory choice in selecting Duperrey. Born in Paris on 21 October, 1786, he entered the Napoleonic navy in 1803, at the age of 17 and soon attracted notice. He served his country well on ships doing convoy work and making hydrographic surveys in difficult warlike circumstances when Britain was blockading France. His abilities and promise in regard to this work was noted on his dossier in 1809 which refers to him as knowledgeable, active, zealous, and a good military sailor. He was made ensign in 1811 and continued on in the navy after the fall of Napoleon. In 1817, he joined Freycinet on the *Uranie* and did impressive scientific work on the voyage of circumnavigation. He also served well, helping Freycinet bring the survivors back to France after the *Uranie* was wrecked.

He was nearly 36 years old when he was selected to command the *Coquille* and take it on its long and serious mission. The ship he was given was small and not a good sailor. It was bluff and could not point too close to the wind. But he was given an enthusiastic crew of young people, some of whom were equipped with skills such as carpentry and surveying which promised success for the mission.

Departure from France

The *Coquille* left Toulon a little later than planned, on 11 August, 1822. The vessel was well equipped with stores of tinned goods and canned water and anti-scorbutics to ensure the health of the crew.

It was from a despatch Duperrey sent home from Teneriffe on the way out that the Naval Ministry first learned that things were not going as they should be.[12] From there, Duperrey reported that he had decided to go by way of Cape Horn instead of by way of the Indian Ocean to the Pacific. He consequently landed in Latin America and made the political reports referred to above. From there he moved belatedly into the Pacific to examine the Carolines, which had always been his main personal aim.[13]

After visiting Port Jackson and then further exploring in the west Pacific, Duperrey put into Ambon on 1 October, 1823, intending to get fresh supplies and refresh his crew before making for western Australia by way of the Sunda Strait as planned. His choice proved to be most unfortunate. There were no stores to be had, and Ambon was badly affected by "Asiatic cholera". The only thing Duperrey managed to pick up was illness which soon affected the bulk of the crew as the ship stayed on at the unhealthy island throughout the month of October. Why Duperrey chose to stay on there although there was nothing for him is not clear. Eventually, he sailed from Ambon on 28 October, intending to call at Dili, in Portuguese Timor, to get the necessary supplies. But near Timor he changed his mind, fearing the crew would contract further diseases if he called there. Therefore he made for Kupang on the western extremity of Timor which he felt might be more suitable and salubrious. But when he reached there, he decided not to put in to port because it seemed too open and his ship could be threatened by westerlies. Consequently he left the East Indies to make the survey of south western Australia with no fresh supplies, a sick crew and only a few dry stores. When he departed from the East Indies he had only enough flour to last a month. The rest of the stores consisted primarily of ship's biscuits, rice and salted lard which were old, weevil or otherwise infected and unpalatable. The journals kept by the members of the expedition subsequently contain information mainly on two subjects, the bad food and the bad weather with occasional references to the mice and cockroaches which battled with the crew for the rotten foodstuffs and water.

It is not fair to solely blame Duperrey for the state of the ship and crew on the eve of his important political mission. He had been ordered to replenish his supplies in the East Indies and then head south into the Indian Ocean, with all speed, to quickly survey "Leeuwin Land" in January 1824. This was a two fold mistake or weakness made by those who drew up Duperrey's itinerary. These planners assumed that the East Indies or Spice Islands were rich in food resources. In fact they were not, and seldom were as far as fresh stores were concerned. There was little fresh meat available near the ports of call, and that which was available was from the local bullocks and was boney and tough. Rice was not always easy to procure, and was often insect infested and could not be kept. There was no wine. There was local arak which was often taken aboard in place of brandy, but it was not easy on the palate, or the stomach. Certainly there was some food available in the islands, but it had to be found and this took time.

Besides this difficulty, Duperrey was presented with the arduous task of sailing a bluff square rigger from the East Indies south to "Leeuwin Land" against contrary winds. That passage is difficult to make at the time of the year he sailed, unless the ship is taken close to hug the coast to pick up the diurnal and other convectional land breezes which can blow on the coast there. But that course is dangerous. The shallow waters of the continental shelf in the north part of western Australia are swept with strong tides and are studded with reefs and banks. Navigation there is consequently often perilous. Duperrey therefore made a correct decision off western Australia. He kept the south winds on his port and made well out from Australia into the Indian Ocean, taking a buffeting all the way which did not make the food more palatable and the trip more comfortable.

Failure to survey western Australia as planned

It is not clear while sailing in these bad conditions, when Duperrey decided to by-pass western Australia. Lesson, the artist who accompanied the expedition, recorded in his journal that as early as 14 November, Duperrey had decided to head straight for Port Jackson; that is not long after he decided to keep clear of Kupang harbour.[14]

In his report to the Naval Ministry from Port Jackson after he arrived there, Duperrey blamed the contrary winds he encountered in the north for preventing him from reaching the Swan River.[15] These, he maintained, held up his progress and caused a supply problem which was certainly the case. The crew from late November were put on salt rations and were engaged in deep sea fishing to try to get fresh supplies. They were still weak from tropical illnesses and suffered from influenza once they reached the colder climate of the south.

But this explanation that he made a decision late in the trip, is not supported by evidence in other journals such as that kept by Lesson which later came into the possession of the French Naval Ministry,

Hunting Kangaroos, by Louis de Sainson, 1826.

Hunting seals, by Louis de Sainson, 1826.

making them aware of his deficiencies. For although Duperrey was given a difficult task to do, and suffered bad luck through sickness at Ambon, he did not display initiative and help himself. He left Ambon with his stores in a dangerously low state and did not call at other close-by noted refreshment ports in the East Indies to rectify the situation. Instead he chose to make the long and arduous trip half way round Australia to Port Jackson with a sick crew and no food reserves which was a dangerous action. It imperilled his crew. There is no reason why, once he had reached the westerly wind belt, he could not have put into King George Sound and taken on the good fresh water there to replace his "oxidized" and putrid water, and get supplies of oysters, fish, birds and other game which abounded there. He could have used the occasion to make at least a partial report on "Leeuwin Land" as ordered. He seemed instead to prefer to sail as quickly as possible to the Pacific, which was his main field of interest. There explorers were denied little. Most of the islands there provided every comfort man needed.

When Duperrey arrived back in France on 24 March, 1825, to be applauded by scientists such as Humboldt for his work in the Pacific, Bougainville the younger had been gone on his circumnavigation for almost a year. He was despatched at last, after long delays, with the *Thetis* which sailed in company with the *Esperance* under Nourquer du Camper who was experienced in eastern waters. Du Camper had sailed with the earlier Kergariou expedition to South and East Asia in 1817.

Bougainville the younger

Hyacinthe Bougainville, who had been a member of the Baudin expedition to Australia and subsequently had a distinguished career as a naval warrior in the Napoleonic wars, was directed to survey the Asian regions from India eastward to China. The reasons for sending him there were many. But prime among these was the French Restoration government's plan to re-establish a French presence eastward of the new British base at Singapore, in the area that was later to become French Indo China.[16]

Bougainville sailed with his expedition from France on 2 March, 1824, and headed for Cochin China by way of the Cape of Good Hope, Reunion Island, which was still a French possession then named Bourbon, and then via India and the Moluccas.[17] After making a report on Cochin China, Bougainville visited Manila and China to assess the trade and international situation there. From there he returned to the East Indies where like Duperrey he was expected to replenish his supplies and make for "Leeuwin Land" to

Albany township viewed from the heights near King Point, looking eastwards, with an aboriginal group. This was settled by the British in 1826 because of fear of the French during Dumont D'Urville's expedition. The artist with the French expedition, Louis de Sainson, made sketches from the same heights, mainly looking westwards. (See plates after page 224). For an eastward view made by de Sainson from Possession Point opposite the heights, looking towards present Albany, see Plate 3.

make an independent survey there. He was given identical orders which called for him to closely examine the resources of Rottnest Island and Garden Island in particular.[18] But Bougainville also ran into trouble in the East Indies. He chose to get his fresh supplies in Sourabaya. But when he arrived there on 25 March, 1825, the town proved unhealthy and food costly and difficult to get. His crew consequently were placed in the same position as Duperrey's had been at Ambon nearly 18 months previously. Thus when he sailed south to make his survey of western Australia, his crew also were ill with dysentery which resulted in six deaths, and his stores were low. He consequently also by-passed south western Australia and made for Port Jackson.

Failure to survey western Australia as planned

In his journal Boungainville blames delays and the weather for preventing him from carrying out his orders. In his explanation to his superiors, Bougainville used a somewhat unsophisticated argument which badly reflects on him as a scientific, experienced navigator. He claimed that he kept clear of the coast because of the enormous swells out at sea, which perturbed him to think what they would be like close to shore.[19]

He was correct in his assessment of the size of the swells off the western Australian coast. These are most impressive out in deep water. They are often borne with the westerlies, from as far away as Cape Horn without interruption and give cause for many a good, experienced sailor to stand in awe. But the continental shelf off the coast is for the most part broad and slopes gently and the swells soon dissipate as they approach the coast which offers dangers of that kind only in cyclonic storms or in deep winter depressions. There was therefore no sound reason for Bougainville to hold the fears he had.

Bougainville added to his explanation for by-passing western Australia by indicating he thought the move for France to establish a colony at the Swan River might upset the British, and therefore might not be well advised and conceived.[20] In fact the evidence in the Naval Ministry in Paris, including Bougainville's own reports home, indicates the Restoration government very much had the situation in hand, and had reason to be confident about the possibility of establishing a settlement in western Australia.

France misses an opportune time to establish a base in western Australia

It is through examining these records in their entirety that the extent of Duperrey's and Bougainville's failures can be appreciated. Both men were asked to report on the different colonial establishments in the east Indian Ocean — west Pacific area. Bougainville in particular reported on the state of the foreign forces in the area.[21] His reports indicated that there were five flags present in the region he surveyed — the French, the British, the Dutch, the Spanish and the Portuguese. Among these Britain was as isolated as it was in Europe. Besides this Britain was seen to be militarily weak in the region. Bougainville assessed there were only 7,000 men and four batteries defending Britain's eastern Australian colony. To make matters worse for Britain at the time France was considering taking action to establish a colony, Britain was further stretching its defence resources by extending to north Australia, at Melville Island.

Other reports coming from Europe and the Americas indicated that Canning and the British were preoccupied with the Americas and the problem of the Bourbons there. Given the fact that Britain was opposed by the Holy Alliance powers in Europe, that it was occupied with what was going on in Spain and Latin America and with what it felt France was up to there, and that it had few military resources in Australia, and that the British were war weary and that they showed no liking or high regard for western Australia which they appeared to feel was unfit for human habitation, then Bougainville was wrong. The French had good reason to believe they could successfully establish a settlement in western Australia or elsewhere if need be, and that the mid 1820's was the opportune time to do this, when British eyes had been led to concentrate on what was going on in Latin America.

Dumont D'Urville

The thing that is most bedevilling in histories of the settlement of western Australia is that while Duperrey and Bougainville the younger by rights should be talked about and never have been, Jules Sebastion Cesar Dumont D'Urville by rights should not be mentioned but often features prominently.

DUMONT D'URVILLE

Dumont D'Urville was not ordered to visit western Australia. Certainly while plans were being formulated for his voyage to the Pacific Ocean, Jules Blosseville was requested to draw up his plan for a settlement at King George Sound, but by the time Dumont D'Urville left France, official opinion had shifted away from establishing a French colony in "Leeuwin Land" in the Indian Ocean, to establishing one in the more salubrious islands of New Zealand. Jules Blosseville consequently in December 1828, drew up a second proposal for a planned penal settlement there.[22] This idea at the same time was being promoted by the adventurer Charles Baron de Thierry who was getting official support for his ambitious personal plans to establish a private settlement for the French in New Zealand.[23] Dumont D'Urville thus was primarily given the task of sailing to the south and west Pacific to survey and examine the resources of New Zealand to see if they could serve as a colony. He is therefore of more significance in the history of New Zealand and the Pacific where France eventually established the settlements it had planned since early in the Restoration period, than in western Australian history.

He has been featured in histories of settlement in western Australia without sound justification primarily because these have been written almost wholly from British documents which are not the most informative and reliable sources about French government intentions, nor about international relations involving powers other than Britain.

The main cause of misconceptions about the part played by D'Urville in the British acquisition of western Australia is that historians of settlement of western Australia have confused British expressions of fear of the French, with official French moves and motives. When they write to explain why Albany was suddenly occupied by Major Lockyer and the 39th Foot regiment in December 1826, they claim that British officials in London got wind of French interests in western Australia, and feared what Dumont D'Urville would do when he was despatched. Governor Darling in New South Wales was consequently alerted by London about the matter.[24] Rumours about the French raising flags in "Leeuwin Land" were circulated, and British forces in consequence were sent to occupy the region to thwart the French.

The only thing true about this explanation is that the British were moved to sudden action by fear. In reality there was then no basis for that fear. Major Lockyer's action in coming to Albany did not thwart the French. The French then no longer had serious intentions of challenging the British there. The time for Britain to have felt real concern was years before, from 1819 to just after the Spanish intervention in 1823 when real French proposals existed and when real moves were made by the French in regard to "Leeuwin Land". The

EXPLANATORY MAP D'URVILLE EXPEDITION 1826

THE EXPEDITION OF DUMONT D'URVILLE ON THE ASTROLABE 1826

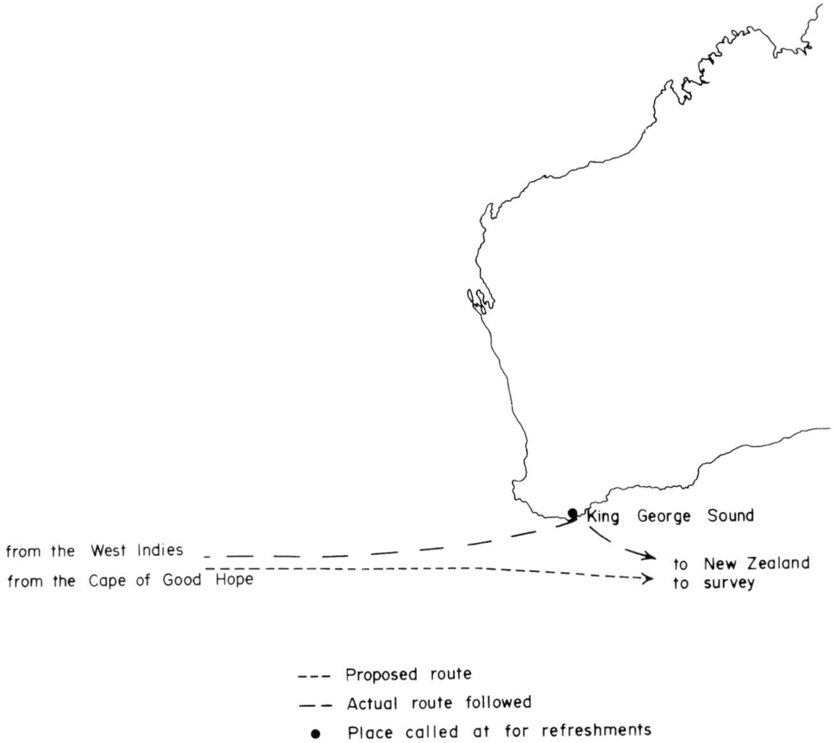

British never got wind of this close kept secret.[25] Their attention was concentrated on the Iberian peninsula and Latin America at that time.

French manuscript sources indicate that Dumont D'Urville landed in western Australia by accident and not by design. He was no stranger to Australia and Australian waters. He had previously visited the region on the *Coquille* as second-in-charge of the Duperrey expedition. The Norman scientific explorer Dumont D'Urville and the Paris born scientist-sailor Duperrey in fact were close colleagues during their early naval careers and jointly planned that expedition which led to them falling out with each other.

Dumont D'Urville was four years younger than Duperrey. Born on

23 May, 1790, at Conde-sur-Noireau in Calvados, Normandy, Dumont D'Urville joined the Napoleonic navy as a novice at the age of 17, in 1807. Like other aspiring officers at the time, he took up scientific studies as a sideline to his naval training. This sort of professional training was a direct result of the Baudin expedition which was troubled by the inclusion of civilian scientists who were not amenable to naval discipline. Thereafter the French navy provided opportunities and incentives for their own officers who were intent on explorations, to qualify in branches of science so the navy would not be forced to take civilians. Dumont D'Urville elected to specialize in botany and entomology in which fields he made a name for himself.

ABORIGINES WITH FISH, KING GEORGE SOUND

This resulted in him becoming a good complementary officer to Duperrey who specialized in hydrography and cartography. Their interests matched. But Dumont D'Urville early on gained the edge on Duperrey. He added to his qualifications a knowledge of English, German, Italian, Spanish, Hebrew and Greek. It was in the latter field of classical studies, combined with his naval profession that he scored his first triumph and honours by securing historical monuments for France from classical areas in the Mediterranean.

NATIVE HUT, KING GEORGE SOUND 1826

He rose gradually from novice to a command of his own after years of service in the Napoleonic and then Restoration navies, in an early career that was similar to Duperrey's. He served France well in the difficult war years when the ports were blockaded, making a contribution to hydrographic work in difficult circumstances. After the war, like Duperrey, he applied to go on the round the world voyage with Freycinet on the *Uranie,* but unlike Duperrey, was not selected. Instead, among other places, he served in the Aegean. There he achieved fame by recognizing the significance of and helping France acquire the famous statue of the Venus de Milo, now in the Louvre for which he was given the Legion of Honour and achieved the notice of scientists and intellectuals thus achieving fame that equalled Freycinet's.

On Freycinet's and Duperrey's return to France, Dumont D'Urville and Duperrey co-operatively drew up and submitted a plan for a further voyage of circumnavigation which resulted in Duperrey being selected as leader and commander of the *Coquille.* It was while they were together on that voyage that the two fell out. Subsequently Dumont D'Urville became Duperrey's most bitter, personal critic.

After the return of the *Coquille* to France on 24 March, 1825, Dumont D'Urville was promoted to the rank of captain-of-frigate. He drew up a plan for a new voyage to the Pacific, to complete the work done sketchily there by Duperrey, and was given command of Duperrey's former vessel, the *Coquille* to take an expedition to the Pacific. The vessel was renamed the *Astrolobe* for the voyage.

Planned Itinerary

The original plans for Dumont D'Urville's journey were drawn up at the end of 1825. The plan was to send the *Coquille* to the Pacific by way of the Cape of Good Hope, then by way of Australia to New Zealand. After this a survey was to be made in the West Pacific, especially of New Guinea, New Britain and the Louisades.

This plan was discussed by Rosily, Rossell and Jules Blosseville at the end of the year. Consequently, early in 1826 Blosseville officially drew up and presented his plan for a convict colony in western Australia. But in March a new and fresh itinerary was drawn up. This called for Dumont D'Urville to go by way of the Cape of Good Hope direct to Port Jackson, refresh and then commence his work of surveying in the Pacific and examining New Zealand.[26]

Political motives and orders

Dumont D'Urville like other restoration explorers in the region was given secret political orders. Specifically he was asked in the minutes which accompanied the itinerary, to report on two matters.[27] Firstly, he was to search for suitable anchorages in southern waters for large ships of war, which France needed and did not possess. Dumont D'Urville was specifically asked to map these and assess the nature of the resources available, in particular the supplies of fresh water and firewood. Secondly he was asked to report on a place where the French government could send convicts like the British did in eastern Australia. Such a place, Dumont D'Urville was advised, should have resources and a good climate.

Growing French interest in New Zealand

The specific place Dumont D'Urville was urged to examine in this regard was the north part of the north island of New Zealand. In the minutes which accompanied his itinerary, Dumont D'Urville was advised to examine the area on the north-east side of the island with the view of taking possession. He was informed that a vast area there was believed to have all of the conditions required for a naval base and penal colony. Charles Baron de Thierry, he was further informed, claimed to be the proprietor of the land in question, and he was urging the navy to possess it in the name of the French monarch. The area in mind was from 35° 32′ south between 173° and 173°45′ east, that is the region between North Cape and Doubtless Bay in the North Island. This region, it was averred, contained excellent anchorages for warships and could be made into a "colony of war like Botany Bay"[28].

FRENCH MAP OF AUSTRALIA AFTER THE COMPLETION OF THEIR SURVEYS

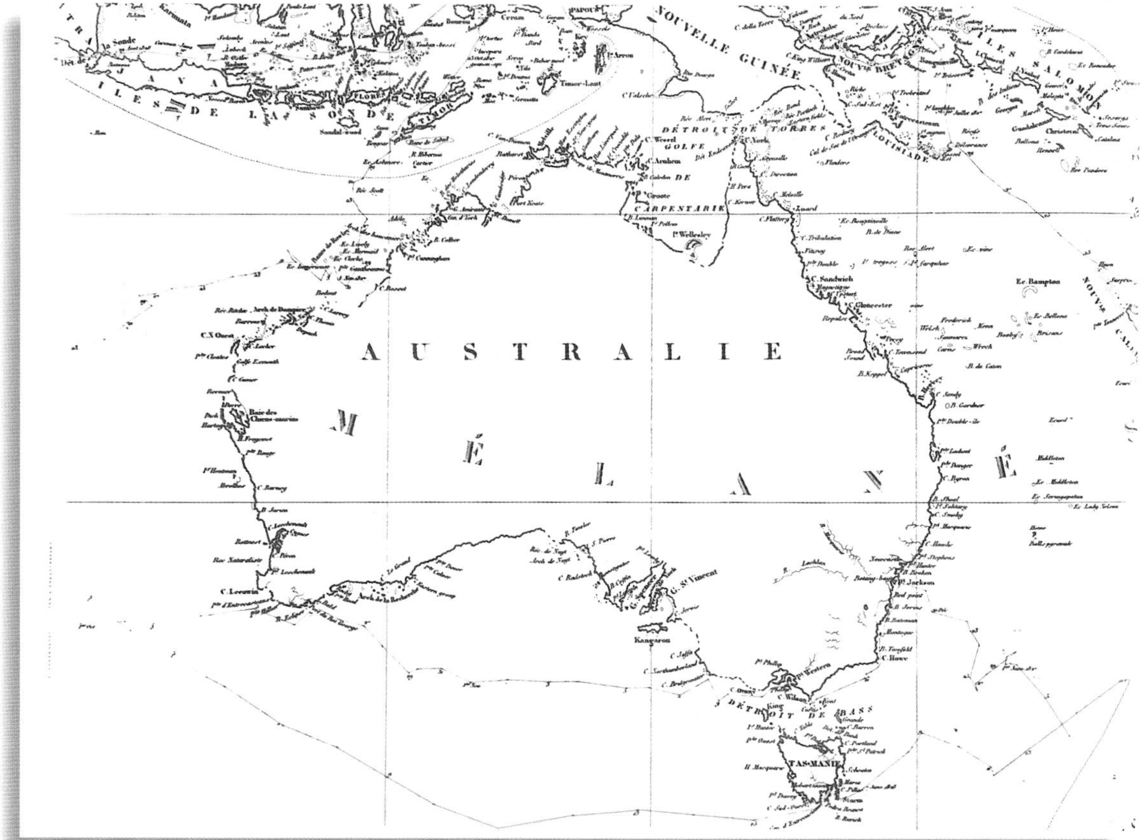

Neither the final official itinerary nor the secret orders included a proposal for Dumont D'Urville to examine western Australia. France by then had become more interested in the Pacific. New Zealand for France was an ideal spot for a colony. There were no European powers established there. Britain did not move to annex land there until 1838. New Zealand was not only 'vacant', but was also convenient; much more-convenient than western Australia for France. It was easy to get to. Ships from France, once they reached the Cape of Good Hope, could ride the westerlies there. Moreover as methods of food preservation improved and ship building techniques improved and the speed of ships increased, a base close to Africa was not needed. Besides this New Zealand was in easy reach of the rich Pacific and of East Asia which could be sailed to with little difficulty from New Zealand by using the Pacific trade winds. Western Australia, conversely, proved virtually impossible to reach from the East Indies, as Duperrey and Bougainville had found out, which made "Leeuwin Land" of limited use strategically. Military reserves could not easily be sent there from the north if needed. New Zealand did not have this problem, and was more isolated and therefore more easy to defend.

Dumont D'Urville kept to his plan, if not to his time table. After being held up by contrary winds and bad weather in the Mediterranean, he reached the Atlantic Ocean and sailed south.[29] After calling at Teneriffe and other ports, he went to Trinidad to survey there. He then set sail south on the long voyage to go direct from there to Port Jackson, keeping well south of Cape Town, sailing along the high thirty parallels.

He had been given the choice of coasting along the south of Australia to check some of the longitudes determined in 1792 by D'Entrecasteaux. But Dumont D'Urville showed no intention of making a landfall in the western part of Australia until he was close to the coast. He left Trinidad on 31 July, 1826, and made south at a bad time for sailing there. He ran into the bad winter gales which blow in the 30^0 parallels of latitude and higher from May. On September 12, when in the southern Indian Ocean in gale conditions which developed after a wind shift north, an able seamen, Binot (or Benoit) fell overboard in the early morning, chicken crates were thrown overboard for the man to use to save himself. The ship was rounded up into the winds and waves with great difficulty, and a whale boat was launched. But the man disappeared before he could be saved. In the meantime, the ship and rigging were both severely strained as a result of rounding up and shortening sail in exceedingly bad weather conditions. With the crew dismal and despondent and the ship in need of repairs, Dumont D'Urville decided to turn north to give his crew a short rest at King George Sound before going on to

New Zealand and the Pacific to do the proper work set for the expedition to do.

Unplanned arrival at King George Sound

"Cape Leeuwin or Cape Hamelin", for the French still did not know which was which, was raised on 5 October and anchor was dropped in King George Sound on 7 October, in foul conditions. There was more than a feeling of relief when the ship was brought into the calmer waters, despite the rain. Dumont D'Urville recorded that they had been at sea for 108 days, in hideous weather and with overwhelming seas. In fact, it took the crew some time to find their land legs.

They were given ample opportunity to do this and refresh. Dumont D'Urville stayed at King George Sound, anchored near the entrance to Oyster Harbour from 7 October until 25 October, 1826.

Surveys of the area

The main centre of attraction for the French on this visit was the aboriginal people. Smoke was seen rising near the shore not long after the ship was anchored. Much to the surprise and interest of the crew the whale boat, which had been sent to sound the waters near Oyster Harbour entrance and to visit the spring to the west, returned with an aboriginal who spent the night on the ship. The next day he was put ashore in French clothes, loaded with presents. This and other scenes were carefully painted and recorded by the official artist with the expedition, Louis Auguste de Sainson.

An observatory was set up ashore near where Major Lockyer landed later in the year, where Albany now stands. The scientists spent the time at their disposal making the astronomical calculations that they found impossible to do on the rolling ship. The rest of the crew went exploring and shooting. Princess Royal Harbour was visited and several expeditions were made to Oyster Harbour and Frenchman's River, now the Kalgan River.

Favourably impressed by the region

The whole area there impressed Dumont D'Urville. He found it green and fertile. "If there is a place suited for a colony in this region then this is the place", he noted after his visit.

He considered the best site for a town was where Albany now is, near a small stream by Mount Clarence. Frenchman's River he believed, would be most suited for "plantations" to be developed in view of its fertility.

But there is no evidence that he had Frenchmen in mind as the colonists there. While he was anchored, English sealers and whalers were in the bay at a permanent camp established on Break Sea Island. For France to establish itself there, and oust these people who already seemed to be in occupation, exploiting its resources, would be fraught with danger. He therefore merely commented he could not understand why Britain had not already occupied such a fine place.

The British act of possession

Preparations were already being made by the British in New South Wales to do this by sending Major Lockyer and take possession of all western Australia which was finally legally done at the mouth of the Swan River on 2 May, 1829. By then France was at work in the Pacific, laying the foundations for its later empire there, which at last saw the old Bourbon dream of having French territories in the temperate part of the southern hemisphere fulfilled by the settlement of New Zealand which did not last long.[29] Both islands were taken over by the British not long after the French settled at Akaroa.

Rottnest Island (background) and the Swan River were put on the map by the Dutch in 1697, but first scientifically explored by Hamelin in 1801. As a result of his reports, France later planned to establish a convict settlement and fort on Rottnest. The British Swan River settlement (foreground) was made in 1829. A convict establishment and fort were later built at Rottnest.

APPENDICES

The maps, sketches and pages of documents contained in appendices 1 to 28 have been included in this volume to show the reader the nature of the archival material listed in the bibliography below, and to show the quality of the work done in western Australia by French explorers from 1772 to 1826.

These few examples have been selected not only to show how the French worked with their sketch books and their day books and on their final drafts, but also to provide the reader with opportunities to personally see and check on the work of the French mappers by visiting the easily accessible places mapped in the charts in the appendices. Calculations made by the French when they were surveying have been included for the reader who wishes to make more detailed checks.

The sources of the documents are printed under the title of each appendix.

LIST OF APPENDICES

With the Géographe *(Baudin) sailing independently in 1801*

Appendix No. 11 Boullanger's map of Shark Bay, June-July, 1801.

With the Naturaliste *(Hamelin) sailing independently in 1801*

Appendix No. 12 Freycinet's sketch map of Rottnest Island and the approaches to Fremantle, June, 1801.

Appendix No. 13 Collas' map of the mouth of the Swan River with bar, June 1801.

Appendix No. 14 Freycinet's sketch map of Rottnest Island salt lakes, Carnac Island and Garden Island, June 1801.

Appendix No. 15 Heirisson's maps of the Swan River, June 1801. (a) Sketch map with soundings. (b) Final draft map of the river.

Appendix No. 16 Map of the course taken by Commander Milius to and from Garden Island, and the site where his boat was wrecked and his crew marooned at Cottesloe, June 1801.

Appendix No. 17 Faure's map of Shark Bay, July 1801.

With the Géographe *(Baudin) and* Casuarina *(Freycinet) in 1803*

Appendix No. 18 Freycinet's map of part of King George Sound from Mistaken Island to Limestone Head, now Frenchman Bay, February 1803 on the *Casuarina.*

Appendix No. 19 Freycinet's map of Princess Royal Harbour, Albany, February 1803 on the *Casuarina.*

Appendix No. 20 Map of Two People Bay and Cheyne Beach east of Albany, February 1803 on the *Géographe.*

Appendix No. 21 Bonnefoi's map of Koombana Bay and Leschenault Inlet, now Bunbury, March 1803 on the *Géographe.*

Appendix No. 22 Freycinet's map from Becher Point, Warnbro Sound to Rottnest Island, March 1803 on the *Casuarina.*

Appendix No. 23 Freycinet's map of the islands from Rottnest Island south to Garden Island, March 1803 on the *Casuarina.*

Appendix No. 24 Freycinet's map of the north coast of Rottnest Island, March 1803 on the *Casuarina.*

FREYCINET'S EXPEDITION 1818

Appendix No. 25 Fabré's map of the north coast of Dirk Hartog Island, showing where the French found the Vlaming plate, September 1818.

Appendix No. 26 Raillard's map of the north end of Peron Peninsula, Shark Bay, with lakes, September 1818.

BLOSSEVILLE AND COLONIZATION

Appendix No. 27 The opening page of Jules de Blosseville's plan for a penal colony in south western Australia, January 1826.

DUMONT D'URVILLE'S EXPEDITION 1826

Appendix No. 28 Map of Oyster Harbour, Albany, October 1826.

LIST OF FRENCH EXPLORERS IN WESTERN AUSTRALIA FROM 1772 to 1826

Appendix No. 29 The leaders of the French missions of exploration, the ships they commanded and the artists who accompanied them.

FRENCH NOMENCLATURE

Appendix No. 30 French place names currently used on the coast of western Australia with a note on their origins.

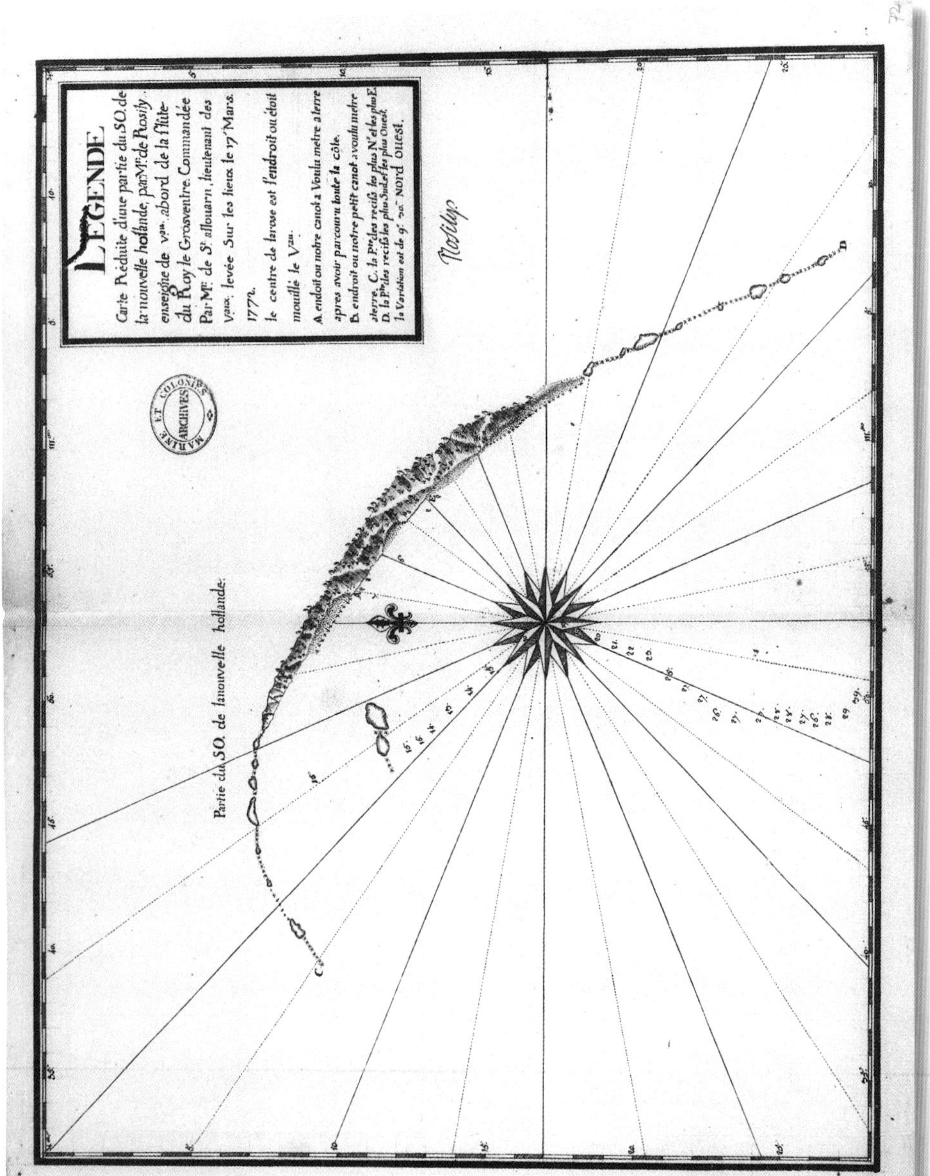
LEGENDE
Carte Réduite d'une partie du S.O. de la nouvelle hollande, par Mr. de Rosily, enseigne de Vau. abord de la flûte du Roy le Grosventre. Commandée Par Mr. de St. allouarn, lieutenant des Vaux. levée Sur les lieux le 17 Mars. 1772.
le centre de la rose est l'endroit ou étoit mouillé le Vau.
A. endoit ou notre canot a Voulu metre a terre apres avoir parcouru toute la côte.
B. endroit ou notre petit canot a voulu metre a terre, C. la Pte. des recifs les plus N. et les plus E.
D. la Pte. des recifs les plus Sud et les plus Ouest.
la Variation est de 9d. 30m. Nord Ouest.
Partie du S.O. de la nouvelle hollande.
ARCHIVES MARINE ET COLONIES

Appendix 1

ROSILY'S MAP OF FLINDERS BAY — MARCH 1772

Archives Nationale. Série marine B4:317

This map of Flinders Bay was drawn by a young naval officer Ensign François Rosily on board the *Gros Ventre,* under the command of St. Allouarn.

François Rosily's visit to western Australia was the result of an accident. He was a member of the crew of Kerguelen's ship the *Fortune.* After Kerguelen and St. Allouarn discovered Kerguelen Island, Rosily was sent from the *Fortune* in a long boat to make an inshore survey of the south coast. Foul weather set in and Kerguelen sailed out to sea on the *Fortune* leaving Rosily behind. Rosily was subsequently picked up by Kerguelen's companion, St. Allouarn, who waited behind on the *Gros Ventre.* The two ships did not meet again. Ensign Rosily consequently remained with St. Allouarn and visited western Australia where he took the opportunity to exercise his cartographic skills.

This was the only visit Rosily made to western Australia. He subsequently rose to the high position as Director of the Hydrographic Office in Paris. While there he was primarily responsible for planning and despatching expeditions to survey western Australia as a site for a French penal colony, which he personally favoured and worked hard to achieve after the fall of Napoleon.

On the map, the ship's position is marked by the compass rose. This position is approximately three miles from the coast. Cape Leeuwin is shown to the west, protecting the bay. The soundings indicate the course taken by the vessel and by the long boat which went ahead of it.

A survey was made of the coast north of the anchored ship, but the party was prevented from landing on the beach by the heavy surf.

There have been some differences expressed about the date St. Allouarn arrived in Flinders Bay. The ship's log book indicates clearly that the *Gros Ventre* entered the bay on the morning of Tuesday, 17th March, 1772 and anchored at midday.

This is the first known map of Flinders Bay which offers protection from the westerly gales and swells which prevail in the region.

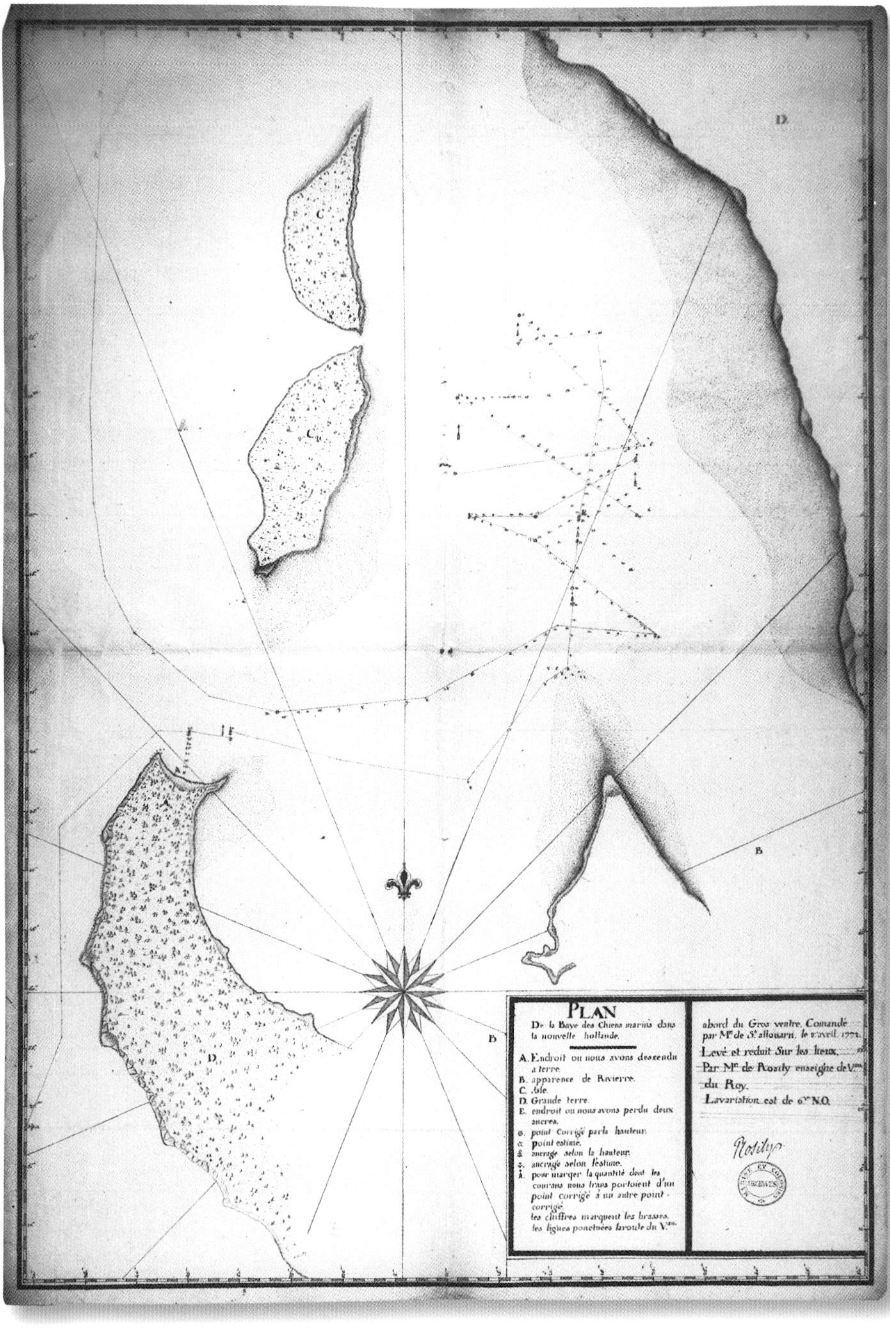
PLAN
De la Baye des Chiens marins dans la nouvelle hollande.
A. Endroit ou nous avons descendu a terre.
B. apparence de Rivierre.
C. Isle.
D. Grande terre.
E. endroit ou nous avons perdu deux ancres.
point Corrigé par la hauteur.
point estimé.
ancrage selon la hauteur.
ancrage selon l'estime.
pour marquer la quantité dont les courans nous trans portoient d'un point corrigé à un autre point corrigé.
les chiffres marquent les brasses.
les lignes ponctuées la route du V.au
abord du Gros ventre, Comandé par Mr de St allouarn, le avril 1772.
Levé et reduit Sur les lieux.
Par Mr de Rosily enseigne de V.au du Roy.
La variation est de 6d N.O.
Rosily
MARINE ET COLONIES
ARCHIVES
A
B
C
D
E

Appendix 2

ROSILY'S MAP OF SHARK BAY — APRIL 1772

Archives Nationale. Série marine B4:317

After visiting and surveying previously undiscovered Flinders Bay, St. Allouarn sailed the *Gros Ventre* to Shark Bay which had been made known by early Dutch explorers and later by William Dampier.

This map of the northern part of the bay was drawn by François Rosily who surveyed and sketched Flinders Bay. His plan compares more than favourably with Dampier's map, offering a good representation of the bay and the navigable entrance to it.

On the map the letter A marks the spot where the French shore party landed on 30th March, 1772 and took possession of western Australia for France. The letter B marks what appeared to the French to be a large river, which in fact is Denham Sound. The letter C marks Dorre and Bernier Islands. The letter D marks Dirk Hartog Island which Rosily calls the mainland. The letters E in the middle of the bay mark the places where the *Gros Ventre* lost two anchors, which have not yet been located.

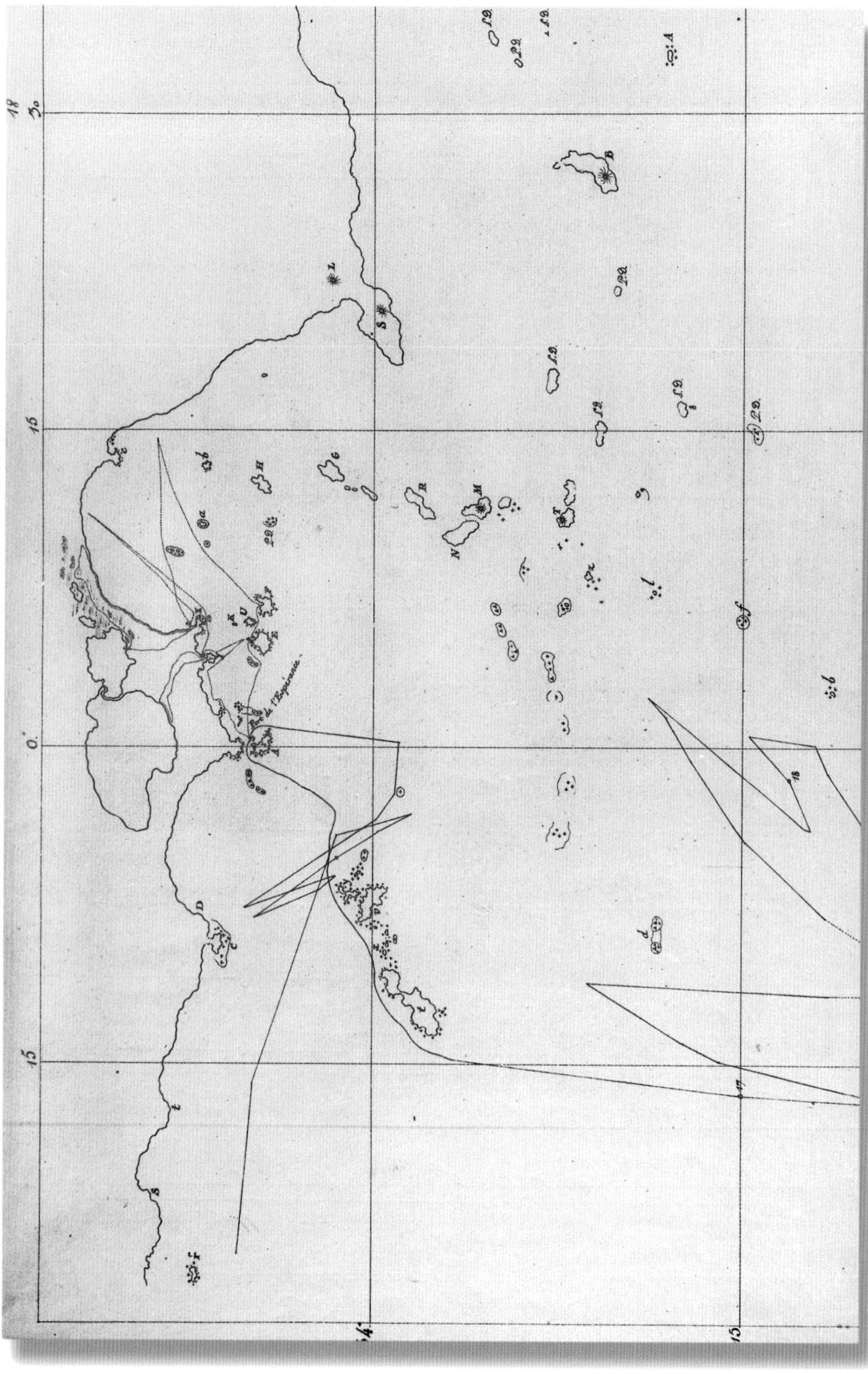

Appendix 3

MAP OF ESPERANCE BAY AND THE RECHERCHE ARCHIPELAGO — DECEMBER 1792

Archives Nationale. Série marine 6JJ3, pl. 18

This map shows the eastward route taken by the *Espérance* and *Recherche* as they moved into Esperance Bay, and their anchorage place behind Observatory Island which D'Entrecasteaux called Port de l'Espérance.

The two vessels approached Esperance on an eastward course, running parallel to the coast as shown by the track commencing at the upper left of the map. When just past Capps and Boner Islands the vessels commenced to tack to the west, to try to get out of the bay by the way they had come in. Failing to do this the ships ran east with the wind for a short distance, and then made north to Observatory Island to find shelter in Port de l'Espérance as marked on the map.

The map also shows the route taken by the shore parties from Blue Haven Beach across the dunes to Lake Spencer (Pink Lake) and along the coast towards Bundy Creek and the head of the bay.

The route taken by Willaumez in the long boat, when he explored the inner islands of the archipelago and the northern coast of Esperance Bay is also shown.

The place where the botanist Riche was lost was in an area east of the lakes near the present townsite of Esperance.

D'Entrecasteaux stayed with his two ships in Port Esperance from 9th to 17th December, 1792, repairing his vessels, using a forge which was established on Observatory Island. No trace of this remains.

Failing to find fresh water he cut short his survey further eastwards and made for Tasmania where plentiful supplies of food and water could be found.

When leaving, the ships sailed between Observatory Island and the coast of the mainland, and sailed back through West Channel, close to the islands before bearing south to the open sea.

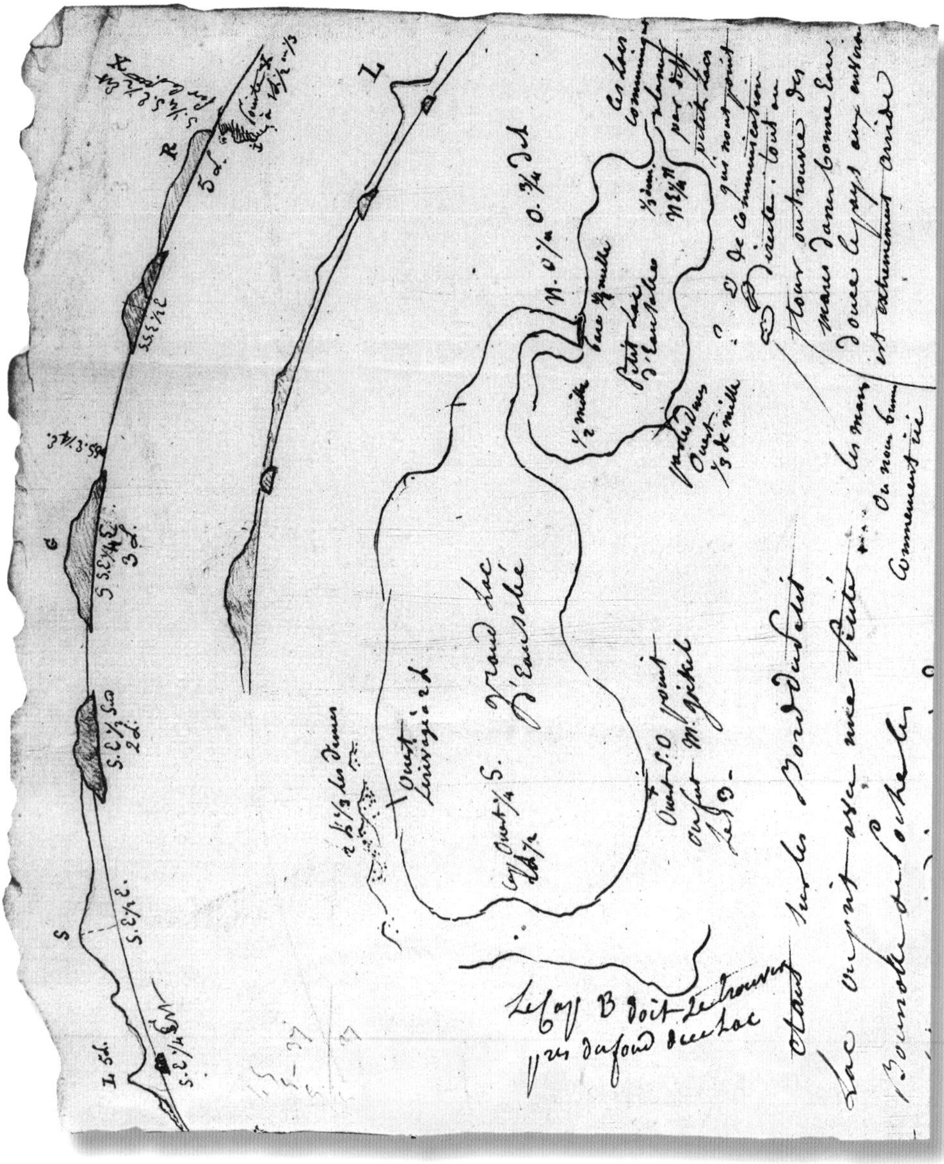
Grand Lac d'Eau salée
Petit Lac d'Eau salée
Le Cap B doit se trouver près du fond de ce Lac

Appendix 4

BEAUTEMPS-BEAUPRE'S SKETCHES OF ESPERANCE BAY AND SALT LAKES — DECEMBER 1792

Archives Nationale. Série marine 5JJ 19

These rough sketches made at Esperance Bay by the hydrographic engineer Beautemps-Beaupré show at the top the profile of the eastern part of Esperance Bay as viewed from Dempster Head close to the townsite of Esperance; and at the bottom the large salt lakes discovered by D'Entrecasteaux's explorers when they walked northwards from the coast at Blue Haven Beach, east of Dempster Head.

The panoramic view shows the profile of the coast from near the head of Esperance Bay, southwards to Cape le Grand and the islands lying to the west of the cape.

The left hand side of the bottom, shorter, sketch is the profile near the head of the bay, close to Rossiter Bay. This short sketch of nearly all of the mainland at the east end of the bay, ends just south of Frenchman Peak (marked L). The profile is continued in the top sketch which begins with the profile just to the north of Frenchman Peak (marked L), thus providing some overlap. The profile is continued beyond Cape le Grand (marked S) to the furthest group of large islands lying to the west of Cape le Grand. Woody Island is marked G and Sandy Hook Island is marked R.

The lower sketch provides a birdseye view of Spencer (Pink) Lake and Lake Warden, which are shown to be separated by a small bar. The small lakes or marshes shown at the bottom of the sketch are those lying eastward from Lake Warden, towards Woody Lake. Sand dunes are noted lying westward of Spencer Lake.

Chapitre Huitième.

Arrivée sur les Côtes Occidentales de la Nouvelle Hollande.

Terre de Leuwin.

(Le Sept Prairéal an 9e. de la République française.)

Vue des Côtes de la Nouvelle Hollande. Erreur sur la position que lui donne nos Cartes Marines. Contrariétés pour rallier la Terre. Sonde et Qualité du Fond. Direction des Courants. Pointe prise pour le Cap Leuwin suivant nos Cartes. Conjecture sur la partie de la Côte où Mr. de St. Allouarn a mis à l'ancre. Changement des Vents de Terre et de Mer. aspect que présente la Côte que nous visitons. Découverte d'une grande Baye en remontant au Nord. Rapports de ceux qui furent envoyés à terre quand nous eûmes mis à l'ancre. Cours inutile fait par la Chaloupe. Départ de notre premier ancrage. Effet du Mirage. Prolongement de la Côte. Vue de différents Feux. Entrée et sortie de l'anse des Peignes. Deuxième Débarquement sur la Côte. Observations Astronomiques faites à terre. Rencontre des Naturels; ce qui se passa. Différents rapports de ceux qui furent à terre. Remarque sur l'Élévation des Marées et formation des Dunes. Présents faits aux Naturels et à la Terre. arrivée à Bord du Capitaine Hamelin. Rapport d'un de ses Officiers. parti auquel je m'arrête. Départ de ma Chaloupe sous le Commandement de Mr. Lebas. Changement de Temps. Retour du Capitaine Hamelin. perte de ma Chaloupe. Situation dans laquelle furent ceux qui étoient dedans. Dangers auxquels le Capitaine Hamelin fut exposé. Moyens employés pour secourir ceux restés à Terre. abandon de la Chaloupe. Retour des Embarcations. Départ. Coup de Vent. Séparation d'avec le Naturaliste. Incident. Retour dans la Baye du Géographe. Rocher à Fleur d'Eau. Position donnée à la Pointe Leuwin. Danger qui se trouve sur la Route que nous faisions. Autre Coup de Vent. Route pour la Baye des Chiens Marins. Table de Loch, Relevements, Sondes &c. &c. &c.)

---

an 9e. de la Rép.

Prairéal 7. Le Sept Floréal à huit heures du Matin nous eûmes Connoissance des Côtes Occidentales de la Nouvelle Hollande, et par l'observation de la hauteur du Soleil à Midi nous reconnûmes que nous étions au Sud des terres qui forment le Cap ou la Pointe de Leuwin à l'ouest, que toutes les Cartes qu'on nous avoit donné au Dépôt de la Marine à l'époque de notre Départ placent assez généralement par 34. Degrés 10. Minutes de Latitude Sud et que nous étions alors par 34°. 35'. 58".

La Longitude indiquée par nos Gardes temps différoit aussi de celle que lui donne ces mêmes Cartes en le fixant par 111°. 25' tandis que nous nous trouvions d'après la Marche de nos Montres, dont l'écart ne pouvoit être Considérable, depuis le Départ de l'Isle de France, par 111°. 43'. 45" en sorte que cette partie de la Côte nous y sembla portée plus à l'Est qu'elle ne l'est

Effectivement

Appendix 5

A PAGE FROM BAUDIN'S UNPUBLISHED HISTORY OF HIS VOYAGE

Archives Nationale. Série marine 5JJ 40

This reproduction is of the opening page of Chapter 8 of the history of the expedition written by Baudin. It records the landfall made by the expedition near Hamelin Bay on 27th May, 1801 (7 Prairial year 9).

Liberté | Egalité

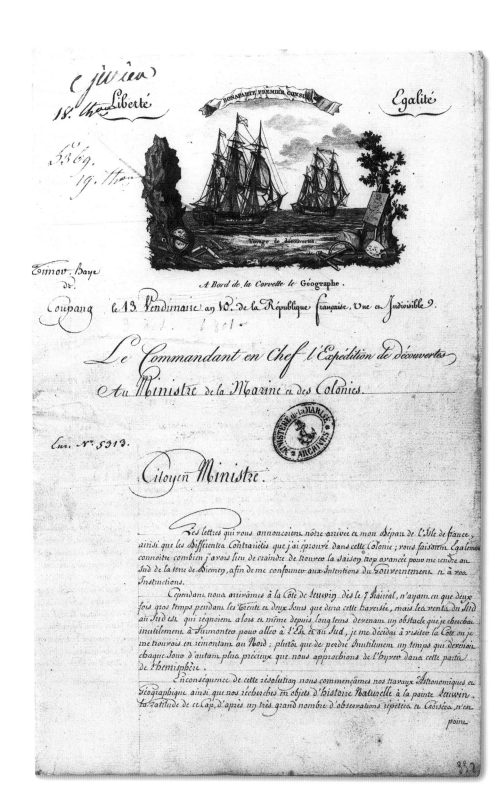

A Bord de la Corvette le Géographe.

Timor. Baye de Coupang le 13. Vendimaire an 10. de la République française, Une et Indivisible.

Le Commandant en Chef l'Expédition de découvertes

Au Ministre de la Marine et des Colonies.

Enr. N° 5313.

Citoyen Ministre.

Les lettres qui vous annoncoient nôtre arrivée et mon départ de l'Isle de france, ainsi que les differentes contrariétés que j'ai éprouvé dans cette Colonie ; vous faisoient Egalement connoitre combien j'avois lieu de craindre de trouver la saison trop avancée pour me rendre au Sud de la terre de Diemen, afin de me conformer aux Intentions du Gouvernement et à vos Instructions.

Cependant nous arrivâmes à la Côte de Leuwin dès le 7 Prairéal, n'ayant eu que deux fois gros temps pendant les trente et deux Jours que dura cette traversée, mais les vents du Sud au Sud est qui régnoient alors et même depuis long tems devenant un obstacle que je cherchai inutilement à surmonter pour aller à l'Est et au Sud, je me décidai à visiter la Côte ou je me trouvois en rémontant au Nord ; plutôt que de perdre inutilement un temps qui devenoit chaque Jour d'autant plus précieux que nous approchions de l'hyver dans cette partie de l'hemisphère.

En conséquence de cette résolution nous commençâmes nos travaux Astronomiques et Géographique ainsi que nos récherches en objets d'histoire Naturelle à la pointe Leuwin. La latitude de ce Cap, d'après un très grand nombre d'observations répetées et Croisées, n'est

point

332

Appendix 6

LETTERHEAD OF THE BAUDIN EXPEDITION — 1800-1803

Archives Nationale. Série marine 5JJ:7

In this letter written at Timor and dated 13 Vendemiaire year 10 (5th October, 1801), Baudin reports on the findings his mission made in western Australia.

Baudin arrived at Coupang in Timor on 21st August, 1801 (3 Fructidor year 9). Hamelin, who remained in western Australia waiting for Baudin to show up at the rendezvous set, surveying during this period, did not arrive in Timor until 21st September, 1801. Baudin left writing his report until after Hamelin arrived.

Hamelin's account of the extensive surveys he made in western Australia with his small crew while Baudin was resting in Timor added to the resentment of the scientists and their open hostility to Baudin who increasingly lost their respect.

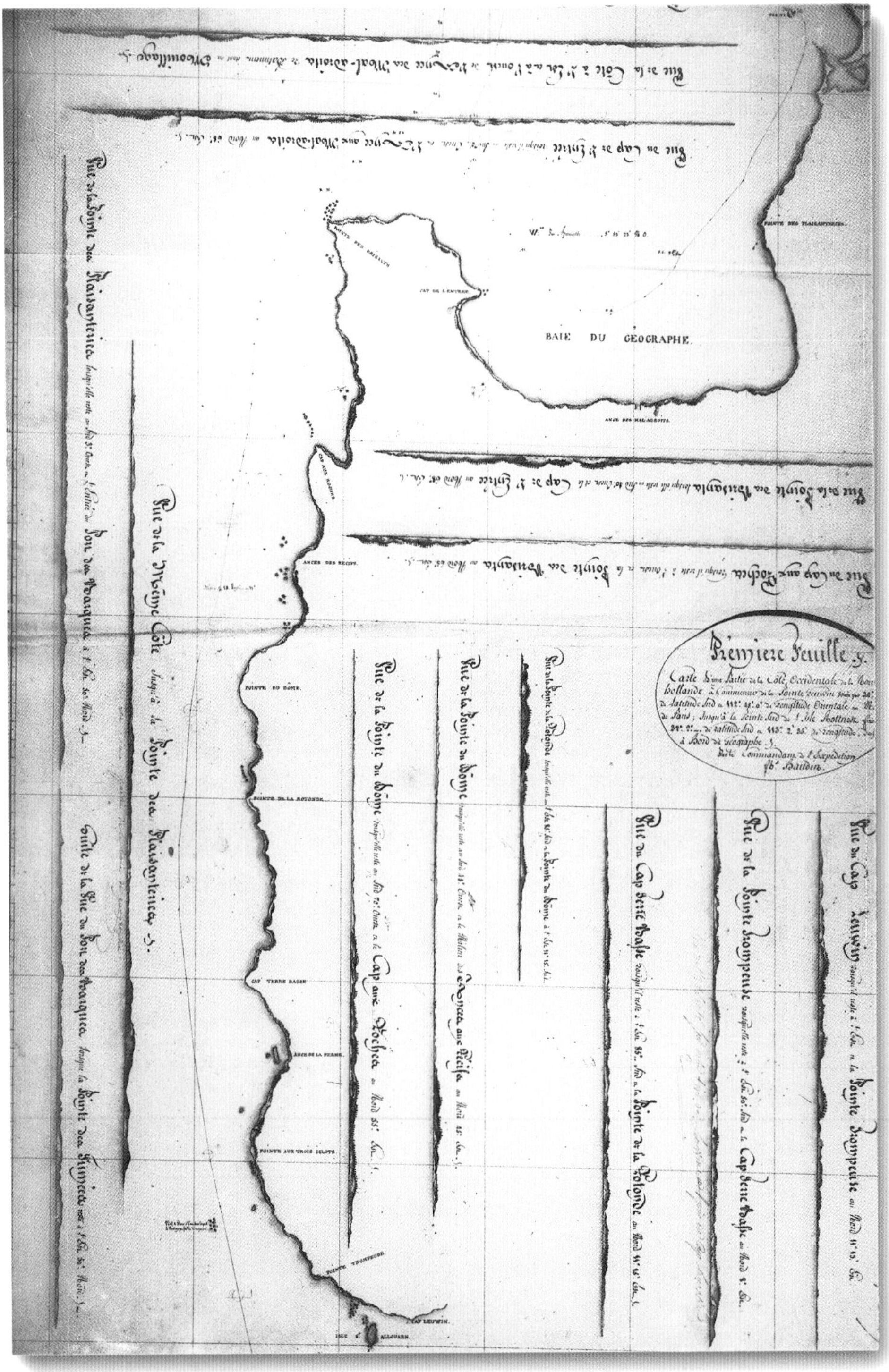
Premiere Feuille
BAIE DU GEOGRAPHE
POINTE DU DOME
POINTE DE LA ROTONDE
ANCE DE LA FERME
POINTE AUX TROIS ILOTS
POINTE TROMPEUSE
CAP LEUWIN
ANCE DES RECIFS
ANCE DES MAL-ADROITS
POINTE DES PLAISANTERIES
CAP DE L'ENTREE

Appendix 7

SKETCH MAP FROM CAPE LEEUWIN TO BUNBURY WITH LAND PROFILES, 1801-1803

Archives Nationale. Série marine 6JJ4:74

This is the first map made of Geographe Bay which was discovered by Baudin on the *Géographe* and Hamelin on the *Naturaliste* on 30th May, 1801.

The ships made a landfall north of Cape Leeuwin and then coasted north observing the land until they came to Cape Naturaliste. This was rounded and the ships sailed into Geographe Bay which was not expected to exist. At the time it was believed that the coast of western Australia from Cape Leeuwin to the Swan River was a straight line with no indentations suitable to be used as anchorages.

Baudin's expedition stayed in the newly discovered bay surveying and collecting specimens from 30th May until 9th June, 1801.

As Baudin's chronometers were not accurate and his assessment of longitude was consequently wrong, and as he did not set up an observatory on shore to make more accurate observations the land on the map is wrongly positioned. Inaccurate compasses and bad visibility at the time, together with inaccurate assessments of longitude account for the distortions of the land drawn on the map.

The point marked Point de l'Entrée, which was taken to be the place where Geographe Bay proper commenced is now called Castle Rock near Dunsborough. The mouth of the present Vasse River and Wonnerup Estuary is called the Anse des Maladroits (Blunder Bay) because of the loss of the *Géographe's* long boat and other disasters there, which Baudin blamed on the scientists and officers.

The place marked Pointe des Plaisanteries (Joke Point) in reality is not a point, but the place where the sand dunes rise north of the Capel River at approximately 33° 27′ south latitude. The harbour later called Port Leschenault, which was not surveyed until March 1803, is marked as Port des Barques (Small Boat Harbour).

Baudin missed seeing South Passage which separates Garden Island from the mainland. This is shown as a bay. The map ends at Rottnest with the Straggler Reefs being sighted and called Les Espions (The Spies).

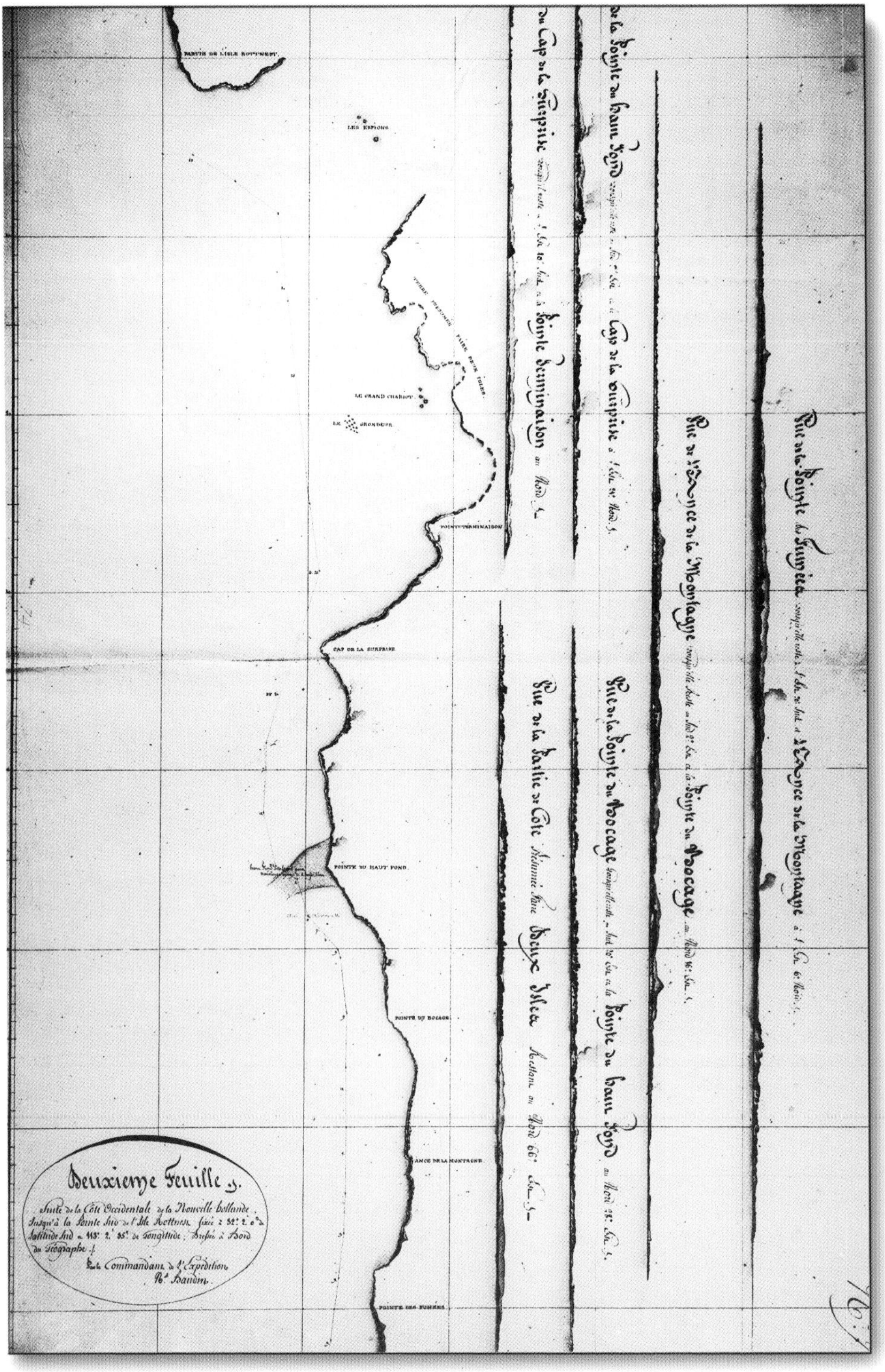
Deuxieme Feuille
LES ESPIONS
LE GRAND CHARIOT
POINTE TERMINAISON
CAP DE LA SURPRISE
POINTE DU HAUT FOND
POINTE DU BOCAGE

Appendix 8

SKETCH MAP FROM THE NORTH OF BUNBURY TO ROTTNEST ISLAND, 1801-1803

Archives Nationale. Série marine 6JJ4:75

This map which shows non existent capes and bays, was drawn from visual observations made on Baudin's ships as they sailed north from Geographe Bay, keeping well out to sea in June 1801 and later in 1803.

The distortions to the land shape were not only the result of distant observations and instrument errors. Illusion also played a part. It is very difficult to determine the real shape of the western Australian coast while sailing along it. The sand dunes which fall away onto the beach often create a false impression of capes and bays.

The cape which was named Pointe des Fumées (Smoke Point) is the dune area near the reservoir between Leschenault Inlet and Lake Preston. The Anse de Montague (Mountain Bay) is the sand dunes two thirds of the way down Lake Preston, at approximately 33° 1′ 30″ south latitude. The Pointe de Bocage (Grove Point) is near the present Preston Beach. The reef area shaded is Bouvard Reefs and the Pointe du Haute Fond (Shallow Point) is the dune area at the north part of Lake Clifton at 32° 45′ south latitude. Baudin obviously did not see the Cape Bouvard which is several miles north of there.

The French-named Cap de la Surpris (Surprise Cape) is near present-day Falcon Townsite, several miles south of Robert Point and Mandurah. The Pointe de Terminaison (Termination Point) is the dune area several miles south of Becher Point.

Present Coventry Reef which can be clearly seen at sea, was sighted and named Le Grandeur (The Splendour). The islands sheltering Port Kennedy and Safety Bay were called Le Grand Chariot (The Big Chariot).

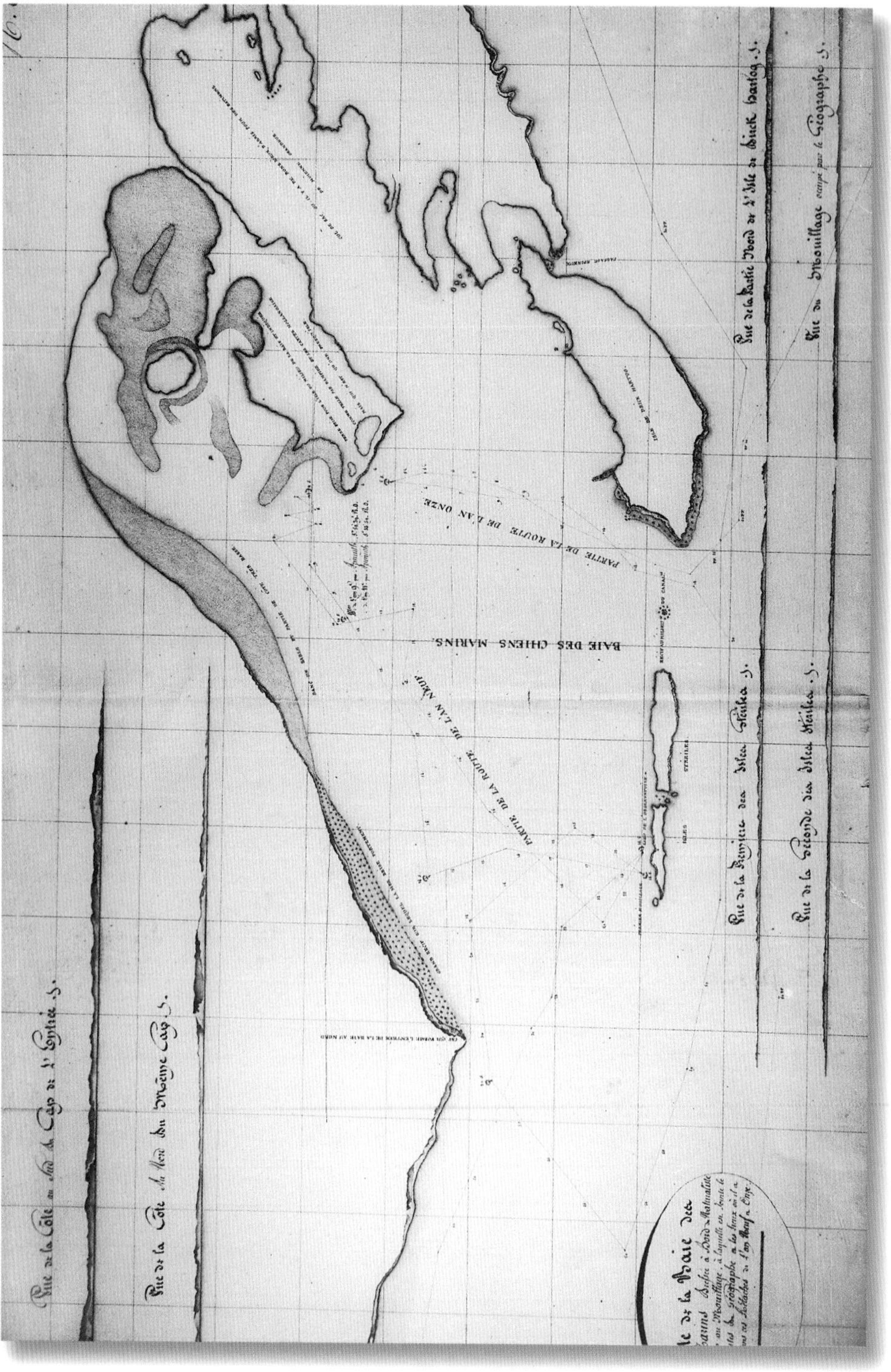
BAIE DES CHIENS MARINS.
PARTIE DE LA ROUTE DE L'AN ONZE
PARTIE DE LA ROUTE DE L'AN NEUF

Appendix 9

SKETCH MAP OF SHARK BAY WITH LAND PROFILES, 1801-1803

Archives Nationale. Série marine 6JJ 86

This is the final sketch map of Shark Bay by Baudin drawn after he revisited the area in March 1803. The routes taken by the *Géographe* and *Naturaliste* in 1801, and the route taken by the *Géographe* and the schooner *Casuarina* are marked on the map together with the position of the anchorages.

nous trouvions, ne tarda pas a trouver l'embouchu
étoit fermée par une barre, il vint ensuite à nous
nous étant embarqués nous remontames la riviere environ
Lieue et demi nous echouames plusieurs fois faute de bie
la passe, enfin a midi et demi nous nous en revinmes l'
nous primes alors le long du bord: ~~avoit perdu de l~~ il étoit
aussi salée que l'eau de mer d'où nous conclumes que si
l'avions ~~remonté~~ cette riviere plus haut nous serions parv
a trouver de l'eau douce

je tracé ici un plan approché de la riviere et du bassin

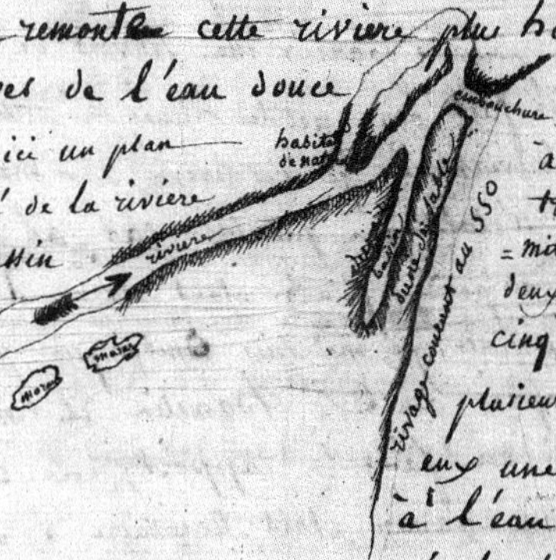

a 1 heure et demi nous n
trouvames par le travers d
=mité d'une isle formée probab
deux bras de la riviere nous
cinq naturels sur l'autre riv
plusieurs d'entre nous desirant
eux une entrevue amicale, se
à l'eau et passerent de l'autr
gué les sauvages s'enfuirent dan
en jettant differents cris, nous les suivimes (j'étois du nombre) quelque t
ensuite nous revinmes sur nos pas, pendant ce temps les sauv
munirent de leurs Lances et Sagayes et vinrent après nous.
deposâ en nous en revenant sur les branches d'arbres differentes Bagatelles pour l
=ver que nos intentions étoient pacifiques, étant parvenus à la rivie
traversames à gué le bras qui nous separoit de l'isle a la quell
canot avoit abordé et nous nous joignimes aux autres, nous vime
six de ces naturels dont cinq étoient armés et tout nuds sans e
et le sixieme étoit sans armes et avoit une peau ~~de~~ sur le
je presume que c'étoit une femme du reste elle étoit entièrement
ces hommes nous parloient, nous leur repondions aussi quelques mot
lesquels le mot pourats que Mr le bas nous disoit signifier ami
entendu et distinctement repeté par un des sauvages aux au
avec le quel celui ci qui paroissoit le chef repeta le mot avoit
contentement il paroissoit dire entendez vous ils disent pourats
leur fimes differents signes pour engager a venir de notre coté en
montrant differents objets enfin le chef et deux autres le

Nota, avant de descendre la riviere, Mr Hamelin voulant savoir si l'endroit ou nous nous trouvions étoit une riviere ou non a pris nos opinions particulières et commencant par lui il a decidé pour l'affirmative les autres opinants étoient Mr depuch Mineralogiste Mr Lechnau Botaniste Mr Laidon Chirurgien Major et moi B. Enseigne de Vau qui avons decidé que nous etions sur une riviere.

Mr Lebas Capne de fregate et Mr Freycinet enseigne de Vau ont été indecis et n'ont pas prononcé.

je n'ai marqué cette note que parceque Mr Hamelin a dit depuis le contraire, et comme on lui a observé que son opinion n'étoit pas la même que celle qu'il avoit eue sur les lieux, il a repondu qu'alors il n'avoit decidé que par Complaisance, cependant on a vu plus haut qu'il avoit decidé le premier

Note du 15 thermidor an IX

Appendix 10

HEIRISSON'S MAP OF WONNERUP ESTUARY, GEOGRAPHE BAY — 5TH JUNE, 1801

Archives Nationale. Série marine 5JJ 56

This sketch made by sub-lieutenant Heirisson on the *Naturaliste* is the first European map of Wonnerup Estuary in Geographe Bay.

The news of the discovery of this estuary by Heirisson on 4th June, 1801 (15 Prairial year 9) changed the course of the Baudin expedition and resulted in tragedy and near disaster for the ships.

Prior to 4th June no sign had been found of waterways inland in Geographe Bay, although the land was well forested. Baudin who was concerned about the shortage of water and food supplies on the *Géographe* and *Naturaliste*, determined to weigh anchor and sail north for Timor on 5th June, 1801 (16 Prairial year 9).

Heirisson returned to the *Naturaliste* on the evening of 4th June and reported that he had landed on the coast south of the anchored ships and when walking inland had observed from the dunes, a broad expanse of water rich with bird life, which appeared to be a wide river going inland. He followed the coast westward in his dinghy searching for the entrance, but failed to find it. However, he examined the shores of the estuary and found tide marks, which indicated to him that a sea entrance existed.

Having been given strict orders to return that evening, Heirisson had to give up his search after going along the coast for six miles. However, his discovery impressed Commander Hamelin who reported the matter to Baudin. Baudin consequently altered his sailing plan and sent two boats equipped with provisions for two days, to look for the river mouth and explore it. Hamelin, who led the expedition, discovered Wonnerup Inlet on 5th June and rowed past the Deadwater (called *bassin* on the sketch) and along Wonnerup Estuary where they were confronted by Aborigines and retreated. The place where they located an Aboriginal settlement is marked on the sketch as *habitat de naturels.*

When the river survey party returned to the coast, a full gale was blowing, and the party was marooned. Vasse was lost from a rescue dinghy during the subsequent rescue operations, and when this was over the ships got out of Geographe Bay with great difficulty.

Baudin subsequently named the small bay which formed the entrance to Wonnerup Estuary, Blunder Bay (Anse des Malaldroits).

The watercourse itself was named after Timothée Vasse, who was lost near the entrance.

His name is now given to the estuary and river which flows westward from the inlet which was not discovered by the French explorers, who made their inland surveys eastward of the inlet.

Detailed accounts of these explorations are given in particular by Heirisson, Leschenault and Péron in their journals.

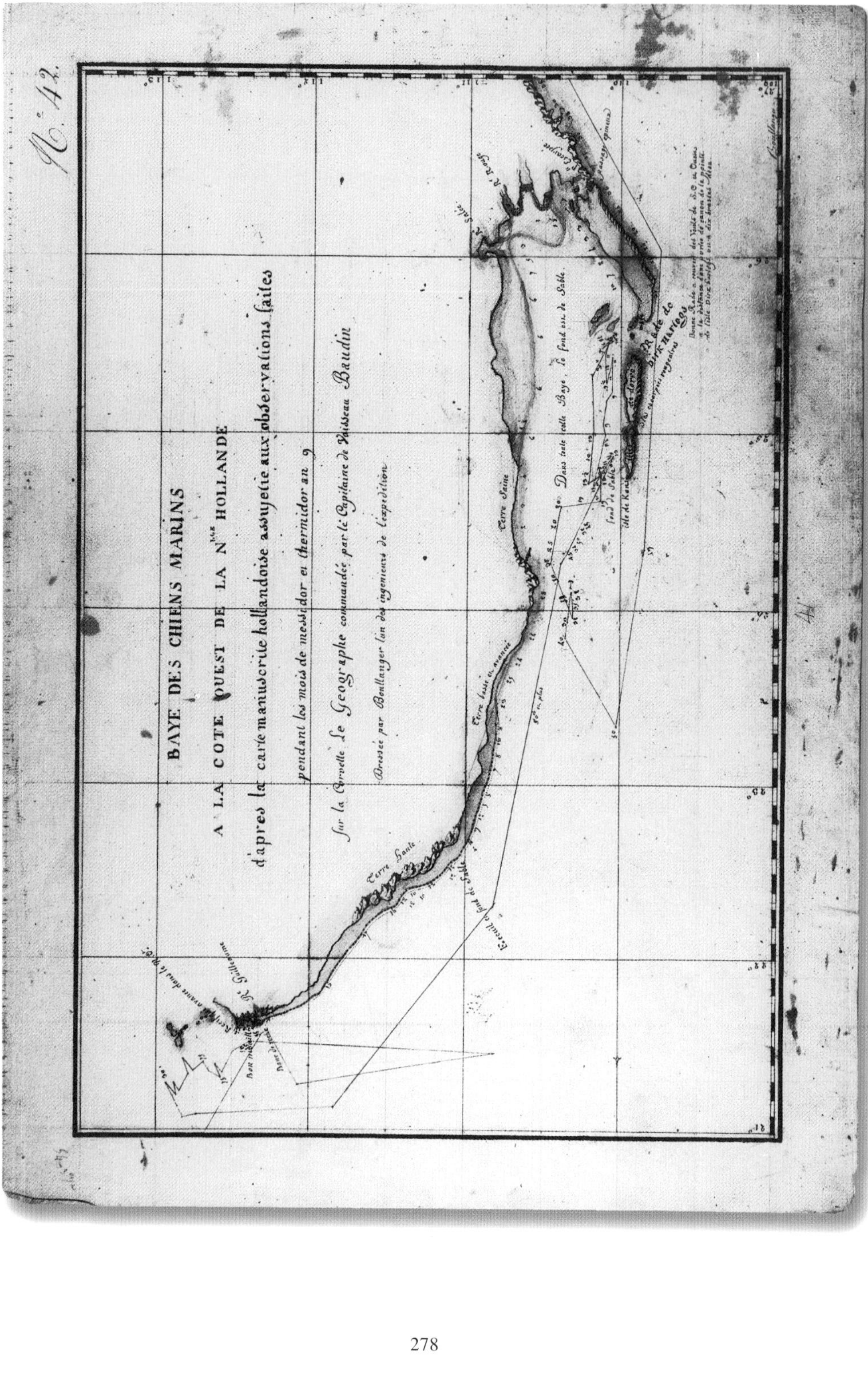
N.° 42.
BAYE DES CHIENS MARINS
A LA COTE OUEST DE LA N.lle HOLLANDE
d'apres la carte manuscrite hollandoise assujetie aux observations faites
pendant les mois de messidor et thermidor an 9
sur la Corvette le Geographe commandée par le Capitaine de Vaisseau Baudin
Dressée par Boullanger l'un des ingenieurs de l'expedition
Terre Haute
Terre basse et avancée
Terre Saine
Dans toute cette Baye, le fond est de Sable.
Rade de Dirk Hartogs

Appendix 11

BOULLANGER'S MAP OF SHARK BAY — JUNE-JULY 1801

Archives Nationale. Série marine 6JJ4:42

This map drawn by Boullanger on the *Géographe* under Baudin shows the northern route Baudin took into Shark Bay, via the Geographe Channel.

The outer islands in the bay are correctly marked but Peron Peninsula is omitted. Only the northern end, previously sketched on the map by Rosily (1772) and others is on the map.

The map consequently is not as important as Faure's detailed survey sketch made on Hamelin's ship the *Naturaliste* which arrived in Shark Bay after Baudin had departed for Timor, and reveals how the scientists on Baudin's vessel were disadvantaged by Baudin's cautiousness which was not the case with Hamelin on the *Naturaliste*.

The mythical William River (R^e Guillaume) is marked by Baudin on the coast north of Shark Bay.

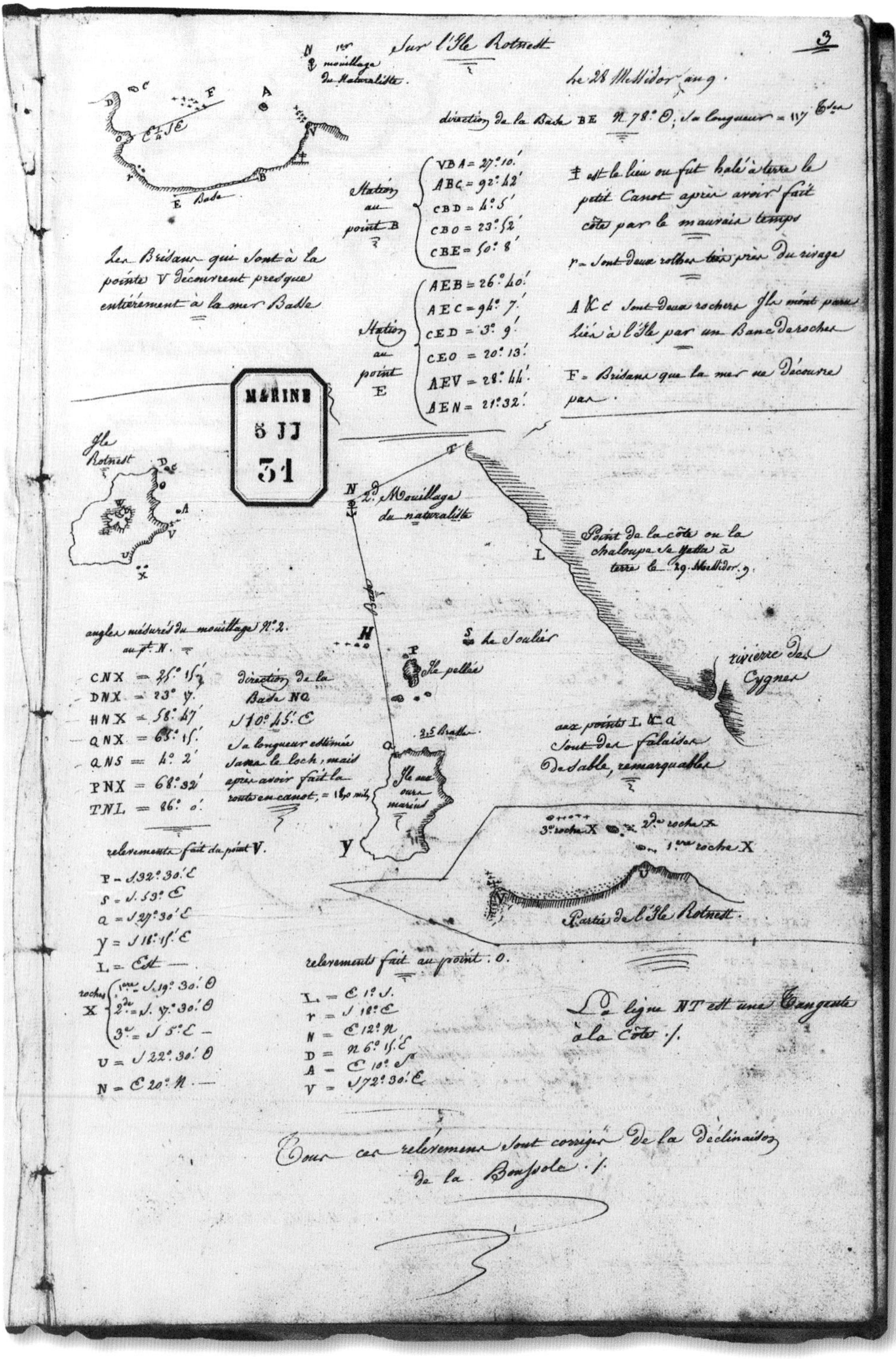

Sur l'Ile Rotnest — 3

N 1er mouillage du Naturaliste.

Le 28 Messidor an 9.

direction de la Base BE N. 78° O. ; sa longueur = 117 Toises

Station au point B

VBA = 27° 10'
ABC = 92° 42'
CBD = 4° 5'
CBO = 23° 52'
CBE = 50° 8'

Station au point E

AEB = 26° 40'
AEC = 94° 7'
CED = 3° 9'
CEO = 20° 13'
AEV = 28° 44'
AEN = 21° 32'

+ est le lieu ou fut halé à terre le petit Canot après avoir fait côte par le mauvais temps

r = sont deux rochers très près du rivage

A & C sont deux rochers qui m'ont paru liés à l'Ile par un Banc de roches

F = Brisans que la mer ne découvre pas.

Les Brisans qui sont à la pointe V découvrent presque entièrement à la mer Basse

angles mesurés du mouillage N° 2. au pt. N.

CNX = 25° 15'
DNX = 23° 7'
HNX = 58° 47'
QNX = 65° 15'
QNS = 4° 2'
PNX = 68° 32'
TNL = 86° 0'

direction de la Base NQ S 10° 45' E

sa longueur estimée sans le loch ; mais après avoir fait la route en canot, = 1,80 mil.

aux points L & Q sont des falaises de sable, remarquables

relevements fait du point V.

P = S 32° 30' E
S = S. 53° E
Q = S 29° 30' E
Y = S 18° 15' E
L = Est
roches X: 1ere = S. 19° 30' O; 2de = S. 7° 30' O; 3e = S 5° E
U = S 22° 30' O
N = E 20° N

relevements fait au point O.

L = E 1° S.
r = S 18° E
N = E 12° N
D = N 6° 15' E
A = E 10° S
V = S 72° 30' E

La ligne NT est une Tangente à la Côte.

Tous ces relevements sont corrigés de la déclinaison de la Boussole.

Appendix 12

FREYCINET'S MAP OF ROTTNEST ISLAND AND THE APPROACHES TO FREMANTLE — JUNE 1801

Archives Nationale. Série marine 5JJ 31

This sketch map by Louis de Freycinet is one of a series he drew of the islands near the Swan River and the approaches to Fremantle.

At the top there is a sketch of his survey of Thomson Bay, Rottnest Island. His base line is marked E,B. This extended from near the present army jetty to near the Rottnest Hotel. Natural Jetty is marked with a V, Philip Rock with an A, and Duck Rock with a C. The first anchorage of the *Naturaliste* in the approaches to Fremantle is marked with an N.

In the centre of the sheet is a map of the approaches to the Swan River, with bearings made of different places.

The blank spot marked L is where the ship's long boat was blown ashore in a gale and the crew marooned. Calculations made from this map and other survey books indicate that the place where the boat was blown ashore was Cottesloe main beach, near the present bathing pavilion.

The small map at the bottom is part of the south coast of Rottnest Island from the Natural Jetty to near Dyers Island.

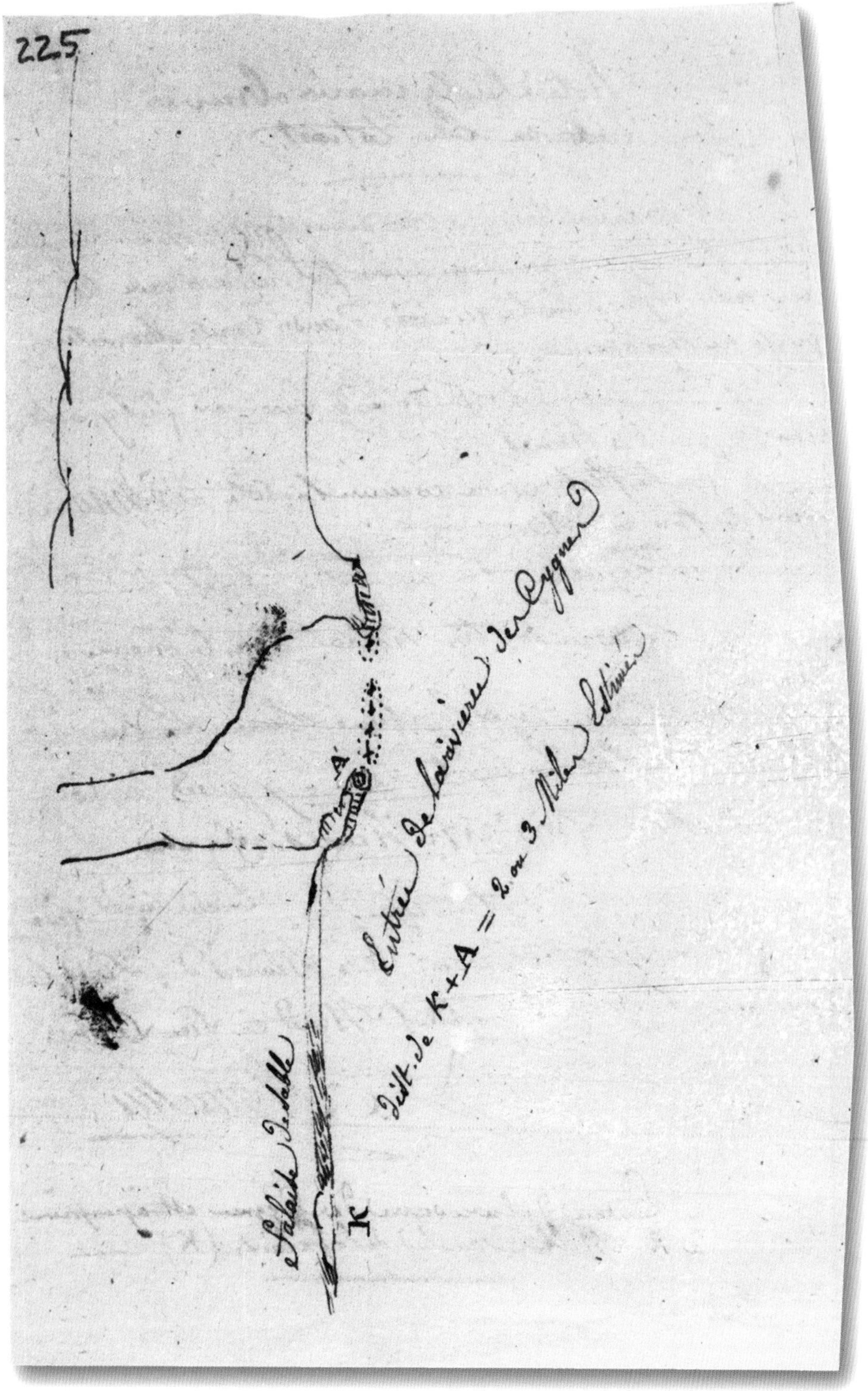
225
Falaise de sable
Entrée de la rivière des Cygnes
Dist. de K à A = 2 ou 3 Miles Estimée
A
K

Appendix 13

COLLAS' MAP OF THE MOUTH OF THE SWAN RIVER WITH BAR — JUNE 1801

Archives Nationale. Série marine 5JJ 41

This sketch map by ship's pharmacist, François Collas, shows the rocky bar across the river entrance and the narrow channel on the south side. Hamelin sent the ship's pharmacist, Collas, to investigate the mouth of the Swan River because he was short of scientists. These were mainly on the *Géographe* which sailed past the Swan River and went to Timor.

The existence of this bar later led Heirisson to conclude that because of the quantity of water in the broad stretches of the river inland, the Swan River had to have a second entrance. He consequently assumed that the Canning River, the mouth of which he saw from the distance, was the beginning of the second entrance to the sea. This was an easy mistake to make. The Canning River, at its mouth, seems to flow westward towards the coast. Further inland the course bends eastwards towards the Darling Ranges.

The conspicuous sand hill marked K on his map is the area later called by the British, the Winding Sheet, and used as a navigation mark for ships approaching Fremantle.

This place is now Cottesloe main beach.

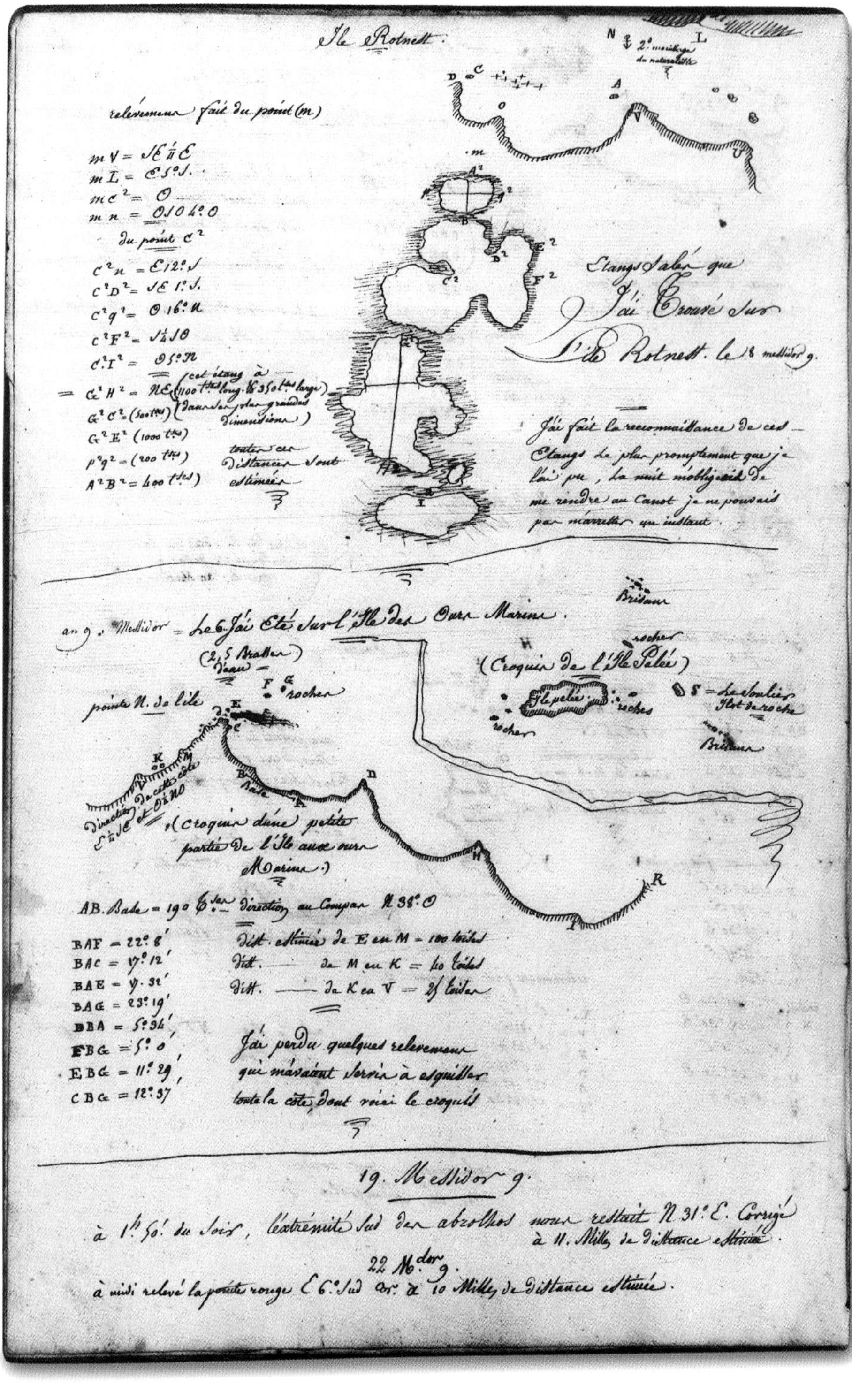

Ile Rotnest.

relèvement fait du point (m)

m V = SE ¼ E
m L = E 5° S.
m c² = O
m n = OSO 4° O

du point c²

c² n = E 12° S
c² D² = SE 1° S.
c² q² = O 16° N
c² F² = S ¼ SO
c² I² = O 5° N
G² H² = NE (cet étang à 1100 toises long. & 350 toises large dans ses plus grandes dimensions)
G² C² = (500 toises)
G² E² (1000 toises)
p² q² = (200 toises)
A² B² = 400 toises)

toutes ces distances sont estimées

Étangs salés que j'ai trouvé sur l'île Rotnest. le 18 messidor 9.

J'ai fait la reconnaissance de ces étangs le plus promptement que je l'ai pu, la nuit m'obligeait de me rendre au Canot je ne pouvais pas m'arrêter un instant.

an 9. Messidor = Le 6. J'ai été sur l'île des Ours Marins.

AB. Baie = 190 toises direction au compas N 38° O

BAF = 22° 8'
BAC = 17° 12'
BAE = 9° 32'
BAG = 23° 19'
DBA = 5° 34'
FBG = 5° 0'
EBG = 11° 29'
CBG = 12° 37

Dist. estimée de E en M = 120 toises
Ditt. de M en K = 40 toises
Ditt. de K en V = 25 toises

J'ai perdu quelques relèvements qui m'avaient servi à esquisser toute la côte, dont voici le croquis

19. Messidor 9.

à 1h 50' du soir, l'extrémité sud des abrolhos nous restait N 31° E. Corrigé à 11. milles de distance estimée.

22 Messidor 9.

à midi relevé la pointe rouge E 6° Sud dr. à 10 milles de distance estimée.

Appendix 14

FREYCINET'S SKETCH MAP OF THE ROTTNEST ISLAND SALT LAKES, CARNAC ISLAND AND GARDEN ISLAND — JUNE 1801

Archives Nationale. Série marine 5JJ 31

This page from Louis de Freycinet's survey sketch book shows at the top the first map drawn of the salt lakes at Rottnest, inland from Thomson Bay which lies between V and D on the map.

These lakes were named Duvaldailly Ponds by the French after a junior officer on the *Naturaliste* who landed on Rottnest with the survey party.

The sketch at the centre is of Carnac Island, called by the French Ile Pelée (Bald Island).

At the bottom is a survey sketch of the north part of Garden Island. On the left is the coast from Entrance Point to Beacon Head, with the places marked where Milius who visited the island made his observations.

Milius survey base line in the Pig Trough (Bay) is marked A B. This extends from near the present jetty northwards to the small point of land near the last of the holiday cabins. South of this Sulphur Bay is marked.

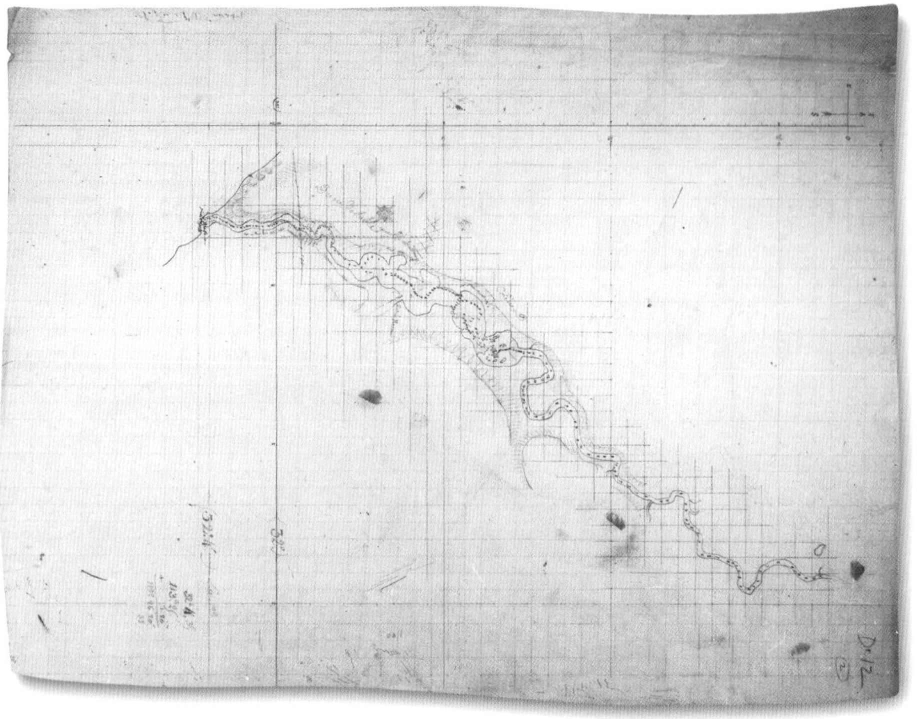

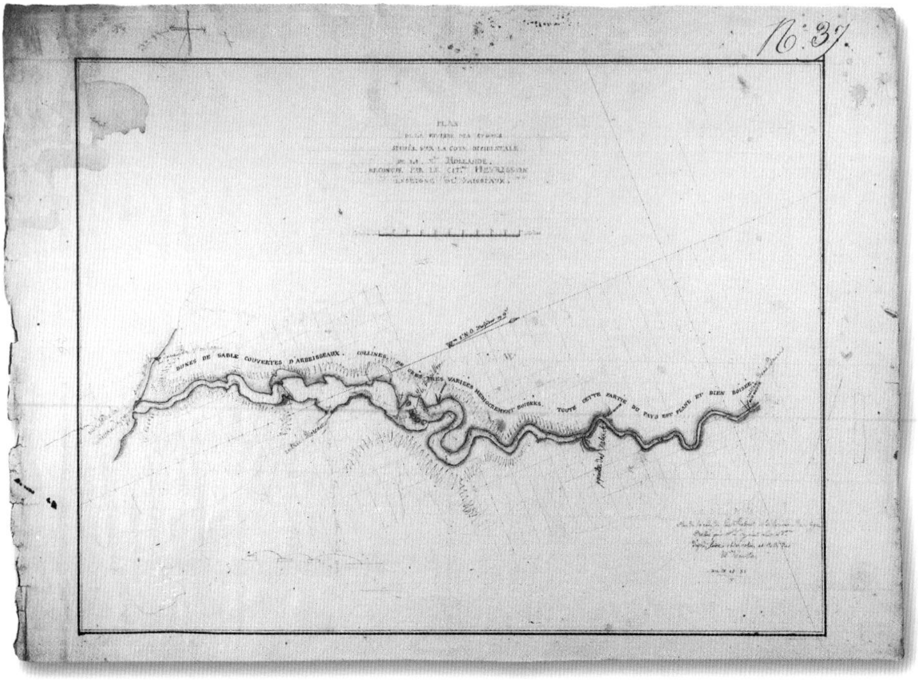
PLAN
DUNES DE SABLE COUVERTES D'ARBRISSEAUX
TOUTE CETTE PARTIE DU PAYS EST PLANE ET BIEN BOISÉE

Appendix 15

HEIRISSON'S MAP OF THE SWAN RIVER — JUNE 1801

Archives Nationale. Série marine 6JJ4:37
Archives Nationale. Série marine 6JJ6:12(2)

These maps of the Swan River were drawn by Sub-lieutenant Heirisson of the *Naturaliste* after he made his survey between 17th and 22nd June, 1801.

The first map (6JJ4:37) is his rough sketch showing the route he took in the long boat. The second map (6JJ6:12(2))is a final draft showing the river and its branches, and the countryside seen.

The bar across the Swan River mouth is shown. Further up the river Point Walter Spit is marked in. Beyond this the entrance to the Canning River is shown as Entrée Moreau. This river entrance was seen from the top of Mount Eliza, on the north bank of the river and because of the bar across the Swan River mouth and the vast amount of water found behind the bar, was assumed to be another mouth of the river. Heirisson could have found fresh water not far up the Canning. Instead he headed inland along the Swan searching for it.

Heirisson Islands are marked on the map, stretching across the river making a further barrier to progress. Just beyond these Claise Brook is marked as a fresh water stream. The Helena River which Heirisson walked along for nearly two miles is shown on the left bank.

The fresh water stream, Bennett Brook, is shown on the right bank beyond the Helena. The map continues to the vicinity of Henley Brook on the upper Swan which is the furthest point Heirisson reached. He estimated he travelled 18 leagues (approximately 55 miles) along the river.

223

Voir cy contre Le Rapport du Lieut. Milius, sur son Excursion dans la Chaloupe, et accidents subséquents. En rade de l'île Rottnest le 3 messidor an 9. pr. l'intelligence duquel j'ai jetté sans dimensions ces idées sur le papier, en attendant un plan exact.

N.ta Dans toute l'étendue de cette Rade le fond est de Sable mouvant, je conseillerai toujours d'empenneler ses ancres.

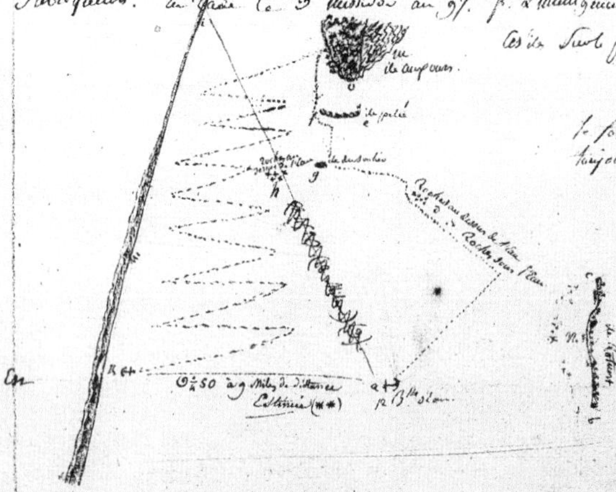

Ouest

Distances Estimées
a f = 14 milles *
a n, 6 milles
a d, 5 milles
a g, 11 milles
a o, 12 milles
a h, 10,5 milles
a i, 18 milles
a k, 9 milles
n x, 2 milles

La Relache à l'île Rottnest n'est pas à dédaigner pr. les Batiments Naviguants dans les mers qui baignent la Côte occ. de la Nouvelle Hollande.

La Rade est poissonneuse et il n'est pas difficile de se procurer des vins marins dont la chair est bonne et salutaire. Sur les bords de la Riv. Swan on trouve beaucoup de Pélicans, Cygnes etc. et de la Casse-pierre. probablement on trouverait aussi du Cellery, si on cherchait bien.

Le Bois est aisé à faire.

S'il y a des Habitants ordinairement sur cette Côte, il paroit que leur éloignement pr. les européens ou plutot leur gout prononcé et inné pr. l'indépendance les portera encore long-tems à s'éloigner à l'approche des batiments, donc on peut envoier sans crainte son monde se promener et se raffraichir à terre.

La grande question est s'il est facile ou difficile de faire de l'eau. c'est possible, mais malaisé, pr. entrer dans la riv. il faut beau tems à cause de la Barre et il faudroit que la Chaloupe remontât le Swan et eut avec elle un pt. canot pr. aller faire son eau à une lieue plus haut qu'une chaine de hauts fonds qu'elle ne pourroit passer. Elle mettroit ce pt. canot lui ramèneroit son eau en drome, elle s'embarqueroit et saisiroit le moment où les vents ne seroient pas du Large pr. tenter sa sortie de la riv. pendant que le pt. canot serviroit à faire son eau, les gens de la Chaloupe pourroient tuer beaucoup de gibier.

D'après le rapport du Lieut. Milius, il paroit cependant qu'on pourroit s'en procurer une petite quantité en creusant çà et là des puits sur les bords de la riv.

Je crois que sur l'île Rottnest dont moitié de la surface est couverte de Pierres. l'autre moitié de Lacs salés (excepté un peu de terre cultivable, où est ramassée une pte. quantité d'eau pluie corrompue, par les canards sauvages.) j'ai du soupçonner que les pluies devoient s'infiltrer dans la terre et de l'eau, tant ou peu, j'ai fait sur le bord (c.à.d. à 10 à 15 pieds) des Lacs d'eau salée creuser des trous d'un pied de profondeur et j'ai en peu de tems puisé de l'eau douce dont le Cit. Colas et moi avons bu.

Appendix 16

MAP OF THE COURSE TAKEN BY COMMANDER MILIUS TO AND FROM GARDEN ISLAND, AND THE SITE OF THE PLACE WHERE HIS BOAT WAS WRECKED AND HIS CREW MAROONED AT COTTESLOE — JUNE 1801

Archives Nationale. Série marine 5JJ 41

Lieutenant-Commander Milius left the *Naturaliste* on 18th June to explore Carnac and Garden Islands which lay south of the anchored ship.

He sailed along the Straggler Reefs, westward of the *Naturaliste* and visited Carnac and then made for Garden Island where he landed in the Pig Trough, a small bay south of Beacon Head on the north east corner of the island.

There he conducted surveys and gained a good impression of Garden Island which he considered to be very fertile.

When he returned to the *Naturaliste* he ran into a north west gale, which prevail in these waters at that time of the year.

He was subsequently driven ashore on the morning of 19th June at the point marked K on the map, and marooned until the party was rescued and returned to the *Naturaliste* on 22nd June. The site of this is Cottesloe main beach, in the vicinity of the present bathing pavilion.

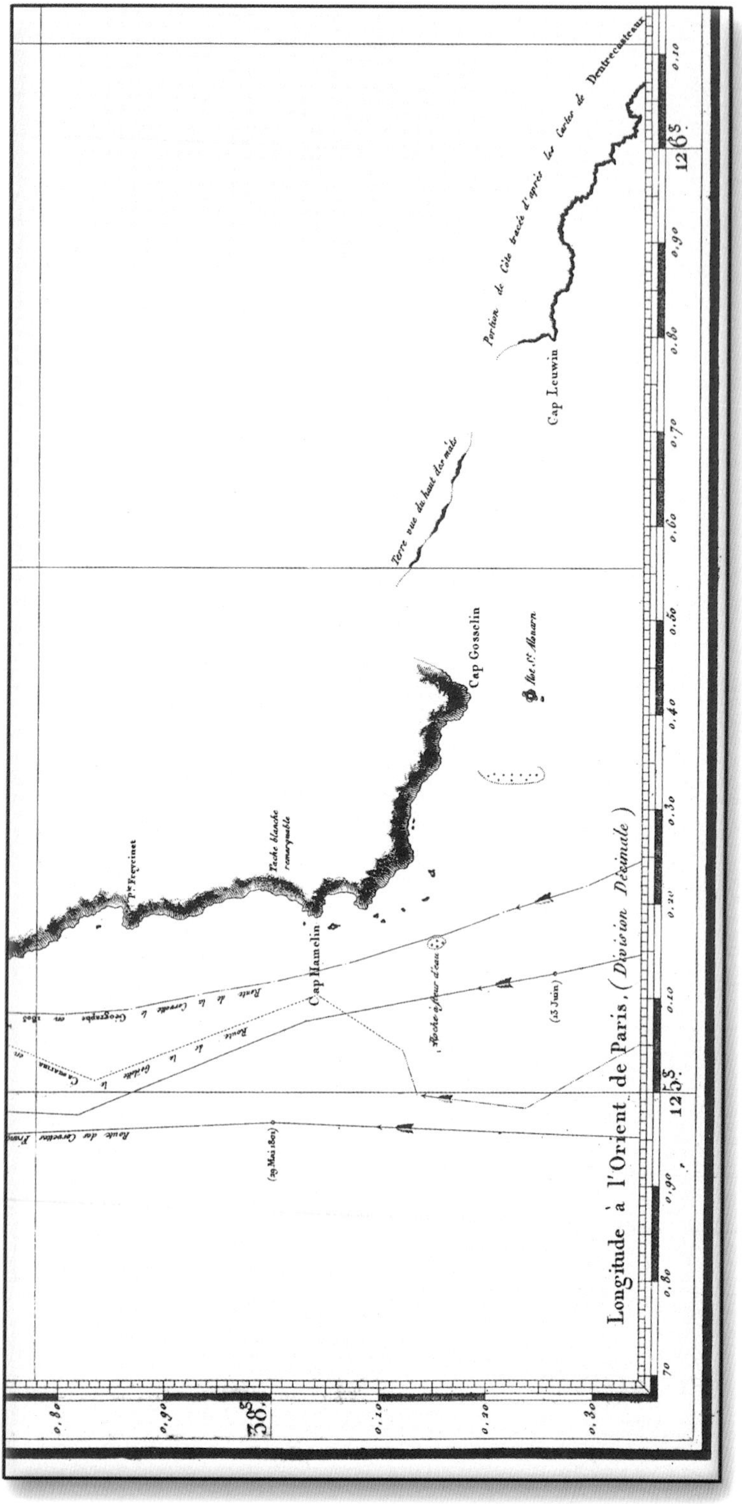
Portion de Côte tracée d'après les Cartes de Dentrecasteaux
Cap Leuwin
Terre vue du haut des mâts
Cap Gosselin
Ilot St. Alouarn
Tache blanche remarquable
Pte. Freycinet
Cap Hamelin
Roche à fleur d'eau
(13 Juin)
(29 Mai 1801)
Route de la Corvette le Géographe en 1803
Route de la Goëlette la Casuarina en
Longitude à l'Orient de Paris, (Division Décimale)
126.
125.
38.

Appendix 17

BAUDIN'S IDENTIFICATION OF "CAPE LEEUWIN"

At the beginning of the 19th century, the increase in trade across the Indian Ocean and the expansion of strategic interests made the identification of the capes at the northern and southern ends of the west coast of Australia essential. Baudin was given the task of locating and charting both.

Fifteenth century French charts by the Dieppe School showed the coast ending near 35° S. This was confirmed by the Dutch on the *Leeuwin* in 1622. In 1627 the *Gulden Zeepard* rounded the corner and surveyed the southern coast to near Ceduna. The Dutch charts drawn from this information showed that the corner was not sharp. It seemed like the Cape of Good Hope which took time to round, which in fact is the case. For ships which sail south along the Western Australian coast hold that course until they reach Cape Hamelin, about 8 miles north of the present Cape Leeuwin. They then bear south east for near 90 miles until they reach Point Nuyts. It is not until then that an eastward course can be set.

The part of the coast from Cape Hamelin to Point Nuyts was generally featured on charts as Leeuwin Land, not Cape Leeuwin.

What Baudin and Flinders who followed him did, was to locate some outstanding feature in Leeuwin Land which could be used as navigational way point that could be fixed precisely on the charts for navigators to identify and use.

Baudin did this in a thoughtful manner, revealing a good knowledge of earlier surveys and sources, and the problem. After rejecting the idea that low lying Cape Hamelin could be Cape Leeuwin, Baudin surveyed further south on the *Géographe* between 11 and 13 June 1801.

He then observed the number of headlands and bays which exist along the coast of Leeuwin Land. He correctly identified, for the first time, the small islands shown on Rosily's chart of Flinders Bay drawn in 1772 (see appendix no 1). He named the prominent headland which marks the southern end of the Cape Naturaliste – Cape Leeuwin ridge, (the present Cape Leeuwin), Cape Gosselin in honour of the noted French scholar of classical geography. He gave the name Cape Leeuwin, with reason, to the next prominent feature he saw further east. This was probably Dickson Peak which lies above Cape Beaufort. When observed from some distance to the west where the *Geographe* was positioned, this and other close by features blend in with the southern end of the high hills where they fall into the sea, creating the appearance of a prominent headland.

The British surveyor Flinders who saw Leeuwin Land from a different angle, from where he could not see Cape Hamelin, named the heights which appeared to form the western side of Flinders Bay, Cape Leeuwin. This was observed from a considerable distance off to the east. He could not see from there the low lying point which forms the present Cape Leeuwin. The position he gives for the Cape therefore is 2 miles too far north.

The independent surveys by Baudin and Flinders poses a problem. As Baudin was the first to name the headland which forms the western entrance to Flinders Bay, in accordance with international convention it should bear the name Cape Gosselin which was given first, instead of Cape Leeuwin which was given later.

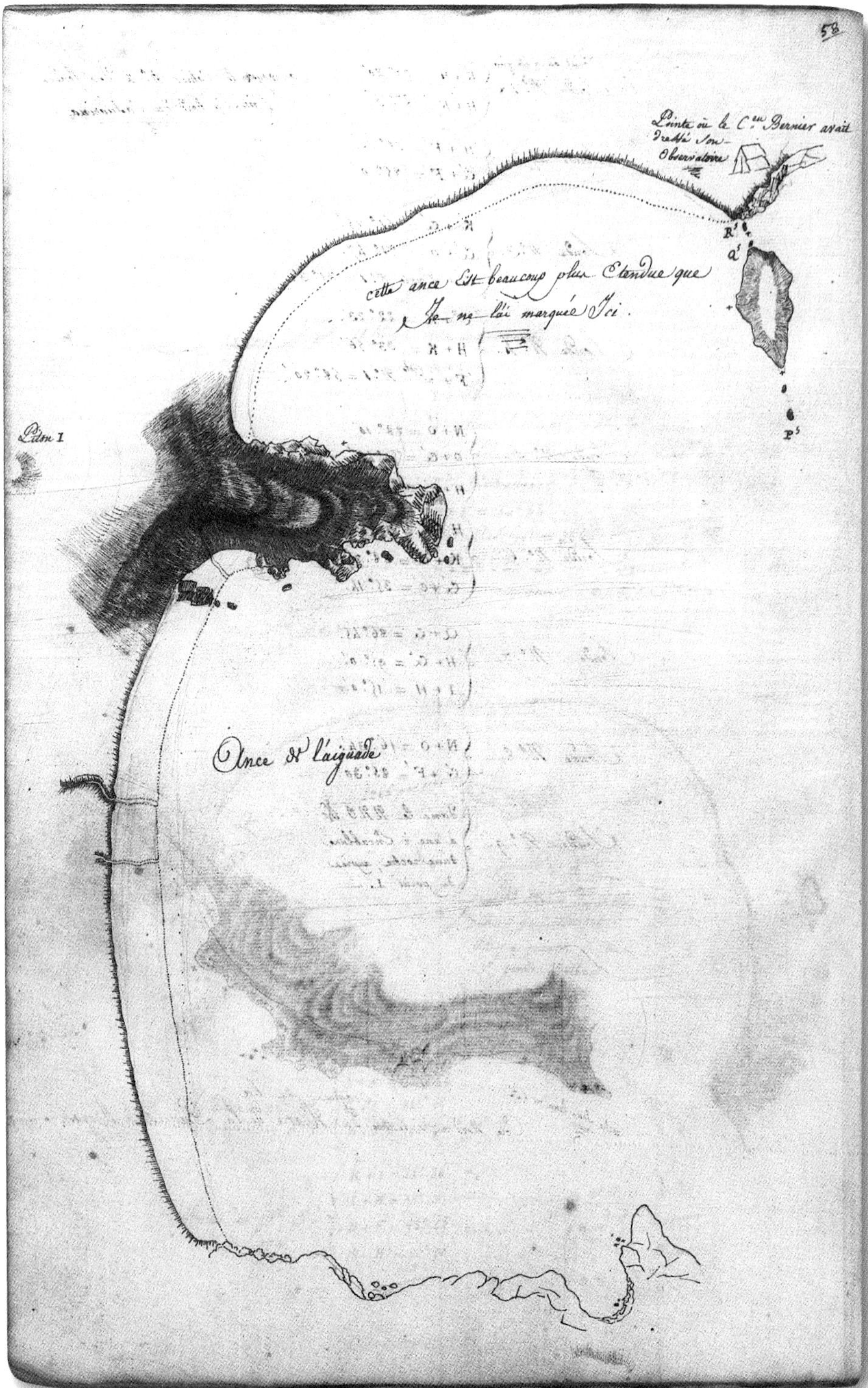
58
Pointe où le Cen Bernier avait
dressé son
Observatoire
cette ance est beaucoup plus Etendue que
Je ne l'ai marquée Ici
Plan I
Ance d'l'aiguade

Appendix 18

PART OF KING GEORGE SOUND FROM MISTAKEN ISLAND TO LIMESTONE HEAD — FEBRUARY 1803

Archives Nationale. Série marine 5JJ 32

This survey sketch made by Louis de Freycinet is of the watering streams used by Baudin and other visitors to replenish the ships' fresh water tanks.

At the top is Mistaken Island, called Observatory Island by the French.

The second site of the observatory set up by Baudin is marked by the sketch of a tent. The first site was on Mistaken Island, but this was dismantled after a fire caused by the stove swept the area.

The watering streams used by Baudin are marked as the Anse de l'Aiguade (Stream Bay) and are near the former whaling station at Albany.

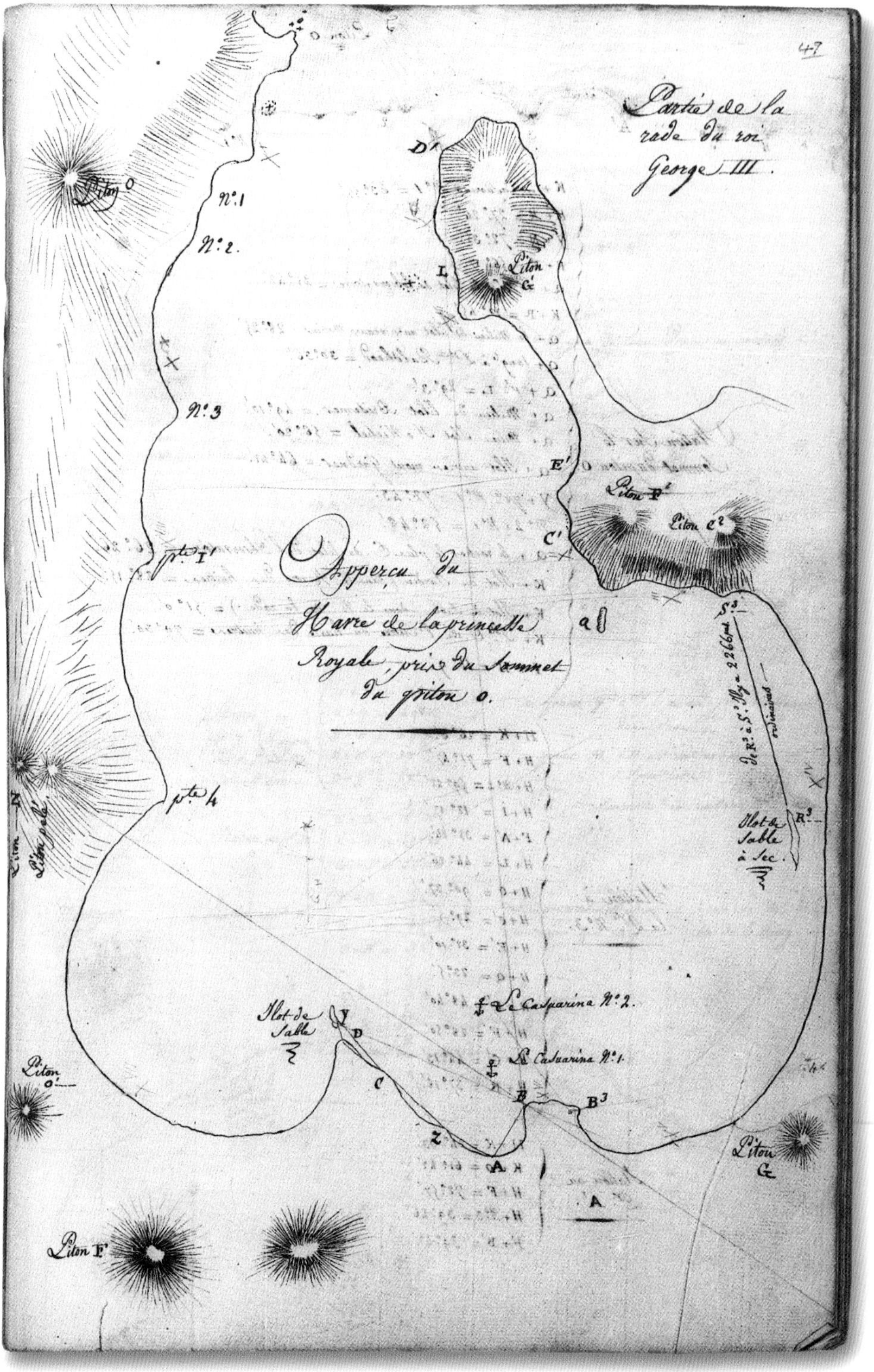
Partie de la rade du roi George III
Apperçu du Havre de la princesse Royale, pris du sommet du piton O.
Piton O
N°.1
N°.2
N°.3
pte I'
pte 4
Piton N
Piton pelé
D'
L
Piton G'
E'
Piton F'
Piton C²
C'
S³
R³
Ilot de Sable à Sec.
Ilot de Sable
Y
D
C
Z
A
B
B³
La Casuarina N°.2
La Casuarina N°.1
Piton O'
Piton G
Piton F'
A

Appendix 19

FREYCINET'S MAP OF PRINCESS ROYAL HARBOUR, ALBANY. FEBRUARY 1803 ON THE *CASUARINA*

Archives Nationale. Série marine 5JJ 31

This survey sketch map of Princess Royal Harbour was made by Louis de Freycinet on the schooner *Casuarina*. The anchorages of the *Casuarina* near Pagoda Point are shown.

Freycinet selected these positions so as to be near fresh water and game, and to careen his boat.

Baudin anchored in the outer harbour of King George Sound when he arrived and immediately fell out with Freycinet who he believed was deliberately isolating himself and his crew from the main body of the expedition.

In consequence Freycinet was not selected to make official surveys of the coast. However he used the occasion to make this survey of the inner harbour before he was called out to King George Sound and given the task of counting and checking store supplies.

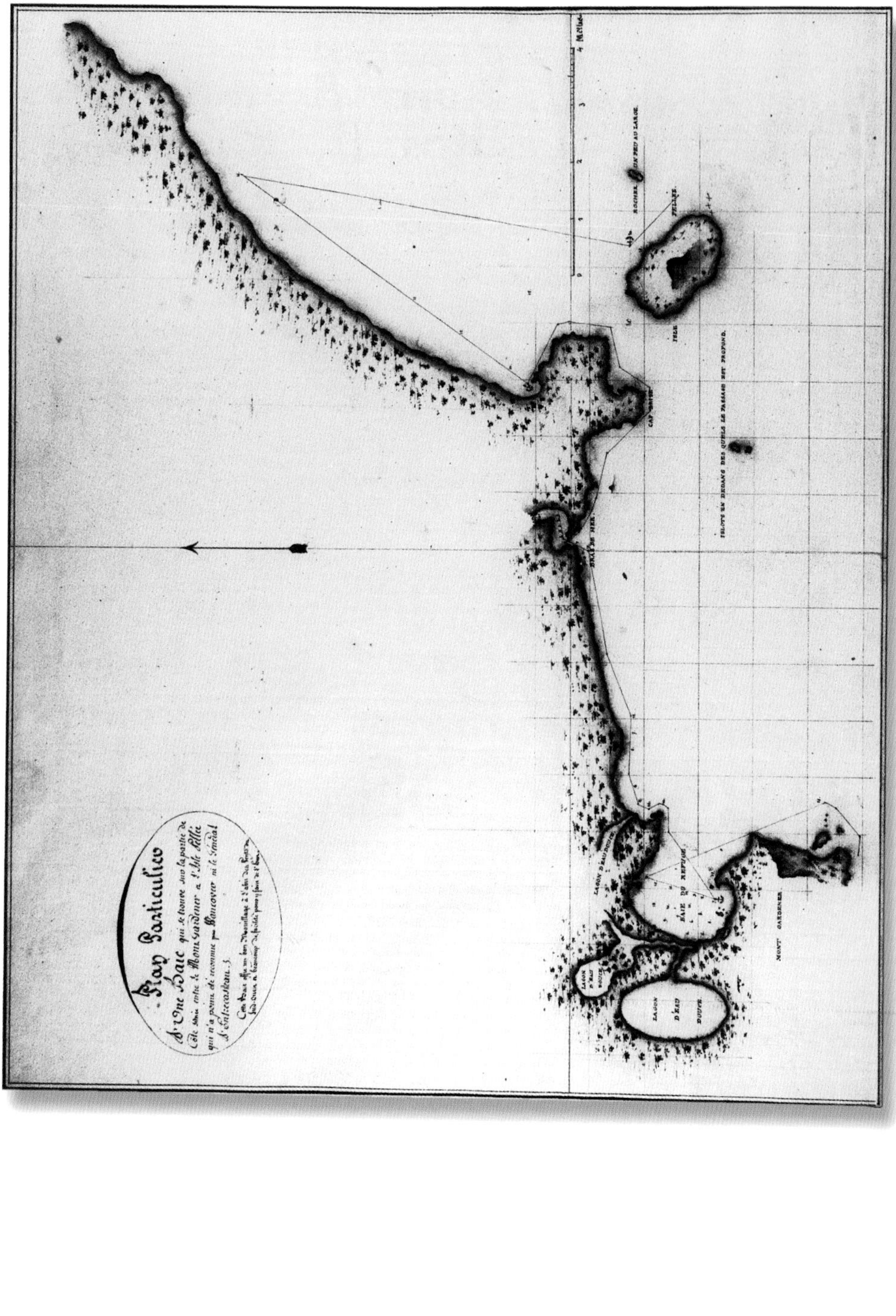
ROCHER UN PEU AU LARGE
ISLE PELLEE
ISLOTS EN DEDANS DES QUELS LE PASSAGE EST PROFOND
LAGON D'EAU DOUCE
BAIE DU REFUGE
MONT GARDENER
Plan Particulier

Appendix 20

MAP OF TWO PEOPLES BAY AND CHEYNE BEACH EAST OF ALBANY — FEBRUARY 1803

Archives Nationale. Série marine 6JJ4:34

On the way into King George Sound in February 1803, Baudin observed what appeared to be some good harbours. Survey parties were consequently despatched to map places where vessels could anchor with safety. This map is of Two People Bay and the region east which impressed the French as a safe anchorage area.

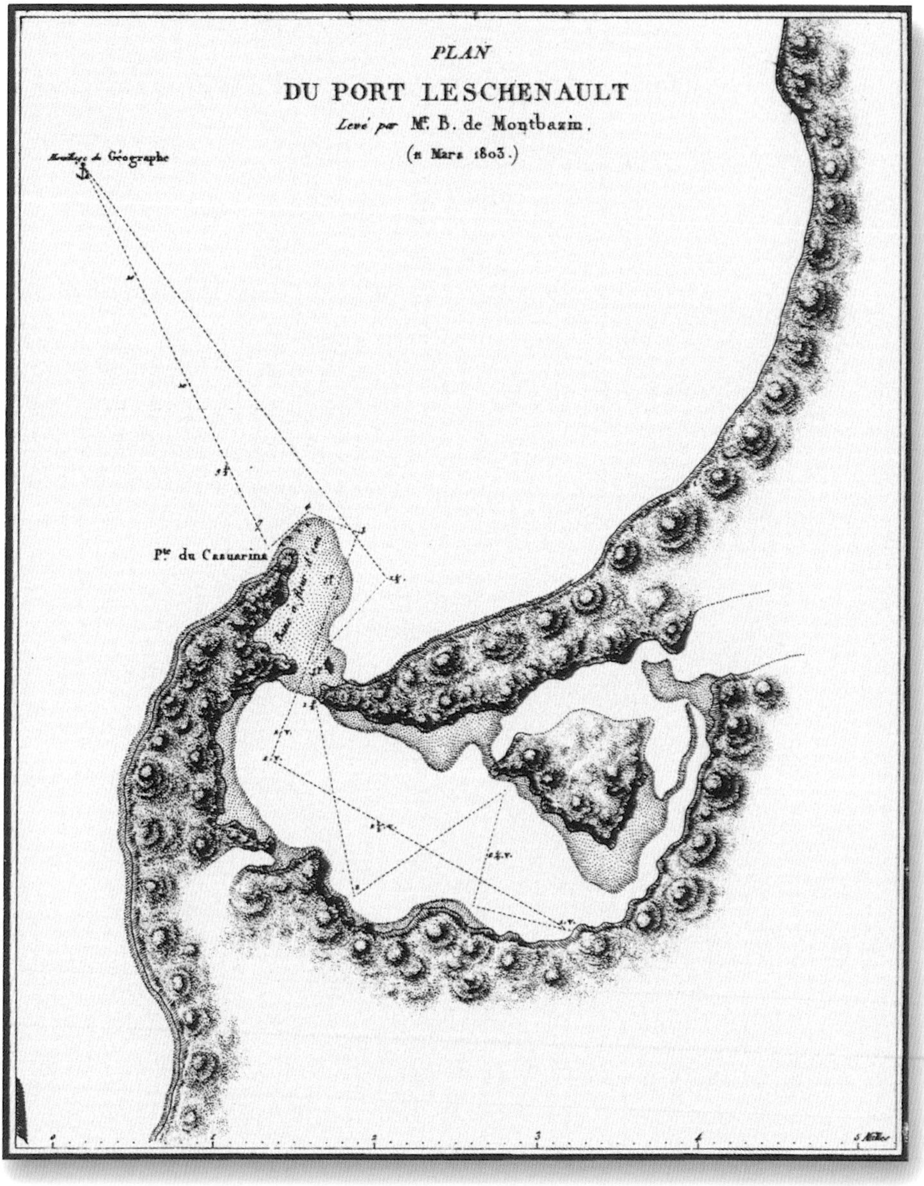
PLAN
DU PORT LESCHENAULT
Levé par Mr. B. de Montbazin.
(11 Mars 1803.)
Géographe
Pte. du Casuarina

Appendix 21

BONNEFOI'S MAP OF KOOMBANA BAY AND LESCHENAULT INLET FROM BUNBURY. MARCH 1803 ON THE *GEOGRAPHE*

Archives Nationale. Série marine 6JJ81

This harbour, now Bunbury, was discovered and mapped by Midshipman Bonnefoi, often referred to as Montbazin, of the *Géographe.*

The bay was viewed as a suitable small boat harbour and was called the Port des Barques (Small Boat or Barque Harbour).

The inlet was not fully explored. Bonnefoi stopped his survey just north of Middle Island which is of botanical significance as it is the furthest point south where mangroves grow.

The inlet was viewed most attractively by the French explorers not only because of the safe refuge it offered small boats, but because of the rich birdlife and other wildlife seen at the inlet.

However, Bonnefoi noted at the bottom of his caption that fresh water was not found in the bay.

In the final chart of south western Australia produced in the published report of the Baudin Expedition, the Point protecting the harbour bears the name Casuarina Point. This was named after Freycinet's Australian built survey schooner the *Casuarina.* This name has been retained. On the same chart, the name Boat Harbour was replaced with Port Leschenault, in honour of the botanist on the *Geographe.*

After the British settled Western Australia, Port Leschenault was renamed Bunbury, and the protected water sheltered by the Point, was given the aboriginal name Koombana Bay. The inland waterway which leads to the island on the map reproduced on page 298, now bears the name Leschenault Inlet. A large stretch of water not seen by the French explorers, now bears the name Leschenault Estuary.

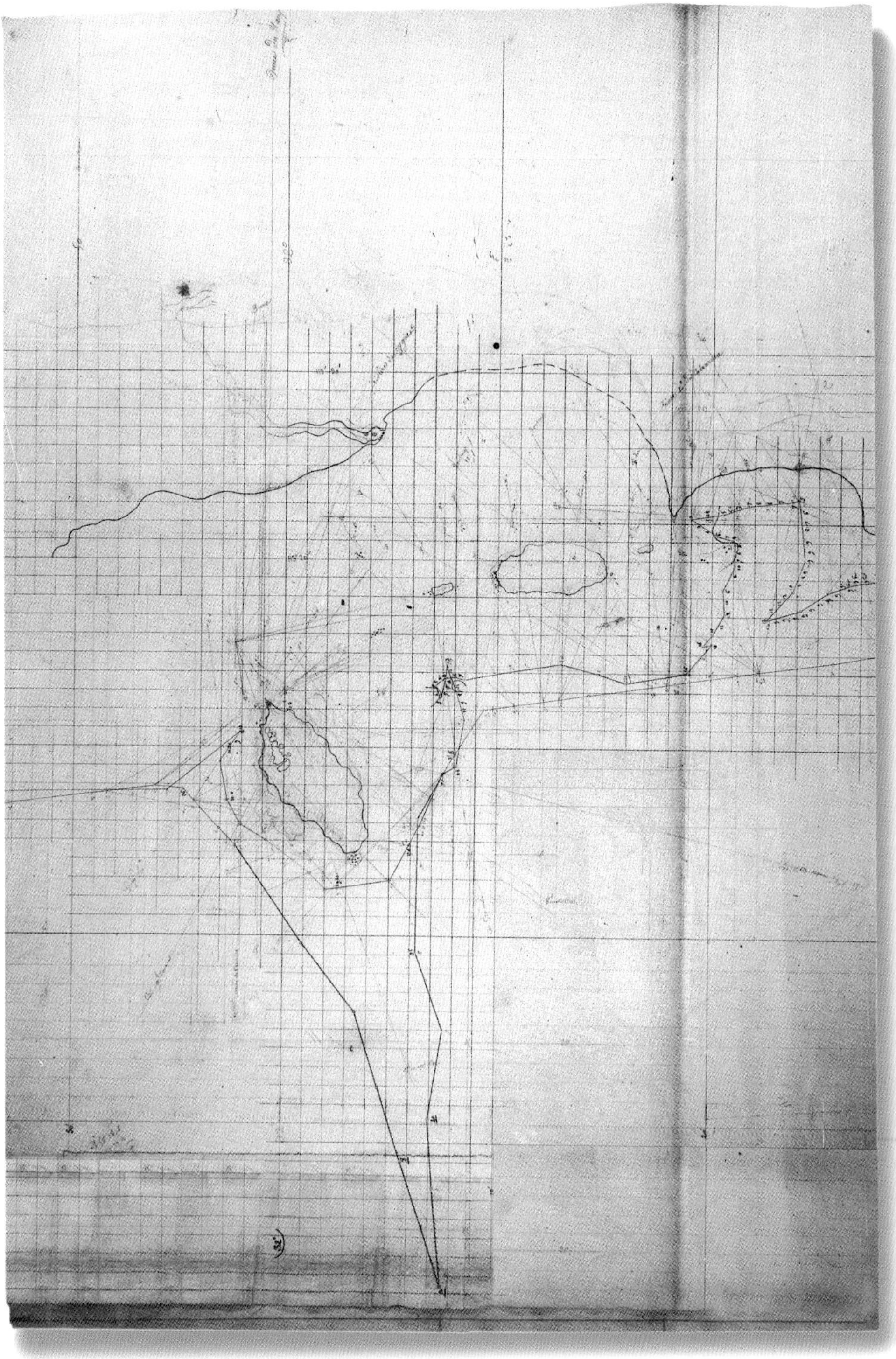

Appendix 22

FREYCINET'S MAP FROM BECHER POINT, WARNBRO SOUND TO ROTTNEST ISLAND. MARCH 1803 ON THE *CASUARINA*

Archives Nationale. Série marine 6JJ6:12(3)

This map of the approaches to the Swan River was made by Louis de Freycinet during Baudin's second visit to western Australia in 1803 (year 11).

The map shows the soundings made in the Port Kennedy, Safety Bay area in the south approaches at the bottom of the map; and surveys and soundings between there and Rottnest Island.

The anchorage of the *Casuarina*, the small schooner on which Freycinet made the close inshore surveys, is shown north of Duck Rock and Thomson Bay at Rottnest. Details of the lakes on Rottnest Island, sketched as a result of the earlier survey of 1801 are shown.

69

Suite du 19. Ventose XI.

La pointe L² (Cap Péron) est très Basse & très Saillante, Elle a un Prolongement formé je crois par un Banc de Sable, sur lequel la mer Brise. ce Banc va se joindre aux rochers M² et il ne m'a pas paru qu'il y eut passage entr'eux et cette pointe. dans tous les cas; ce passage ne serait qu'entre des Brisans. – Je n'ai pas cru devoir le tenter plus avant, Lorsque la Sonde m'a rapporté que deux Brasses d'eau. /.

Après les rochers M² s'étend un Banc qui Brise de O² en M² presque sans interruption. Je crois que ce Banc est presque en Entier de Sable. J'ai essayé de passer entre N² & O² mais le fond m'ayant paru diminuer rappidement et ne trouvant plus que deux Brasses d'eau, J'ai changé de route.

Le 20 Ventose XI.

5h 30' matin (a^2 = Est)

6h 10' { V^2 = S 47° E ; $V^2 + X^2$ = 25° 0' }

6h 30' {
T^2 = E 22° N
$T^2 + R^2$ = 22° 10'
$T^2 + a^2$ = 26° 30'
$T^2 + S^2$ = 32° 10'
$T^2 + V^2$ = 51° 10'
V^2 = (6° 0')
$T^2 + X^2$ = 79° 25' }

7h 25' matin (A^3 = N 23° E)

7h 30' { B^3 = N 24° E ; $B^3 + C^3$ = 65° 0'. } N 37 O

8h 0' {
A^3 = E 50° N
$A^3 + C^3$ = 54° 0'
$A^3 + B^3$ = 7° 30'
$A^3 + V^2$ = 4° 15' }

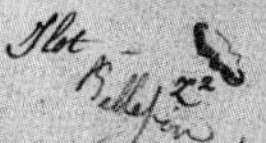

Appendix 23

FREYCINET'S MAP OF THE ISLANDS FROM ROTTNEST ISLAND SOUTH TO GARDEN ISLAND. MARCH 1803 ON THE CASUARINA

Archives Nationale. Série marine 5JJ 32

This final survey map of the islands off the Swan River shows the completed survey of Rottnest Island, with the south coast and West End being shown in heavy lines.

In this chart Carnac Island, called Ile Pelé (Balad Island) is marked with a further name, Ile Levilian, and Garden Island is named Ile aux Ours-Marins (Sea Bear of Seal Island). A further name given to Garden Island on this map is Ile Bertholet.

Partie Nord de l'île Rotnest /

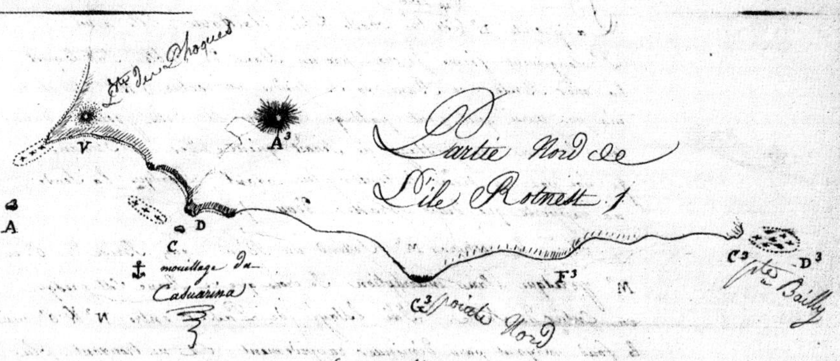

9h 10' matin
- C³ = 147° E
- C³+D³ = 5°0'
- C³+F³ = 34°0'
- C³+A³ = 55°40'
- C³+E³ = 62°40'

9h 30'
- F³ = S 5° O.
- F³+C³ = 23°15'
- F³+A³ = 58°30'
- F³+E³ = 79°0'

E³ Est un point d'avan l'Est de F³

Au Mouillage lat. 31°–58'–18",8

- ☉+L = 93°48'30"
- ☉ 61°54'
- L+A = 50°10'
- A+V = 59°24' plutôt 9°24
- V+D = 31°0'
- V+G³ = 97°40'

(L Est à gauche de A)

L Est une grande tache de sable sur la côte du Continent & dans le Nord de la rivière des Cygnes /

c'est à peu près le point où échoua la chaloupe du Naturaliste ~~Géographe~~ en l'an 9. (par 32° Latt. Sud.)

Quoiqu'il y ait des Brisans auprès de La Côte de L'Ile Rotnest, J'ai cru remarquer cependant qu'il y en a Beaucoup moins, que l'on en a marqué sur la Carte Hollandaise qui M'a été remise de Cette Partie de la Nouvelle Hollande. / carte N° 28 remise par le Dépôt de la Marine

Appendix 24

FREYCINET'S MAP OF THE NORTH COAST OF ROTTNEST ISLAND. MARCH 1803 ON THE *CASUARINA*

Archives Nationale. Série marine 5JJ 32

This sketch of the north coast of Rottnest Island was made by Freycinet in the *Casuarina* while he was waiting for Baudin from whom he was separated near Denmark on the south coast.

West End in the map is named Point Bailly. Pointe Nord, named by Freycinet is now called North Point.

The anchorage of the *Casuarina,* off Duck Rock and the Transit Reefs is marked with an anchor.

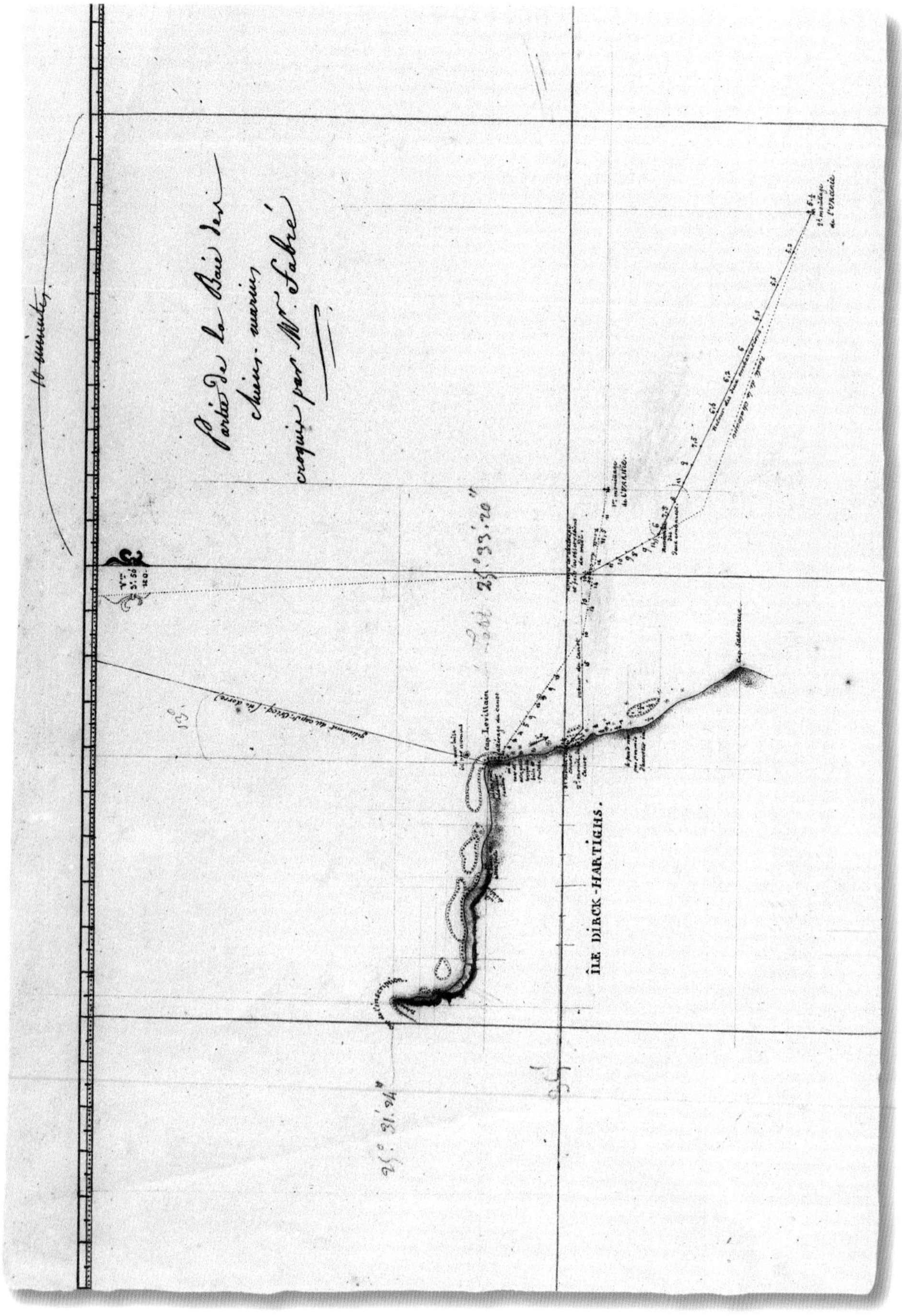
Partie de la Baie des chiens-marins croquis par Mr. Fabré
10 minutes
ÎLE DIRCK-HARTIGHS
Cap Levillain
25° 31' 24"
Lat 25° 33' 20"

Appendix 25

FABRE'S MAP OF THE NORTH COAST OF DIRK HARTOG ISLAND SHOWING WHERE THE FRENCH FOUND THE VLAMING PLATE — SEPTEMBER 1818

Archives Nationale. Série marine 6JJ6:2

This map was drawn by Fabré who accompanied Louis de Freycinet on the *Uranie* on his visit to Shark Bay in September 1818.

The map shows the place where Fabré landed to get the Vlaming plate in order to take it to France for safe-keeping. The location of the plate at Inscription Point is marked and the latitude of the point noted.

Appendix 26

RAILLARD'S MAP OF THE NORTH END OF PERON PENINSULA, SHARK BAY WITH LAKES — SEPTEMBER 1818

Archives Nationale. Série marine 6JJ6

This sketch map by Cadet Raillard of the *Uranie* shows the area around Dampier Road where Louis de Freycinet set up his camp and observatory in 1818.

Freycinet stayed in Shark Bay from 13th September to 26th September 1818 and took the opportunity to make a detailed survey of the north end of Peron Peninsula, and the native life there. Raillard's sketch shows the physical features including the extensive salt lakes found there at that time. The great lake (grand étang) was regarded by the French party more as a small harbour than a lake.

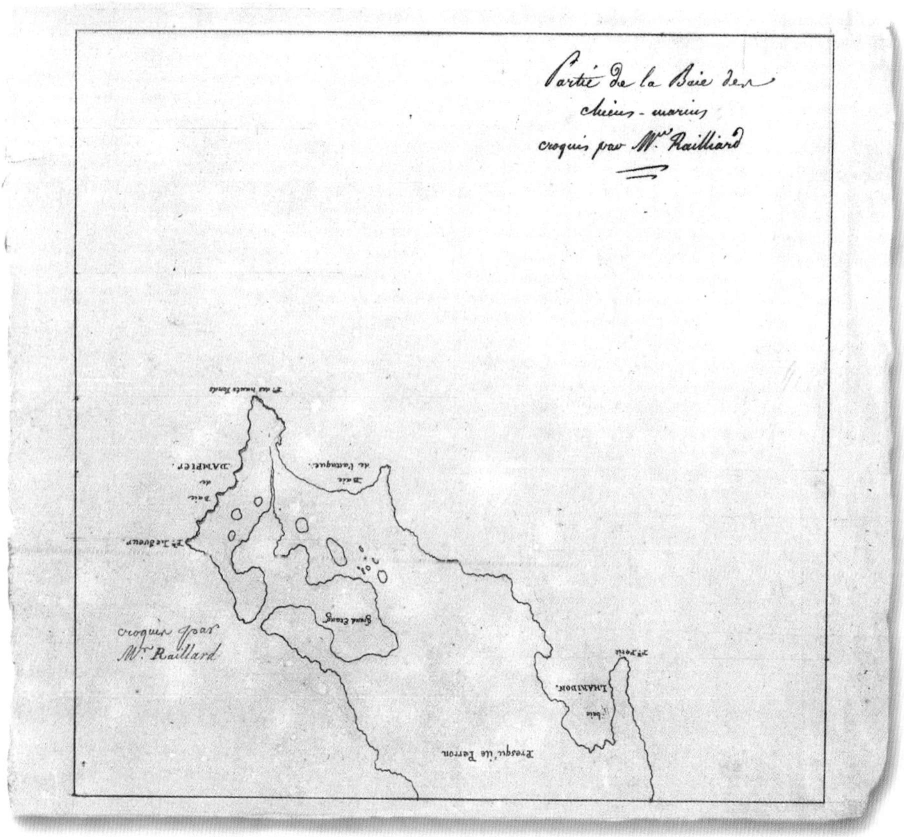

Projet d'une colonie pénale
Sur la côte S.O. de la nouvelle Hollande

La Russie envoie ses exilés dans les déserts de la Sibérie; l'Espagne, envoie ses galériens, à ses presidios d'Afrique; le Portugal faisait passer autrefois ses malfaiteurs à Mozambique et dans les Indes, et la Hollande a versé en Asie, l'écume de sa population. En adoptant aussi le système de déportation, sur des plans plus ou moins vastes, le but principal s'est trouvé rempli sous le rapport de la peine, mais en choisissant mal le lieu où les coupables devraient la subir, ces gouvernements n'ont obtenu aucun résultat remarquable de l'établissement de leurs colonies pénales. L'excès du froid ne permet pas aux climats

Appendix 27

THE OPENING PAGE OF JULES DE BLOSSEVILLE'S PLAN FOR A PENAL COLONY IN SOUTH WESTERN AUSTRALIA — JANUARY 1826

Bibliothèque Nationale. Nouvelles acquisitions françaises 6785

This plan was prepared by the explorer Jules Blosseville who visited Australia with the Duperrey expedition (1822-1825). It was requested by the police in France who were concerned about the number of criminals in gaols, and who supported the idea of transportation as used by Britain in the process of its colonization.

The report was followed by a specific plan to explore south western Australia in detail, so as to select the best possible site for a French colony. Albany was suggested as the best place for the headquarters of the exploring party because of the water and other supplies which were readily available there.

The site of the settlement was left to be determined. High on the list of suggestions were Rottnest Island and Garden Island which were regarded as good places for a prison.

The plan was not proceeded with. By 1826 French officials had decided to acquire and settle New Zealand instead of western Australia.

Appendix 28

MAP OF OYSTER HARBOUR, ALBANY — OCTOBER 1826

Archives Nationale. Série marine 6JJ7:8

This survey map of the Rivière des Françaises, now the Kalgan River, shows the region Dumont D'Urville viewed as a favourable site for plantations.

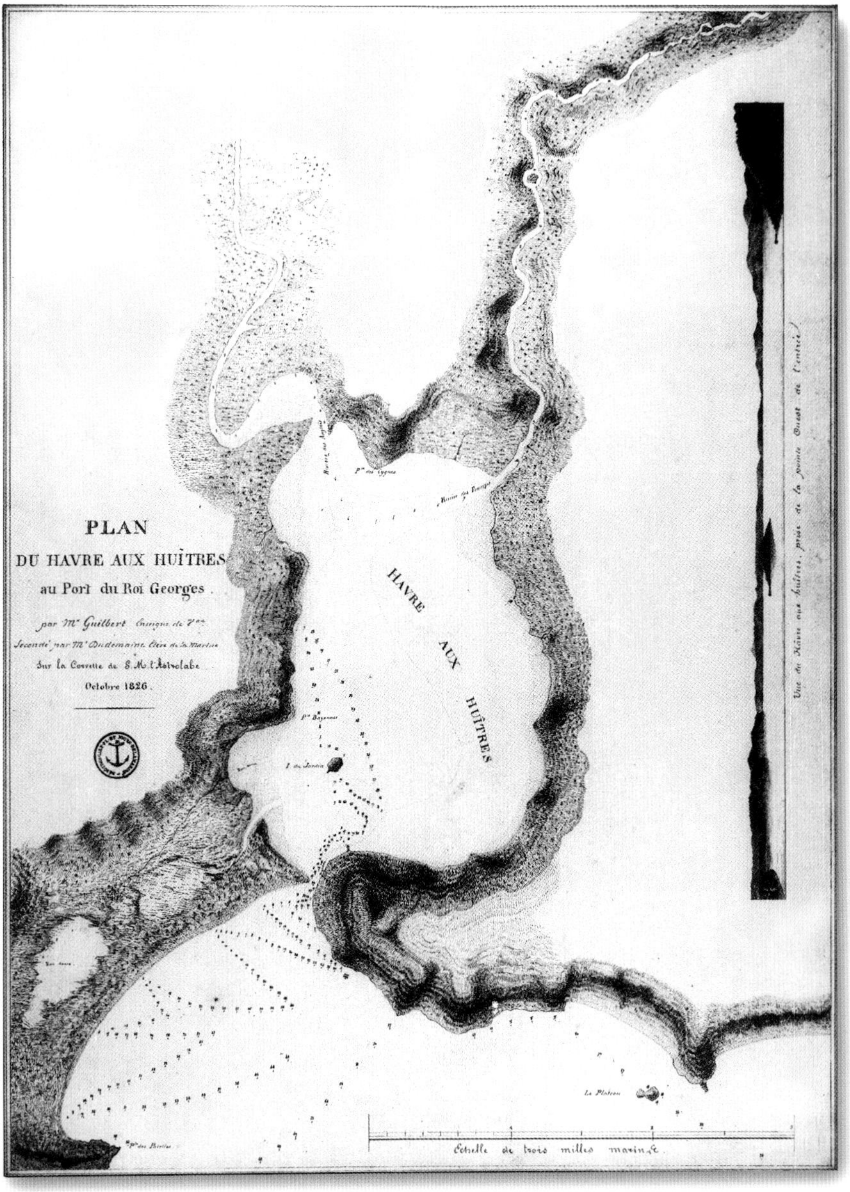

Appendix 29

THE LEADERS OF THE FRENCH MISSIONS OF EXPLORATION, THE SHIPS THEY COMMANDED AND THE ARTISTS WHO ACCOMPANIED THEM

Date Of the Expedition	Date When in western Australia	Name of the Expedition Leader	Name of the Ship	Name of the ship's Commander	Name of the Artist
1772	March-May 1772	François Alesno, comte de Saint Allouarn. NOTE: Saint Allouarn was in command of the second vessel belonging to the Kerguelen expedition. The lead ship, the *Fortune,* under the command of the leader of the expedition, Yves Joseph de Kerguelen-Tremarec, did not reach western Australia.	*Gros Ventre*	Saint Allouarn	—
1791-1794	December 1792-January 1793	Joseph Antoine Bruny D'Entrecasteaux	*Recherche*	D'Entrecasteaux	Piron
			Espérance	Huon de Kermadec	
1800-1804	May-August 1801	[Thomas] Nicolas Baudin	*Géographe*	T.N. Baudin	Charles Alexandre Lesueur
			Naturaliste	Jacques Felix Emmanuel, baron Hamelin.	Nicolas Martin Petit

Date Of the Expedition	When in western Australia	Name of the Expedition Leader	Name of the Ship	Name of the ship's Commander	Name of the Artist
	February-June 1803	Baudin *(continued)*	*Géographe*	T.N. Baudin	as above
			Casuarina	Louis Claude Desaules de Freycinet.	
1817-1820	September 1818	Louis Claude Desaules de Freycinet	*Uranie**	L.C.D. de Freycinet	Jacques Etienne Victor Arago
1826-1829	October 1826	Jules Sebastien Cesar Dumont D'Urville	*Astrolabe*	J.S.C. Dumont D'Urville	Louis Auguste de Sainson

*NOTE: The *Uranie* was wrecked at the Falkland Islands in 1820 and was replaced by the *Physicienne*.

Appendix 30

FRENCH PLACE NAMES CURRENTLY USED ON THE COAST OF WESTERN AUSTRALIA WITH A NOTE ON THEIR ORIGIN

The Difficulty of Compiling a List

It is not easy to make a definitive list of places on the coast of western Australia, which bear names given by the French explorers. The British colonial practice of choosing their own names for landmarks in western Australia after they acquired the legal right of nomenclature by their act of annexation on 2 May, 1829, makes it difficult to identify and list French given names. For British surveyors and cartographers, when they drew their charts of the colony, used their power of nomenclature not only to discard some French name places[1] and to replace others with British names,[2] but also to translate French names into English thus making them appear English in origin,[3] to Anglicize the spelling of other French names,[4] and to add French names of their own choosing. Baudin Island in the Bonaparte Archipelago and Baudin Island in Shark Bay, which commemorate the memory of the leader of the greatest French scientific expedition to visit western Australia, for example, were both named by the British and not by the French.

To add to the difficulty, changes of names are still being made in western Australia, primarily from British given names to earlier French given names. The foundations for this policy were laid in 1911 as a result of the visit made to Western Australia and other states by the Comte de Fleurieu, whose family had played a prominent part in sending out the French scientific missions of exploration. He successfully urged state and federal governments to preserve the French names on the maps of Australia, supporting his plea with the argument that such a policy of retention was in accordance with decisions made at pre World War I international conferences on geography where the international use of the first name given to a place was recommended. In Western Australia, Fleurieu's appeal resulted in research efforts being made, and in the subsequent publication of a volume containing the maps of the French explorers showing which names had been and would be preserved on the Western Australian maps.[5] More recently, the Western Australian Lands and Surveys Department, conforming to the earlier determined policy, has been renaming features, reverting to the first given name. Bessieres Island, which the British had renamed Anchor Island, for example, was renamed Bessieres Island in 1968. More recently still, Freycinet Estuary and Freycinet Reach in Shark Bay have been renamed Henri Freycinet Harbour, as it was earlier called by the French. And the lagoon at the northern end of Peron Peninsula in Shark Bay has once again been given the French name Montbazin Lagoon. Thus the process of naming and renaming is continuing, making the compilation of a definitive list as yet impossible.

Types of French Names Contained in the List

This list is therefore a preliminary one which provides a register of French given place names on the coast of western Australia in 1982, and a summary of what is known about the origins of the names. The list is arranged alphabetically to make the identification and location of places easy. It is not a simple list of places named by the French explorers, it is a comprehensive register of:-

1. Places named by the French which still appear on maps of western Australia.
2. French named places which have been translated into English such as Turtle Is.
3. Anglicized names which were originally French such as Hermite Island.

4. French names given by the British, but only when these are related to the French explorers. For example, Adieu Points on Augustus Island in the north west has nothing to do with the French explorers and is therefore not included.
5. Landmarks or features with French names, which were not seen by the French but were later given names because of their proximity to French named places. Near Adèle Island which was named by the French, for example, there is Adèle Island Reef and Adèle Island Shoal which were not discovered and named by the French explorers.
6. Corruptions or possible corruptions of French names, Quoin Bluff on the east coast of Dirk Hartog Island, which was called Coin de Mire by the French, for example, could be an English corruption of the French name.

Description of the Form of the List

Information available about each entry in the list is given in four columns:

I. *The first column* contains the place name. Variations of spelling and corruptions are noted here where applicable.

II. *The second column* describes the location of the place. The first phrase indicates if the place is on the north west, west or south coast and is followed by more detailed information to help with location.
Those requiring more precise information can refer to the extensive, detailed maps issued by the Western Australian Department of Lands and Surveys, and volumes 1 and 5 of the *Australian Pilot* published by the Hydrographic Department of the British Admiralty.

III. *The third column* lists the name of the expedition in whose charts the place name first appears. It is not possible to be more precise than this. The final French charts of western Australia were not drawn until years after the expeditions ended. Names in these sometimes vary from those given in earlier made sketch charts for varieties of reasons. Thus it is not possible to say that the place names were given by leaders of the missions of exploration. Those suggested were or were not accepted by French naval cartographers. Consequently it is less misleading to say which of the missions of exploration discovered the place and was therefore responsible for it being named.

IV. *The fourth column* describes after whom or what the place was named. This part of the list is far from being definitive and complete. This is because the log books and ship's journals give little information on nomenclature. Varieties of records and charts made and kept by the French have to be searched to get information. For example, Point Picquet in Geographe Bay was spontaneously named by the scientists and crew who accompanied Baudin's mission. It was named by them to honour a comrade. Sub-lieutenant Picquet whom they believed had been unjustly maligned and treated by Baudin. The naming of the place consequently is not to be found in the ship's journals. It is contained in the diaries of the men opposed to Baudin.
Where uncertainties exist, the alternative names which could have been commemorated by the French are given. For example, in the case of Muiron Island, it is not clear if this was named after Napoleon's close friend and aide-de-camp who died at Napoleon's side at the Battle of Arcole, or if it commemorates Napoleon's pseudonym which he used in his correspondence, or if it commemorates the name of the French naval vessel which brought Napoleon from Egypt back to France — or indeed all three.
The compilation of a final definitive list will take much time and research effort. In the meantime what is at present known is presented to readers for their information.

Types of Names used by the French Explorers and Cartographers

A survey of the types of names given by the French indicates that French nomenclators left a rich and complex series of monuments to French culture, French notables, French revolutionary and restoration politics, and French revolutionary and Bourbon restoration values. Eight different types of names were used. There are places:

1. named after those who participated in the expeditions to western Australia such as Hamelin, Vasse and Depuch;
2. named after the ships which made the surveys such as the *Géographe* and the *Naturaliste;*
3. named after earlier notable French maritime explorers such as Bougainville and Latouche-Treville;
4. named after notable figures in French cultural history, especially in the fields of science, literature and war, such as Molière, Racine, Lamarck and Suffren;
5. named after then contemporary notable figures in France, especially in the fields of politics, administration, war and science, such as Bonaparte, Bouvard and Gantheaume;
6. named after French revolutionary and Napoleonic military victories in Europe, such as Montebello;
7. named to describe the physical appearance of a place, such as Useless Inlet;
8. named to describe a happening at a place, such as Attack Bay;
9. named after zoological or other species of life such as Baleine Bank and Holothuria Bank.

Fortunately for the researcher and others interested in the origins and meanings of the names used, French nomenclators tended to use the types of names in groups or in chain like forms. On the south and west coasts, for example, the names used tend to commemorate the survey ships, the explorers who sailed on them and some notable French contemporaries such as Mentelle. The north west coast, which was the last place reached and surveyed by Baudin, conversely bears for the most part, memorials to the French revolution, its wars and its notables. A journey along the coast there eastwards from Cape Murat and the Montebello Islands to the Joseph Bonaparte Gulf, in fact, is rather like making a journey through French revolutionary history at the time of Napoleon. But as the French charts took decades to produce, the names along the north west coast portray not only the values and heros idolized by Napoleonic Frenchmen, but also those which were still highly regarded in the later Bourbon restoration period after the fall of Napoleon, when final drafts of the maps were produced. The names of the islands in the Institut Group,[6] for example, exist as memorials of most eminent Frenchmen such as Molière and Racine who were highly regarded throughout the *ancien regime,* the revolution, the Napoleonic period and the later restoration period.

The political revolutionary names used by the French are very select and limited. The names of Jacobins noted for extremism and other extremists are not commemorated in western Australia. Their victim, the scientist Lavoisier is. For the most part the names of eminent political, naval and military figures who held office under Napoleon and later under the restored Bourbons, that is those acceptable to both regimes have been used by the French for places names in western Australia.

Interestingly only two French revolutionary wars, the Battle of Montebello and the Battle of Arcole, both marking victories against the Austrians, were commemorated. The curious thing is that there were other great battles which were household names in France — Valmy where the French revolutionary forces secured a victory for the revolution against an allied coalition — Jemappes which marked the French advance into Germany and other later victories of note. None of these were commemorated.

Lack of a Place Name to Commemorate Baudin

Certainly the battles of Montebello and Arcole were highlights and memorable events for Napoleon and could be considered as fitting memorials made to him by the explorers he sent out and supported. However, if we look at the naming of the Montebello and Arcole Islands in the context of the history of the Baudin mission, which was responsible for the discovering and naming of these groups, it does not take much use of the imagination to conclude from the records of the mission that the naming of these could be a further slight given to Baudin by his enemies who were responsible for later drawing the French maps. At the beginning of the French revolution, the fiery revolutionary François Péron, who later sailed on the Baudin mission and later still wrote the official history of the mission, fought in the French revolutionary armies against the Austrians. He was wounded, lost an eye, was captured by the Austrians and imprisoned. He was later repatriated, invalided out of the revolutionary army and took up the study of science which secured him a position on the scientific mission led by Baudin.

Péron's future commander, in contrast, was in the service of the Austrian monarch when the revolution in France broke out. He was leading a scientific mission to the Indian Ocean area to collect specimens for Austrian scientists. Baudin certainly tried to get out of this when he heard of the troubles in France, but on advice continued to serve the Austrians. It is therefore not surprising to see deep and continued conflicts existing between Péron and Baudin on the voyage. They not only had little in common. They were bitterly divided. Unfortunately for Baudin, Péron and his colleague Louis Freycinet proved to be vindictive and gained the power to use this. When they returned to France they were appointed to write the official history of the voyage of exploration and to issue the maps and atlases. Baudin's name, in consequence, is not apparent in the official history of the mission he led, and his name does not appear on the maps of the region he surveyed. Instead the region bears the name of his assistants Hamelin, Péron, Freycinet, Lesueur and others who united to oppose him. The depth of feeling held by scientists and others who opposed Baudin during the mission is easily seen from the Picquet incident. Point Picquet in Geographe Bay, mentioned above, was named by Baudin's enemies to show him up. Sub-lieutenant Picquet's sole claim to fame was that he was some hours late returning from a short survey trip from the *Géographe* to Cape Naturaliste, made just after Baudin first anchored in Geographe Bay. For this Baudin removed Picquet from his command of a long boat. In retaliation the men on board named a landmark after him.

Whether or not Baudin deserved to be left off the maps of western Australia while others of junior, insignificant rank were commemorated is a complex question. Baudin's survey work was unimpressive. He worked fast and from far off the coast. He saw exposed reefs which he thought were islands, such as in the case of Adele Island which he believed was two islands. He saw islands when they were in fact capes on the mainland. And he concluded, by seeing the land from a distance, that parts of the mainland were islands. Baudin and those with him on the *Géographe* consequently made little contribution to the survey knowledge of western Australia. His sketchy work cannot be compared in quality with the excellent, detailed inshore surveys made by Hamelin on the *Naturaliste* and Freycinet on the *Casuarina* and their crews.

As an excuse it could be claimed that Baudin, who had earlier inspired confidence, was too old and too ill in 1800 to lead a mission. He died of his illness before reaching home. This could account for his bad judgements and his excessive caution at sea. He also had a difficult crew of revolutionaries who were bigoted, intolerant and difficult to control. Although these enemies of Baudin secured the power to denigrate him by getting positions as official historians, it is nevertheless a fact that Baudin led the greatest French scientific mission which sailed from France, and the greatest scientific mission to visit Australia and collect specimens.

NOTES

1. Louis Napoleon Archipelago, the name given to the group of islands off Fremantle, for example, was disregarded by British cartographers who left the group without a name. They named only the individual islands.
2. Bessieres Island off Onslow, for example, was renamed Anchor Island.
3. Attack Bay in Shark Bay, for example, was translated from the French Baie d'Attacque.
4. In the Montebello Islands, for example, the French named L'Hermite Island has become Hermite Island, changing the meaning; and La Trimouille Island, named after a famous French family, has become Trimouille Island which is meaningless as far as French nomenclature is concerned.
5. West Australian Department of Lands and Surveys file 15255/11 *French names on the plans of Western Australia to be retained, A de Fleurieru, comte.* 2 vols.
6. The Institut Islands named by the French consisted of:

Berthoud Island	Lagrange Island
Borda Island	Laplace Island
Bougainville Island	Lavoisier Island
Cassini Island	Moliere Island
Condillac Island	Monge Island
Corneille Island	Montesquieu Island
Descartes Island	Pascal Island
Lafontaine Island	Racine Island

Appendix 30 *(Continued)*

FRENCH PLACE NAMES CURRENTLY USED ON THE COAST OF WESTERN AUSTRALIA WITH A NOTE ON THEIR ORIGINS

Place Name	Coastal Location	Named by the Expedition under:	Named for:
Adele Island	North west coast — north of King Sound	Baudin	A species of insect, Adèle
Adele Island Reef	North west coast — off Adele Island	—	ditto
Adele Island Shoal	North west coast — south of Adele Island	—	ditto
Amphinome Shoals	North west coast — off Poissonier Point	Baudin	A zoological species
Attack Bay	West coast — Shark Bay, east of Cape Peron	Baudin named it Baie d'Attacque	Translated from the French. The place was named as a result of a clash with the Aborigines
Augereau Island	North west coast — the northern extremity of King Sound	Baudin	Pierre François Charles Augereau, duc de Castiglione, 1757-1816, military commander who fought for Napoleon at the Battle of Arcole, later a peer
Baleine Bank	North west coast — near Beagle Bay	Baudin	French for Whale Bank
Barthelemy Hills	North west coast — on the east side of Joseph Bonaparte Gulf	Baudin named it Barthélemy Island	François Barthélemy, 1747-1830, diplomat and member of the Directory, later marquis *or* Nicolas Martin baron Barthélemy, 1765-1835 distinguished general who fought in Spain

Place Name	Coastal Location	Named by the Expedition under:	Named for:
Baudin Island	North west coast — Institut Islands, Bonaparte Archipelago	Named in 1890 by the British surveyor Osborne Moore	Thomas Nicolas Baudin, 1754-1803, leader of the French expedition to western Australia 1800-1804
Baudin Island	West coast — Shark Bay, in the lower part of Henri Freycinet Harbour	Named in 1858 by the British surveyor Denham	ditto
Bedout Island	North west coast — off Larry Point, north of Amphinome Shoals	Baudin	Rear Admiral Bedout, 1751-1818, a distinguished naval officer in the revolutionary wars
Bellefin, Cape	West coast — Shark Bay, the northern extremity of Bellefin Prong near Useless Inlet	Baudin	Jérôme Bellefin, surgeon on the *Naturaliste* commanded by Hamelin
Bellefin Flats	West coast — Shark Bay, north of Cape Bellefin	—	ditto
Bellefin Prong	West coast — Shark Bay, between South Passage (Blind Strait) and Useless Inlet	—	ditto
Bernier, Cape	North west cape — on the west coast of Joseph Bonaparte Gulf	Baudin	Pierre François Bernier, astronomer on the *Géographe* commanded by Baudin
Bernier Island	West coast — Shark Bay, the northernmost island	Baudin	ditto

Place Name	Coastal Location	Named by the Expedition under:	Named for:
Bernouilli Island	North west coast — D'Arcole Islands, off Brunswick Bay	Baudin	A distinguished family of scientists including: Jacques Bernouilli, 1654-1705; mathematician Jean Bernouilli, 1667-1748; mathematician Nicolas Bernouilli, 1687-1759; mathematician Daniel Bernouilli, 1701-1782; mathematician and hydrodynamics researcher Jean Bernouilli, 1710-1790; mathematician Jean Bernouilli, 1744-1807; mathematician and astronomer Jérôme Bernouilli, 1715-1829; celebrated naturalist Jacques Bernouilli, 1759-1789; physicist
Berthier Island	North west coast — Bonaparte Archipelago, York Sound off Bigge Island	Baudin	Jean Baptiste Berthier, 1721-1804, engineer and military administrator who constructed government offices at Versailles *or* Alexandre Berthier, 1753-1815, military commander who served with Lafayette and Napoleon
Bertholet, Cape	North west coast — off Carnot Bay near Cape Leveque	Baudin spelt it Bertholet	Claude Louis, comte Berthollet, 1749-1822, celebrated chemist
Berthoud Island	North west coast — Admiralty Gulf, east of Bigge Point	Baudin	Ferdinand Berthoud, 1727-1813, celebrated watchmaker who perfected a chronometer

Place Name	Coastal Location	Named by the Expedition under:	Named for:
Bessieres Island	North west coast — north west of Onslow, formerly Anchor Island, renamed in 1968	Baudin	A celebrated family, members of whom include: Jean Baptiste Bessières, duc d'Istrie, 1768-1813, military commander under Napoleon, who fought in Egypt; Bertrand, baron Bessières, 1773-1854, military commander under Napoleon; Julien Bessières, 1770-1840, administrator and savant, later peer of France
Bezout Island	North west coast — north of Cape Lambert	Baudin	Etienne Bezout, 1730-1783, mathematician, especially applying malthematics to use at sea
Bezout Rock	North west coast — off Bezout Island	—	ditto
Boileau, Cape	North west coast — north of Roebuck Bay	Baudin	Nicolas Boileau-Despréaux, 1636-1711, poet friend of Racine, noted for his anti-English verse *or* Marie Louis Joseph de Boileau, 1741-1817, reform lawyer, *or* Jacques Boileau sometimes Boilleau, 1752-1793, revolutionary politician
Boileau Patches	North west coast — off Cape Boileau	—	ditto
Bonaparte Archipelago	North west coast — off Montague Sound, York Sound and Brunswick Bay	Baudin	Napoleon Bonaparte (Napoleon I) 1769-1821, Consul of France (1799-1804) at the time the Baudin expedition was in western Australia
Borda, Cape	North west coast — the northern extremity of Pender Bay	Baudin	Jean Charles Borda, 1733-1799, French geometrist, mathematician and mariner who wrote on latitudes and longitudes and perfected the reflective circle used by navigators

Place Name	Coastal Location	Named by the Expedition under:	Named for:
Borda Island	North west coast — the eastern part of Admiralty Gulf in the Institut Islands	Baudin	ditto
Bossut, Cape	North west coast — the southern extremity of La Grange Bay	Baudin	Charles, Abbé Bossut, 1730-1814, celebrated mathematician who wrote on navigation and astronomy
Bossut, False Cape	North west coast — the northern extremity of Lagrange Bay near Cape Latouche Treville	—	ditto
Bougainville, Cape	North west coast — the western extremity of Vansittart Bay, on the Bougainville Peninsula	Baudin — named it Bougainville Island	Louis Antoine de Bougainville, 1729-1811, mathematician, soldier and Pacific explorer. Bougainville's son, Hyacinthe Yves Philippe Potentier Bougainville served as midshipman on the *Géographe* commanded by Baudin
Bougainville Peninsula	North west coast — between Vansittart Bay and Admiralty Gulf	Baudin — named it Bougainville Island	ditto
Bougainville Reef	North west — off Cape Bougainville	—	ditto
Bougner Entrance (sometimes incorrectly named as this on charts — see Bouguer Entrance)			
Bouguer Entrance	North west coast — near Cape Lambert	Baudin	Pierre Bouguer, 1698-1758, hydrographer and explorer, author of a treatise on navigation, professor of hydrography at Le Havre
Boullanger, Cape	West coast — Shark Bay, the northern extremity of Dorre Island	—	Charles Pierre Boullanger, engineer-geographer on the *Géographe*, commanded by Baudin.

Place Name	Coastal Location	Named by the Expedition under:	Named for:
Boullanger Island	West coast — off Island Point at the southern extremity of Jurien Bay	Baudin	ditto
Bouvard, Cape	West coast — south of Mandurah	Baudin	Alexis Bouvard, 1787-1843, astronomer, discoverer of comets in 1797 and 1798, just before Baudin sailed to Australia
Bouvard Reefs	West coast — off Cape Bouvard	—	ditto
Breton Bay	West coast — north of Cape Leschenault near Moore River	Baudin	François Désiré Breton, (?)-1801, a Creole from Guadaloupe, midshipman on the *Géographe*
Brue Reef	North west coast — north of Cape Leveque	Baudin	Midshipman Adrien Hubert Brué, 1786-1832, who joined the Baudin expedition at Mauritius; later a noted cartographer and geographer
Brugieres, Cape	North west coast — near Dampier	Baudin	Jean Guillaume Bruguières, 1750-1799, naturalist who accompanied the second Kerguelen expedition to the Indian Ocean, and later made botanical explorations in the Middle East
Buffon Island	North west coast — Bonaparte Archipelago, north of Brunswick Bay	Baudin	Georges Luis Leclerc, comte de Buffon, 1707-1788, French naturalist who attempted to describe all scientific knowledge in his *Histoire Naturelle*

Place Name	Coastal Location	Named by the Expedition under:	Named for:
Caffarelli Island	North west coast — Buccaneer Archipelago, north of King Sound	Baudin spelt it Cafarelly Island	Name of a French family which came from Italy in the reign of Louis XIII (1610-1643). Distinguished members at the time of the Baudin expedition included: Louis Marie Joseph Maximilien Cafarelli du Falga, 1750-1799, republican general killed in Egypt; Charles Ambrose Caffarelli, 1758-1826, economist and administrator under Napoleon François Marie Auguste Caffarelli, 1766-1849, military leader under Napoleon, later peer of France
Carnot Bay	North west coast — at Cape Baskerville, north of Broome	Baudin named it Carnot Island	Lazare Nicolas Marguerite Carnot, 1753-1823, distinguished revolutionary and military official, mathematician and engineer
Carnot Peak	North west coast — inland from Red Bluff, Cape Baskerville	Baudin named it Carnot Island	ditto
Cassini Island	North west coast — Institut Islands, north of Admiralty Gulf	Baudin	Named after the Cassini family of astronomers who for four generations succeeded each other in charge of the observatory in Paris as follows: Giovani Domenico Cassini, 1625-1712; Jacques Cassini, 1677-1754; César François Cassiny de Thury, 1714-1784; Jacques Dominique Cassini, comte, 1748-1845

Place Name	Coastal Location	Named by the Expedition under:	Named for:
Casuarina Islands	South coast — off Walpole and Nornalup Inlet	Baudin	The Australian built schooner *Casuarina* commanded by Louis de Freycinet who discovered these islands. The *Casuarina* was purchased in Sydney to replace the *Naturaliste* which sailed for France in 1803, to take specimens to the Institut
Casuarina, Mount	North west coast — inland from Cape Lambert, Joseph Bonaparte Gulf	Baudin	ditto
Casuarina Point	West coast — Geographe Bay, the north western extremity of Koombana Bay	Baudin	ditto
Casuarina Reef	North west coast — off Cape Bossut, Lagrange Bay	Baudin	ditto
Casuarina Shoal	West coast — Garden Island, west of Entrance Point	Named in 1829 by west Australian surveyor Roe	ditto
Champagny Islands	North west coast — off Camden Sound	Baudin	Jean Baptiste Nompère de Champagny, duc de Cadore (1756-1834) French statesman and diplomat, Foreign Minister under Napoleon after Talleyrand
Champagny Island	North west coast — off Camden Sound	Baudin	ditto
Championet Island	North west coast — off the west coast of Bigge Island, York Sound	Baudin spelt it Championnet Island	Jean Antoine Etienne Championnet, 1762-1800, celebrated general in the French Republican forces
Chateaurenaud Cape	North west coast — the northern extremity of Bigge Island	Baudin marked it as a cape on the mainland	François Louis Rouselet, marquis de Chateaurenaut, 1637-1716, admiral, sometimes Chatearenaud or Chateau Regnaud

Place Name	Coastal Location	Named by the Expedition under:	Named for:
Clairault Cape	West coast — south of Cape Naturaliste	Baudin spelt it Clairault	Alexis Claude Clairaut, 1713-1765, celebrated mathematician, who worked with Maupertius on determining the meridian at Paris
Colbert Island	North west coast — Bonaparte Archipelago, off Brunswick Bay	Baudin	Jean Baptiste Colbert, 1619-1683, French statesman, Minister of Finance to Louis XIV
Commerson Island	North west coast — D'Arcole Islands off Brunswick Bay	Baudin	Philibert Commerson, 1727-1773, naturalist selected to accompany the Bougainville expedition, later resident in Mauritius
Condillac Islands	North west coast — Bonaparte Archipelago, north of Cape Voltaire	Baudin	Etienne Bonnot de Condillac, 1715-1780, French philosopher and encyclopedist who wrote on psychology, logic and economics
Corneille Island	North west coast — Institut Islands north of Admiralty Gulf	Baudin	Pierre Corneille, 1606-1684, French dramatic poet regarded as the creator of French classical tragedy
Corvisart Island	North west coast — Maret Islands off York Sound	Baudin	Jean Nicolas, baron Corvisart, 1755-1821, physician to Napoleon
Cossigny, Cape	North west coast — south west of Cape Thouin	Baudin	The Cossigny family were closely associated with the development of the Ile de France (Mauritius) as a fortified French base in the Indian Ocean. Distinguished members of the family include: Jean François Charpentier de Cossigny (sometimes Coussigny), 1693-1778, engineer who surveyed Mauritius for port facilities and then served there in a military capacity Joseph François Charpentier de Cossigny de Palma, 1730-1806, military engineer on Mauritius who introduced sugar cane from Indonesia to the island

Place Name	Coastal Location	Named by the expedition under:	Named for:
Cossigny Hill	North west coast — near Cape Thouin	—	ditto
Coulomb Point	North west coast — north of Broome	Baudin	Charles Augustin de Coulomb, 1736-1806, distinguished physicist and academic noted for researches on electricity. His unit for electricity, the coulomb, was later renamed the ampere
Couture, Cape	West coast — Shark Bay, the southern extremity of Bernier Island	Freycinet	Joseph Victor Couture, midshipman on the *Naturaliste*
Cuvier, Cape	West coast — north of Shark Bay	Baudin	Georges Léopold Chrétien Frédéric Dagobert, baron Cuvier, 1769-1832, French naturalist
D'Aguesseau Island	North west coast — in the D'Arcole Islands off Brunswick Bay	Baudin	Henri François d'Arguesseau, 1668-1751, noted French jurist and law administrator. D'Aguesseau's statue was erected near the French legislature in 1810
D'Arcole Islands	North west coast — Brunswick Bay	Baudin	The Battle of Arcole (Bataille d'Arcole) where Napoleon defeated the Austrians on 15, 16 and 17 November, 1796
De Freycinet Island (see under Freycinet for other places with this name)	North west coast — Bonaparte Archipelago off Brunswick Bay	Baudin	Henri Desaules de Freycinet, 1777-1840, sub-lieutenant on the *Géographe*, later baron and rear admiral *or* his brother, Louis Claude Desaules de Freycinet, 1779-1842, sub-lieutenant on the *Naturaliste* later commander of the *Casuarina* and then the *Uranie*
Degerando Island	North west coast — Champagny Islands off Camden Sound	Baudin	Joseph Marie, baron de Gerando, 1772-1842, philosopher who worked with the French Interior Minister Champagny (see Champagny)

Place Name	Coastal Location	Named by the Expedition under:	Named for:
Delambre Island	North west coast — Dampier Archipelago near Nickol Bay	Baudin	Jean Baptiste Joseph Delambre, 1749-1822, distinguished French astronomer and mathematician
Delambre Reef	North west coast — near Cape Lambert	—	ditto
D'Entrecasteaux Point	South coast — between Cape Leeuwin and Point Nuyts	d'Entrecasteaux	Joseph Antoine Raymond Bruny d'Entrecasteaux, 1739-1793, admiral and explorer who surveyed the south coast of Australia in December 1792
Depuch Bay (Eagle Bay)	West coast — Geographe Bay near Cape Naturaliste	Baudin	Louis Depuch, mineralogist on the *Géographe* who was excited by finding granite in the creek running into the bay
Depuch Island	North west coast — off Port Hedland	Baudin	ditto
Depuch Loop	West coast — Shark Bay, in the southern part of Henri Freycinet Harbour (Freycinet Estuary)	Baudin named it Depuch Entrée	ditto
Desaix Island	North west coast — off York Sound, north of Coronation Islands	Baudin	Louis Charles Antoine Desaix de Veygoux (or Des Aix), 1768-1800, celebrated French general killed at the Battle of Marengo, 1800
Desault Bay	North west coast — between Cape Jaubert and Cape Missiessy	Baudin	Pierre Joseph Desault, 1744-1795, celebrated surgeon
Descartes Island	North west coast — Institut Group off Admiralty Gulf	Baudin	René Descartes, 1596-1650, French philosopher
Desfontaines Island	North west coast — Brunswick Bay, west of Coronation Island	Baudin	17th century French dramatic author, *or* René Loriche Desfontaines, 1752-1833, noted botanist

Place Name	Coastal Location	Named by the Expedition under:	Named for:
Dombey, Cape	North west coast — the east side of Joseph Bonaparte Gulf	Baudin	Joseph Dombey 1742-1793, French botanist and explorer associated with Buffon
Duguesclin Island	North west coast — D'Arcole Islands off Brunswick Bay	Baudin	Noted French family from Bretagne (Brittany)
Duhamel, Cape (also on the charts as Cape Dahamel)	North west — near Cape Frezier, Lagrange Bay	Baudin	Jean Pierre François Guillot Duhamel, 1730-1816, French metallurgist
Dupuy, Cape	North west coast — northern extremity of Barrow Island	Baudin	André Julien, comte Dupuy, 1753-1832, administrator in Mauritius and France *or* Louis Dupuy, 1709-1795, mathematician, philologist and academician
Dussejour, Cape	North west coast — the west entrance to Cambridge Gulf	Baudin. This was the first named Cape Berquin. Cape Dussejour was previously north of Barthelemy Island on the east side of Joseph Bonaparte Gulf. The French changed the name in their final map	Archille Pierre Diones du Séjour, 1734-1794, astronomer and official
Emeriau Point	North west coast — the western extremity of Pender Bay	Baudin named it Emeriau Island	Maurice Julien, comte Emeriau, 1762-1845, vice admiral

Place Name	Coastal Location	Named by the Expedition under:	Named for:
Epineaux Entrance (alternative name for South Passage)	West coast — Shark Bay, between the south point of Dirk Hartog Island and the mainland	Baudin. Note the Australian pilot incorrectly states Epineaux Entrance is an alternative name for False Entrance. Baudin's explorers discovered South Passage by sailing to it from inside Shark Bay. They therefore could not have confused this with False Entrance which is a blind bay and was not seen by the French from the sea.	French for thorny or prickly
Esperance Bay Formerly called Le Grand Bay by the French	South coast — off Esperance	D'Entrecasteaux	The *Espérance,* commanded by Huon de Kermadec — the second ship in the expedition under d'Entrecasteaux
Faure Island	West coast — Shark Bay near Hopeless Reach, east of Peron Peninsula	Baudin	Pierre Ange François Xavier Faure, astronomer on the *Naturaliste*
Faure Flat	West — Shark Bay, south of Faure Island	Baudin named it Turtle Flats. Named Faure by the British	ditto
Fenelon Island	North west coast — Institut Islands off Admiralty Gulf	Baudin	François de Salignac de la Mothe Fénelon, 1651-1715, French theologian and author
Fenelon Passage	North west — Institut Islands, south of Fenelon Island	—	ditto

Place Name	Coastal Location	Named by the Expedition under:	Named for:
Flacourt Bay	North west coast — west coast of Barrow Island, near Cape Malouet	Baudin	Etienne de Flacourt, 1607-1660, administrator and traveller who took possession of Indian Ocean islands for France, director of the French Compagne de Orient and administrator in Mauritius
Fontanes Island	North west coast — York Sound, north of Coronation Islands	Baudin	Louis de Fontanes, 1757-1821, author, prominent publicist and state figure
Forbin Island	North west coast — D'Arcole Islands off Brunswick Bay	Baudin	A distinguished family from Provence which included: Claude, comte de Forbin, 1656-1733, noted mariner who served in the Indian Ocean; Gaspard François Anne de Forbin, 1718-1780, mathematician; Louis Nicolas Philippe Auguste, comte de Forbin, 1777-1841, painter and archaeologist
Forestier Islands	North west coast — south of Cape Thouin	Baudin	François Louis, baron Forestier, 1776-1814, distinguished French general who fought with Napoleon
Fourcroy, Cape	North west coast — south west side of Bathurst Island	Baudin	Antoine François, comte de Fourcroy, 1755-1809, academician and Counsellor of State, professor of chemistry at the Museum in Paris
Frenchman Bay	South coast — King George Sound, Albany	Named by the British	Where the French explorers called to get fresh water supplies from the streams. Known to the French as Stream (Aiguade Bay)
Freycinet Estuary (now Henri Freycinet Harbour)	West coast — Shark Bay, east of Peron Peninsula	Baudin named it Henri Freycinet Harbour	Henri Desaules de Freycinet, 1777-1840, sub-lieutenant on the *Géographe,* later baron and rear admiral

Place Name	Coastal Location	Named by the Expedition under:	Named for:
Freycinet Island	West coast — Shark Bay, Freycinet Estuary near Cararang Peninsula	Named in 1858 by the British surveyor Denham	Henri Desaules de Freycinet, 1777-1840, sub-lieutenant on the *Géographe*, later baron and rear admiral *or* his brother, Louis Claude Desaules de Freycinet, 1779-1842, sub-lieutenant on the *Naturaliste* later commander of the *Casuarina* and the *Uranie*
Freycinet Island (see De Freycinet Island)	North west coast — Bonaparte Archipelago off Brunswick Bay		
Freycinet Point	West coast — near Hamelin Bay, Cape Leeuwin	Baudin	ditto
Freycinet Reach (now part of Henri Freycinet Harbour)	West coast — Shark Bay, west of Peron Peninsula	Baudin named it Henri Freycinet Harbour	Henri Desaules de Freycinet, 1777-1840, (see above)
Frezier, Cape	North west coast — northern extremity of Geoffroy Bay, near Lagrange Bay	Baudin	Armédée Frezier, 1682-1773, engineer and navigator
Gantheaume Bay	West coast — near the Murchison River	Baudin	Honoré Joseph Antonin, comte Gantheaume, 1775-1818, vice admiral and peer of France who sailed with Suffren and commanded the *Muiron* which took Napoleon from Egypt to France
Gantheaume Bay	North west coast — near Roebuck Bay	—	ditto
Gantheaume Point	North west coast — near Broome	Baudin named it Gantheaume Island	ditto
Geoffroy Bay	North west coast — north of Cape Jaubert	Baudin named it Geoffroy Island	Etienne Louis Géoffroy, 1725-1810, physician and naturalist

Place Name	Coastal Location	Named by the Expedition under:	Named for:
Geographe Bay	West coast — west of Bunbury, east of Cape Naturaliste	Baudin	The *Géographe* commanded by Nicolas Baudin
Geographe Channel	West coat — Shark Bay, between Bernier Island and the mainland	Baudin	ditto
Geographe Reef	West coast — off Cape Leeuwin	—	ditto
Geographe Shoals	North west coast — off Cape Thouin	Baudin	ditto
Giraud Point	West coast — Shark Bay, the east point of Depuch Loop	Baudin	Ete. Stanislas Giraud, midshipman on the *Naturaliste*
Gourdon Bay	North west coast — south of Broome	—	Antoine Louis, comte de Gourdon, 1765-1833, admiral
Gourdon, Cape	North west — south of Broome	Baudin	ditto
Guichen Reef	North west coast — off Cape Bougainville	—	Luc Urbain du Bonexic, comte de Guichen, 1712-1790, naval administrator and commander
Guichenault Point (See Guichenot Point)			
Guichenot Point	West coast — Shark Bay, on north east coast of Peron Peninsula	Baudin, also called it Guichenault	Antoine Guichenot, gardener on the *Géographe*
Hamelin Bay	West coast — north of Cape Leeuwin	—	Jacques Félix Emmanuel, baron Hamelin, commander of the *Naturaliste,* 1768-1839, later rear admiral. After his return to France from Australia, Hamelin was put in charge of the fleet planned to invade England
Hamelin, Cape	West coast — north of Cape Leeuwin	Baudin	ditto
Hamelin Island	West coast — off White Cliff Point, Hamelin Bay	—	ditto

Place Name	Coastal Location	Named by the Expedition under:	Named for:
Hamelin Pool	West coast — Shark Bay, east of Taillefer Isthmus	Baudin, gave the name Hamelin Harbour to the large bay east of Peron Peninsular. Hamelin Pool is part of this area	ditto
Hauy Island	North west coast — Dampier Archipelago	Baudin named it Haüy Island	René Just Haüy, 1743-1822, celebrated mineralogist *or* Ventin Haüy, 1745-1822, educator of the blind
Heirisson, Cape	West coast — Shark Bay, north end of Heirisson Prong	Baudin	Antoine Boniface Heirisson of Madrid, sub-lieutenant on the *Naturaliste*
Heirisson Flats	West coast — Shark Bay, north of Cape Heirisson	—	ditto
Heirisson Island	West coast — Swan River, near Perth	Baudin named it Heirisson Islands	ditto
Heirisson Prong	West coast — Shark Bay, between Useless Inlet and Henri Freycinet Harbour	—	ditto
Helvetius, Cape	North west coast — on the west coast of Bathurst Island	Baudin	Claude Adrien Helvétius, 1715-1771, philosopher and 18th century celebrity associated with the encyclopedists and natural historians
Henri Freycinet Harbour (see also under Freycinet)	West coast — Shark Bay, west of Peron Peninsula	Baudin	Henri Desaules de Freycinet, 1777-1840, sub-lieutenant on the *Géographe*, later baron and rear admiral
Herald Bight (also known as Attack Bay, see Attack Bay)			

Place Name	Coastal Location	Named by the Expedition under:	Named for:
Hermite Island (in French this is L'Hermite Island)	North west coast — in the Montebello Islands	Baudin named it L'Hermite Island	Jean Marthe Adrien L'Hermite, 1766-1826, admiral noted for his bravery when fighting the British; a popular hero at Mauritius
Holothuria Reef	North west coast — north of Cape Bougainville off Admiralty Gulf	Baudin named it Holothuria Banks	Name of a marine zoological species (the sea cucumber)
Holothuria Reef East	North west coast — north of Admiralty Gulf	—	ditto
Holothuria Reef West	North west coast — north of Admiralty Gulf	—	ditto
Huygens, Cape	North west coast — north of Broome	Baudin	Christian Huygens or Huyghens, 1629-1695, Dutch physicist, geometrist, astronomer and maker of marine chronometers who was associated with the French Academy of Sciences
Institut Islands	North west coast — north of Cape Voltaire off Admiralty Gulf	Baudin	The Institut de France, founded in 1795 to encourage research. This Institut took over the functions of the Académie des Sciences
Jaubert, Cape	North west coast — the northern extremity of Desault Bay near La Grange	Baudin	François, comte Jaubert, 1758-1822, administrator and statesman
Joseph Bonaparte Gulf	North west coast — north of Cambridge Gulf	Baudin	Joseph Bonaparte, 1768-1844, Napoleon Bonaparte's elder brother, Councillor of State, later King of Spain (1803)
Jurien Bay	West coast — south of Geraldton near the Hill River	Baudin	Charles Marie, vicomte Jurien, 1763-1836, naval administrator

Place Name	Coastal Location	Named by the Expedition under:	Named for:
Jussieu Island	North west coast — York Sound, west of Bigge Island	Baudin	French family of distinguished botanists including: Antoine de Jussieu, 1699-1777; Joseph de Jussieu, 1704-1779; Adrien Laurent Henri de Jussieu, 1797-1853
Keraudren, Cape	North west — east of Port Hedland	Baudin	Pierre François Kéraudren, 1769-1857, physician in charge of naval medical services at Brest, author of works on diseases of seamen
Keraudren Island	North west — Bonaparte Archipelago off Brunswick Bay	Baudin	ditto
Lacepede Channel	North west coast — Beagle Bay between the Lacepede Islands and the mainland	—	Bernard Germain Etienne de Laville, comte de Lacépède, 1756-1825, natural historian
Lacepede Islands	North west coast — off Beagle Bay	Baudin	ditto
Lacrosse Islands	North west coast — off Cambridge Gulf	Baudin	Jean Baptiste Raymond, baron de Lacrosse, 1765-1829, admiral who fought in the Indian Ocean and led the invasion of Ireland in the revolutionary wars
Lafontaine Island	North west coast — off Admiralty Gulf	Baudin	Jean de la Fontaine or Lafontaine, 1621-1695, poet and academician
Lagrange Bay	North west coast — west of Roebuck Bay	Baudin	Joseph Louis Lagrange, 1736-1813, celebrated mathematician and geometrist
Lagrange Island	North west coast — Institut Islands, north east of Corneille Island	—	ditto

Place Name	Coastal Location	Named by the Expedition under:	Named for:
Lamarck Island	North west coat — off York Sound	Baudin	Jean Baptiste Pierre Antoine de Monet, Chevalier de Lamarck, 1744-1829, naturalist who wrote on evolution
Lancelin Island	West coast — north of Moore River	Baudin	P.F. Lancelin, (?)-1809, scientific writer, author of the World Map of Sciences and works on the planetary system and analyses of science
Laplace Island	North west coast — north of Cape Voltaire	Baudin	Pierre Simon, marquis de Laplace, 1749-1827, celebrated mathematician and geometrist who worked with Clairaut
Laridon Bight (See Lharidon Bight)			
Larrey Point	North west coast — near Poissonier Point, east of Port Hedland	Baudin	Dominique Jean, baron Larrey, 1766-1842, celebrated military surgeon and health officer
Latouche Treville, Cape	North west coast — the northern extremity of Lagrange Bay	Baudin	Louis René Madeleine le Vassor de Latouche Tréville, 1745-1804, French admiral who planned to survey the south coast of Australia
Latreille, Cape	North west coast — north of Gautheaume Point, Broome	Baudin	Pierre André Latreille, 1762-1833, celebrated naturalist and entymologist at the Museum of Natural History
Lavoisier Island	North west coast — Voltaire Passage, Admiralty Gulf	Baudin	Antoine Laurent Lavoisier, 1743-1794, noted chemist and scientific researcher, guillotined in Paris during the revolution
Lefebre Island	West coast — Shark Bay in Henri Freycinet Harbour off the Cararang Peninsula	Baudin	Jean Thomas Lefebre, helmsman on the *Naturaliste*

Place Name	Coastal Location	Named by the Expedition under:	Named for:
Legendre Island	North west coast — Dampier Archipelago off Nickol Bay	Baudin	Adrien Marie Legendre, 1752-1834, mathematician, geometrist and academician
Le Grand, Cape	South coast — the southern extremity of Esperance Bay	D'Entrecasteaux	Ensign Le Grand of the *Espérance* who climbed the mast and guided the *Esperance* and *Recherche* through the reefs into Esperance Bay during a gale, thus saving the ships
Le Grand, Mount	South coast - near Cape Le Grand	—	ditto
Leschenault, Cape	West coast — near Moore River	Baudin	Jean Baptiste Louis Claude Théodore Leschenault de la Tour, 1773-1826, botanist and naturalist on the *Géographe*
Leschenault Inlet	West coast — Bunbury	Baudin named it Port Leschenault	ditto
Leschenault Island (also known as Salutation Island)	West coast — Shark Bay near Fording Point in Henri Freycinet Harbour	—	ditto
Leschenault Reefs	West coast — off Cape Leschenault	—	ditto
Lesueur, Cape	West coast — Shark Bay west coast of Peron Peninsula	Baudin	Charles Alexandre Lesueur, 1778-1846, gunner on the *Geographe* then appointed artist-sketcher, later the conservator of the Museum of Natural History at Le Havre
Lesueur Island	North west coast — off Cape Rulhieres, on the western side of Joseph Bonaparte Gulf	Baudin	ditto
Lesueur Mount	West coast — near Jurien Bay	Baudin	ditto
Leveque, Cape	North west coast — the western extremity of King Sound	Baudin	Pierre Lévèque, 1746-1814, engineer hydrographer
Leveque Island	North west coast — off Cape Leveque	—	ditto

Place Name	Coastal Location	Named by the Expedition under:	Named for:
Levillain, Cape	West coast — Shark Bay, the north east point of Dirk Hartog Island	Baudin	Stanislas Levillain, zoologist on the *Géographe*
Levillain Shoal	West coast — east of Cape Levillain	—	ditto
Lharidon Bight	West coast — Shark Bay on the east side of Peron Peninsula	Baudin	François Etienne Laridon or Lharidon, ship's doctor on the *Géographe*
Lowendal Island	North west coast — Montebello Islands	Baudin	Ulric Frédéric Woldemar, comte de Lowendahl, 1700-1755, marshal of France, linguist, soldier and academician
Lucas Island	North west coast — D'Arcole Group, off Augustus Island	Baudin	Jean Jacques Etienne Lucas, 1764-1819, distinguished naval commander who fought heroically against the English in the Indian Ocean and on the *Redoubtable* at Trafalgar, which opposed Nelson's *Victory*
Mably Island (sometimes incorrectly spelt Malby on charts)	North west coast — off the Coronation Islands	Baudin	Gabriel Bonnot Abbé de Mably, 1709-1785, philosopher and historian
Malouet, Cape	North west coast — west coast off Barrow Island	Baudin	Pierre Victor, baron Malouet, 1740-1814, state official and author, director of the navy
Malus Island	North west coast — Dampier Archipelago, off Regnard Bay	Baudin	Etienne Louis Malus, 1775-1812, physicist and academician noted for his study of optics

Place Name	Coastal Location	Named by the Expedition under:	Named for:
Maret Islands	North west coast — Bonaparte Archipelago, York Sound	Baudin	Hugues Bernard Maret, duc de Bassano, 1763-1839, French diplomat and statesman
Medusa Banks	North west coast — Joseph Bonaparte Gulf at the entrance to Cambridge Gulf	Baudin	Name of a group of zoophytes
Mentelle, Cape	West coast — north of Cape Hamelin and Cape Leeuwin	Baudin	A prominent scientific family including: Edmonde Mentelle, 1730-1815, geographer and historian; François Simon Mentelle, 1731-1799, geographic engineer who drew topographical charts
Middle Island	South coast — Recherche Archipelago	D'Entrecasteaux named it Ile de Milieu	Translated from the French. Named for its location as the central island in the Recherche Archipelago
Missiessy, Cape	North west coast — north of the Eighty Mile Beach, south west of Cape Jaubert	Baudin	Edouard Thomas Burgues, comte de Missiessy, 1754-1832, admiral
Moliere Island	North west coast — Admiralty Gulf	Baudin	Jean Baptiste Poquelin (Molière), 1622-1673, French dramatist
Mollien, Cape	North west — the north point of Adele Island	Baudin	Nicolas François, comte de Mollien, 1758-1850, state official and financier
Mondrain Island	South coast — Recherche Archipelago	D'Entrecasteaux	French for abrupt hillock. In the eighteenth century the word mondrain was used to describe an abrupt hill or cliff near the sea, as recorded by Bougainville. The word is now usually used for a hillock or sandhill

Place Name	Coastal Location	Named by the Expedition under:	Named for:
Monge Island	North west coast — Institut Islands off Cape Voltaire	Baudin	Gaspard Monge, comte de Péluse, 1746-1818, celebrated mathematician and geometrist, founder of the Polytechnic in Paris *or* Louis Monge, 1748-1827, astronomer who sailed with La Pérouse on his expedition in 1785, later a noted hydrographer and mathematician
Montalivet Island	North west coast — Montague Sound	Baudin named the surrounding group the Montalivet Islands	Comte Jean Pierre Bachasson de Montalivet, 1766-1823, statesman and soldier
Montbazin Lagoon	West coast — Shark Bay on the north point of Peron Peninsula	Baudin	Louis Charles Gaspard Bonnefoi de Montbazin, also known as Louis Bonnefoi, midshipman on the *Géographe*
Montebello Islands	North west coast — north of Barrow Island	Baudin	The battle of Montebello where Napoleon defeated the Austrians on 9 June 1800, near Pavia in north Italy. Montebello was subsequently made a Duchy and the victorious French general Lannes was made the Duke of Montebello
Montesquieu Islands	North west coast — Institut Islands, Admiralty Gulf	Baudin	Charles de Secondat, baron de la Bréde et de Montesquieu, 1689-1755, French writer on politics and law
Muiron Islands	North west — off North West Cape	Baudin named one island Muiron Island	Muiron, Napoleon's aide de camp and close friend killed by his side at the battle of Arcole *or* Napoleon's pseudonym "Colonel Muiron" *or* the ship *Muiron* commanded by Gantheaume which took Napoleon from Egypt back to France

Place Name	Coastal Location	Named by the Expedition under:	Named for:
Murat Point	North west coast — on the east coast of North West Cape, Exmouth Gulf	Baudin gave this name to North West Cape	Joachim Murat, 1767-1815, King of Naples and military commander with Napoleon
Naturaliste, Cape	West coast — the western extremity of Geographe Bay	Baudin	The *Naturaliste* commanded by Hammelin
Naturaliste Channel	West coast — Shark Bay, north of Dirk Hartog Island	Baudin	ditto
Naturaliste Reef	West coast — north of Cape Naturaliste	Baudin	ditto
Observatory Island	South coast — Recherche Archipelago	D'Entrecasteaux	Site of d'Entrecasteaux's observatory
Pascal Island	North west coast — Institut Islands, off Cape Voltaire	Baudin	Blaise Pascal, 1623-1662, French religious writer and scientist
Peron, Cape	West coast — the west part of Cockburn Sound	Baudin	François Péron, naturalist on the *Géographe*
Peron, Cape	West coast — Shark Bay, the northern extremity of Peron Peninsula	Baudin named it Pointe des Hauts Fonds	ditto
Peron Flats, Cape	West coast — Shark Bay, north of Cape Peron	—	ditto
Peron Hills	West coast — Shark Bay, Peron Peninsula near Cape Lesueur	—	ditto
Peron Island	North west coat — east coast of Joseph Bonaparte Gulf	Baudin	ditto
Peron Mount	West coast — near Jurien Bay	Baudin	ditto
Peron Peninsula	West coast — Shark Bay, between Henri Freycinet Harbour and Hamelin Pool	Baudin	ditto

Place Name	Coastal Location	Named by the Expedition under:	Named for:
Petit Point	West coast — Shark Bay, opposite Faure Island	Baudin	Nicolas Martin Petit, gunnery petty officer on the *Géographe* who was promoted to artist/sketcher
Picard Island	North west coast — off Cape Lambert	Baudin	Louis François Picard, 1769-1828, dramatic author and novelist
Picquet Point	West coast — the western side of Geographe Bay	Baudin	Antoine Farcy Picquet, sub-lieutenant on the *Naturaliste,* later transferred to the *Géographe,* later sent home in disgrace. The crew of the expedition named this point after Picquet as a protest to Baudin's treatment of him in Geographe Bay, which they considered unwarranted.
Planaires, Bank	North west coast — off the Eighty Mile Beach	Baudin	A species of marine animal
Poissonier, Point	North west coast — the east side of the entrance to the De Grey River	Baudin named it Ile Poissonier	Pierre Poissonier, 1720-1798, physicist and chemist who helped Bougainville get a distiller for use at sea
Poivre, Cape	North west coast — west coast of Barrow Island	Baudin	Pierre Poivre, 1719-1786, traveller and naturalist who introduced spices to Mauritius and other French islands in the Indian Ocean to develop the economy
Poivre Reef	North west coast — off Barrow Island	—	ditto
Quoin Bluff	West coast — Shark Bay in Tetrodon Bay on the east coast of Dirk Hartog Island	Baudin named it Coin de Mire	The French term for measuring point which possibly has been corrupted to the English word Quoin
Racine Island	North west coast — Institut Islands, Admiralty Gulf	Baudin	Jean Racine, 1639-1699, French tragic dramatist

Place Name	Coastal Location	Named by the Expedition under:	Named for:
Ransonnet, Cape	West coast — Shark Bay, the south the south point of Dirk Hartog Island	Baudin	Jacques Joseph Ransonnet, midshipman on the *Naturaliste*
Recherche Archipelago	South coast — off Esperance	D'Entrecasteaux	The *Recherche* under the command of Bruny d'Entrecasteaux
Regnard Bay	North west coast — near Cape Preston, east of Fortescue	Baudin	Jean François Regnard, 1655-1709, noted French poet
Regnard Island	North west coast — off Cape Preston, Regnard Bay	—	ditto
Riche, Cape	South coast — Cheyne Bay	D'Entrecasteaux	Claude Antoine Gaspard Riche, 1762-1797, physician and naturalist on the *Espérance*
Rivoli Islands	North west coast — off Exmouth Gulf	Baudin	Battlefield near Verona, Italy, where Napoleon decisively defeated the Austrians on 14 and 15 January 1797
Ronsard Bay	West coast — near Cervantes Islands	—	François Michel Ronsard, naval engineer on the *Géographe*
Ronsard, Cape	West coast — Shark Bay, the northern extremity of Bernier Island	Baudin	ditto
Ronsard Island	North west coast — Forestier Islands near Cape Thouin	Baudin	ditto
Ronsard Rocks	West coast — off Cervantes Islands	—	ditto
Rose, Cape	West coast — Shark Bay on the east coast of Peron Peninsula.	Freycinet?	Rose de Freycinet, wife of Louis de Freycinet who secretly boarded the *Uranie* and accompanied her husband on his voyage of exploration, later author of a book about this

Place Name	Coastal Location	Named by the Expedition under:	Named for:
Rosily Islands	North west coast — off Onslow	Baudin	François Etienne, comte de Rosily Mesros, 1748-1832, vice admiral who accompanied Saint Allouarn to western Australia in 1772 and later, as Director of the Hydrographic Office of the French navy, helped plan for the establishment of a French convict colony in western Australia
Rulhieres, Cape	North west coast — the west side of Joseph Bonaparte Gulf	Baudin	Claude Carloman de Rulhière, 1735-1791, historian and poet
Sable Island	North west coast — in the Forestier Islands	Baudin	French for Sandy Island
Saint Allouarn (see Saint Alouarn)			
Saint Alouarn Island	South coast — off Cape Leeuwin	D'Entrecasteaux, later verified by Baudin	Louis François Marie Alleno de Saint-Allouarn who made the first scientific survey of western Australia on the *Gros Ventre* in 1772.
Saint Cricq, Cape	West coast — Shark Bay, the southern extremity of Dorre Island	Baudin	Jacques Saint Cricq, sub-lieutenant on the *Naturaliste*
Saint Lambert, Cape	North west coast — the west coast of Joseph Bonaparte Gulf	Baudin	Jean François, Marquis de Saint Lambert 1716-1803, philosopher and poet
Salutation Island (see Leschenault Island)			
Serrurier Island	North west coast — on the east side of Exmouth Gulf	Baudin	Jean Mathieu Philibert Séruier, sometimes Sérurier, 1742-1819, celebrated military commander under Napoleon, marshal of France

Place Name	Coastal Location	Named by the Expedition under:	Named for:
Suffren Island	North west coast — York Sound, west of Bigge Island	Baudin	Pierre André de Suffren de Saint Tropez, 1726-1788, popular French admiral and strategist who sailed in the Indian Ocean
Taillefer Isthmus	West coast — Shark Bay, this joins Peron Peninsula to the mainland	Baudin	Hubert Jules Taillefer, medical officer on *Géographe*
Tetrodon Loop	West coast — Shark Bay, the east coast of Dirk Hartog Island	Baudin, named this area Tetrodon Island and Tetrodon Bay	Name used for a four-toothed or globe fish
Thevenard Island	North west coast — off Onslow	Baudin	Antoine Jean Marie, comte Thévenard, 1733-1815, vice admiral
Thouin, Cape	North west coast — west of Port Hedland	Baudin	André Thoüin, 1747-1824, noted botanist who worked with Jussieu and Buffon
Three Bays Island	West coast — Shark Bay, in the south part of Henri Freycinet Harbour	Baudin, named it Ile aux Trois Baies	Describes the shape of the island
Tournefort Island	North west coast — off York Sound	Baudin	Joseph Pitton de Tournefort, 1656-1708, celebrated botanist
Trimouille Island (in French this should read La Trimouille)	North west coast — Montebello Islands	Baudin	A distinguished family, La Trimouille, sometimes La Tremoille, noted for expelling the English from France in the early struggles between the two nations. Charles Bretagne Marie Joseph, duc de Tarente, prince de la Trimouille, 1764-1839, a prominent military commander at the time Baudin sailed, fought with the allies against the French revolutionary forces and therefore is not likely to have had the island named after him personally

Place Name	Coastal Location	Named by the Expedition under:	Named for:
Turtle Islands	North west coast — near Point Poissonier	Baudin named them Ile des Tortues	Translation from the French
Two Peoples Bay	South coast — east of Albany	Baudin,named it Port des Deux People	Supposedly to commemorate the meeting of the French and American revolutionary people, when an expedition under Ransonnet on the *Géographe* met the American sailing ship the *Union*
Uranie Bank	West coast — Shark Bay, east of Dorre Island	Freycinet	The *Uranie* under the command of Louis de Freycinet
Useless Inlet	West — Shark Bay	Baudin, named it Havre Inutile (Unusable Harbour)	Loose translation of the French
Vasse Estuary	West coast — Geographe Bay, near Busselton, part of Vasse River. See Vasse River		
Vasse River	West coast — Geographe Bay, near Busselton	Baudin. The name was given by the French to what is now the Ludlow River	In memory of Thomas Timothée Vasse of Dieppe, helmsman on the *Naturaliste* drowned near the Vasse Estuary during a gale on 8 June 1801
Victor Island	North west coast — Exmouth Gulf, Rivoli Islands	Baudin	Claude Victor Perrin, duc de Bellune, 1764-1841, celebrated military commander who fought with Napoleon at the battle of Montebello, marshal of France
Villaret, Cape	North west coast — Roebuck Bay	Baudin	Louis Thomas, comte Villaret de Joyeuse, 1750-1812, vice admiral who fought in the Indian Ocean
Voltaire, Cape	North west coast — the east side of Montague Sound	Baudin	François Marie Arouet de Voltaire, 1694-1778, French academician and author
Voltaire Passage	North west coast — Montague Sound, Admiralty Gulf	—	ditto

FOOTNOTES

CHAPTER 1

1 This legal belief which prevailed in France from the eighteenth century was also the general European continental view, and was a product of the age of rationalism in Europe. The most prominent early writer responsible for the widespread acceptance of this legal view was the influential German rationalist philosopher and international jurist Christian Wolff (1679-1754), whose book, *Jus gentium* (1749) was used as the basis for Emerich de Vattel's (1714-1767) treatise entitled *Le droit des gens (The law of nations)* (1758). This book which is subtitled *Principles of natural law applied to the conduct and affairs of nations and sovereigns*, established that exploring nations had the legal right to possess lands discovered by them, if those lands were not in lawful possession.

2 A.N. Série marine BB4. 1000, plaquette 2, itinéraire 20.2.1822.

3 Marchant, L.R. "The French discovery and settlement of New Zealand 1769-1846: a bibliographical essay on naval records in Paris" p. 516 in *Historical Studies, Australia and New Zealand*, Vol. 10, No. 40, May 1963, p. 511-518.

4 Estancelin, L. *Recherches sur les voyages et découvertes des voyageurs normands en Afrique.* Paris, 1832 La Roncière, C. de *La découverte de l'Afrique au moyen age.* Cairo, 1924-25.

5 Estancelin, L. *Dissertation sur les decouvertes faites par les navigateurs Dieppois.* Abbeville, s.d.

6 Muhammed ibn Abdallah Ibn Battutah, 1304-1368. See Trapier, B. *Les voyageurs Arabes au moyen âge.* Paris, 1937.

7 d'Avazec, A. *Note sur le premier expedition de Béthencourt aux Canaries.* Paris, 1846; Bontier. P. *Histoire de la première descouverte et conquests dès Canaries faites des l'an 1402 par messire Jean de Béthencourt, escrite du temps mesme par Pierre Bontier.* Paris, 1630.

8 Anselme d'Isalguier c1370-c1420-22. See Wolff, P. *Une famille de XIII*[e] *au XIV*[e] *siècle: les Ysalguier de Toulouse.* Paris, 1942. Wolff, P. *Commerces et marchands de Toulouse vers 1350 vers 1450.* Paris, 1954.

9 Prégent de Bidoux c1468-c1528, a prominent French admiral who fought in the Atlantic Ocean and the Mediterranean against the English and the Turks.

10. Hervé de Portzmoguer c1470-1512. See for his exploits as an admiral, Bonchart, A. *Les grands chroniques de Bretagne.* Paris, 1514. His fight against the English fleet in the Atlantic is immortalized in G. Brice's poem, *Chordigerae Navis Conflagratio.* Paris, 1513.

11 Anthiaume, A. *Cartes marines, constructions navals voyages de découvertes chez les Normands, 1500-1560.* Paris, 1916. Anthiaume, A. *Evolution et enseignement de la science nautique en France, et principalement chez les Normands.* Paris, 1920.

12 Guillaume de Rubrouk. *Relation des voyages en Tartary.* Paris, 1634. Remusat, A. *Memoire sur les relations politiques des princes chrétiens, et particulièrement des rois de France, avec les empereurs Mongols.* Paris, 1824.

13 Giovanni da Verrazano, 1485-1528, a Florentine scholar and seaman. See Gaffarel, P. *Etude sur les rapports de l'Amérique et de l'ancien continent avant Christophe Colomb.* Paris, 1869. Gaffarel, P. *Jean Ango.* Rouen, 1889.

14 Jacques Cartier, 1491-1557. See d'Avazec A. *Bref récit et succincte narration de la navigation faite en MXXXV et MDXXXVI par le capitaine Jacques Cartier aux Iles de Canada, Hochelaga, Saguenay et autres.* Paris, 1863. La Roncière, C de. *Jacques Cartier,* Paris, 1931.

15 Nicolas Durand de Villegaignon 1510-1571. See Gaffarel, P. *Histoire du Brésil Français.* Paris, 1878. Heulhard, A. *Villegaignon.* Paris, 1897.

16 Louis Antoine de Bougainville, 1729-1811. See B.N., N.A.F. 9407 *Bougainville journal;* N.A.F. 9438 *Mers australes.*

17 d'Avazec, A. *Compagne du navire l'Espoir de Honfleur 1503-1505.* Paris, 1869. This is a published copy of Gonneville's declaration to the Admiralty Court at Rouen.

18 B.N., N.A.F. 9439 *Mers australes . . . relations du voyage de Paulmier de Gonneville.* See also Flacourt, E. de. *Histoire de la grande isle Madagascar.* Paris, 1658. Julien, C.A. *Les Français en Amérique pendant la première moitié du XVI*[e] *siècle.* Paris, 1946.

19 Julien, C.A. *Les Français en Amérique pendant la première moitié du XVIe siècle.* Paris, 1946.

20 See e.g. B.N., N.A.F. 9439, 52, where in a document written in 1767 it is claimed that "that place in Terra Australis where Gonneville landed in 1503 is perhaps 600 to 700 leagues from the Ile de France (Mauritius), nearly south (from there)".

CHAPTER 2

1 Gaffarel, P. *Jean Ango.* Rouen, 1889; Guérin, E. *Ango et ses pilotes.* Paris, 1901; Hellot, A. *Jean Ango et sa famille.* Dieppe, s.d.; Mollat, M. *Le commerce maritime normand à la fin du moyen âge.* Paris, 1952.

2 Asseline, D.L. *Les antiquités et chroniques de la ville de Dieppe.* Dieppe, 1874; Estancelin, L. *Recherches sur les voyages et découvertes des voyageurs normands.* Paris, 1832; Howe, S.E. *Les grands navigateurs à la recherche des épices.* Paris, 1939; Parmentier, J. *Description nouvelle des merveilles de ce monde.* Paris, 1531.

3 "Ango et la liberté des mers" in *Le Correspondant,* Paris, 25.2.1902.

4 Henri Lancelot Voisin de la Popelinière (1540-1608), author of *L'histoire de France enrichée des plus notables occurrances survenues ez provinces de l'Europe et pays voisins depuis l'an 1550 jusques â ces temps.* La Rochelle, 1581; *Les trois mondes,* Paris, 1582. In this work La Popelinière discusses the use of the newly discovered Americas as a place of refuge for Calvinists and other religious groups facing opposition in Europe.

5 Hervé, M.R. "Australia in French geographical documents of the renaissance", *Royal Australian Historical Society Journal and Proceedings,* Vol. 41, 1955, p. 23-38.

6 Little is known about the life of Roze who is presumed to be of Scots origin. See Hervé, M.L. *Op. cit.,* and for his work in England see Taylor, E.G.R. *Tudor Geography,* 1485-1583. London, 1930.

7 Alfonce, J. *Les voyages avanteureux du capitaine Jan Alfonce, sainctogeois.* Portiers, 1559.

8 Gaspard de Châtillon, comte de Coligny 1519-72, admiral and Huguenot leader in France. He supported the proposal for a French colony in Brazil. See Delaborde, L.J.H. *Gaspard de Coligny, amiral de France.* Paris, 1879-82; Gaffarel, P. *Histoire du Brésil Français.* Paris, 1878.

9 Paulmier de Courtonne, Jean, chanoine de Lisieux. *Mémoires touchant l'établissement d'une mission chréstienne dans le troisième monde, autrement appellé la terre Australe, meridionales, antartique et inconnues . . . par un ecclesiastique originaires de cette mesme terre.* Paris, 1663.

10 Intriguing stories about the discovery of the mysterious islands of Solomon and the riches existing there attracted French adventurers and treasure hunters to the south seas after Alvaro de Mendana sailed to colonize the islands for Spain (1595), and Fernandez de Quiros was later sent to take possession of the great southland (1606). See La Roncière, C de "Le premier voyage Francais autour du monde" in *La Revue Hebdomadaire,* September 1907, p. 22 et seq for Malherbe's story of his search for riches in the area. The myth of great riches continued until the exploration of the Pacific and Indian Oceans was completed by Cook and others.

11 Thevenot, M. *Relations de divers voyages curieux.* Paris, 1663-72.

12 See e.g. Fénelon, F. de S. de la Motte. *Télémaque.* Paris, 1699, Lahontan, L.A. de L. d' A. *Dialogues de M. le baron de Lahontan et d'un sauvage dans l'Amérique.* Amsterdam, 1704. *Mémoires de l'Amerique septentrionale.* Amsterdam, 1728.

13 Cyrano de Bergerac, S. de. 1619-1655. *Histoire comique ou voyage dans la lune.* 1656 *Nouvelles ouvres comprenant l'histoire comique des états et empires du soleil.* 1662.

14 Foigny, G. de. *La terre australe connue, c'est à dire la description de ce pays inconnu jusqu' ici, de ses moeurs et de ses coutumes, par Mr. Sadeur, avec les avanteurs qui la conduiserent en ce continent.* Vannes, 1676.

15 Vairasse, D. d'Alais. *L'histoire des Sevarambes, peuples qui habitent une partie du troisième continent communement, appellé la Terre Australe (Australie).* Paris, 1677-79.

16 Jonathon Swift (1667-1745) used the region of south Australia as the setting for Lilliput in his satire, *Gulliver's travels.*

17 Neville, Henry (1620-94) *The Isle of Pines, or a late discovery of a fourth island near Terra Australis Incognita by Henry Cornelius van Sloetten.* London, 1668.

18 John Law (1671-1729), a Scottish monetary reformer and colonial developer who became

widely known after the publication of his book, *Money and trade considered, with a proposal for supplying the nation with money.* Edinburgh, 1705. He was subsequently controller of the French East India Company which then collapsed financially. See Wood, J.P. *Sketch of the life and projects of John Law of Lauriston;* Bonnassieux, L.J.P.M. *Les grandes compagnies de commerce: études pour servir à l'histoire de la colonisation.* Paris, 1892. There is a very large volume of material in France on Law and his monetary system and overseas commerce and colonization.

19 B.N., N.A.F. 9341, N.A.F. 9438; N.A.F. 9439.

20 A.N. Série marine B4, 45; 77. B.N. N.A.F. 9439.

21 Maupertius, P.L.M. de. *Lettre sur le progrès des sciences.* Paris, 1752.

22 Brosses, C. de. *Histoire des navigations aux terres australes.* Paris, 1756.

CHAPTER 3

1 See for an account of his voyage, Crozet, J. *Nouveau voyage à la mer du sud commencé sous les ordres de M. Marion.* Paris, 1783.

2 Alexandre Gui Pingré. (1711-1796). For an account of the stimulus given to maritime scientific expeditions by the need to widely observe the eight year intervalled transit of Venus see Woolf, H. *The transits of Venus.* Princeton, 1959. In the eighteenth century the transits were in 1761 and 1769.

3 Bougainville, L.A. de. *Voyage autour du monde.* Paris, 1771.

4 Pierre Poivre (1719-1786). See Madeleine Ly Tiuo Fane. *Mauritius and the spice trade: the odyssey of Pierre Poivre.* Port Louis, 1958; Poivre, P. *Voyages d'un philosophe.* Yverdun, 1768.

5 B.N., N.A.F. 9439 project presenté p. 66.

6 Alexis Marie de Rochon (1741-1817), scientific explorer who promoted the establishment of the French bureau of longitudes for navigational research. He wrote accounts of explorations to the south seas. See *Nouveau voyage à la Mer de Sud.* Paris, 1783. *Voyage aux Indes orientales et en Afrique, avec une dissertation sur les iles Solomen.* Paris, 1807.

7 A.N. Série marine 4JJ 142 (18). *Journal du vaisseau le Mascarin et de la flûte le Marquis de Castries . . .* par Jean Roux; Série marine 5JJ 142 (19). *Journal du voyage de Marion et du Clesmeur* par Jean Roux; B.S. Hyd. Marine. A 5708. Duclesmeur. *Relation du voyage.*

8 A.N. Série marine 4JJ 142 (18) Roux. Journal; B.N., N.A.F. 9437, Roux, Journal.

9 A.N. Série marine 4JJ 143. Journal de la flûte Gros Ventre; Série marine B4, 317, de Ternay 19.10.1772. Buffon later cautioned French authorities that for some inexplicable reason the southern hemisphere was colder than the northern one, B4, 317, Buffon 26.12.1772.

10 See for an account of his life, Dupouy, A. *Le Breton Yves de Kerguelen.* Paris, 1929.

11 Jean Baptiste Nicolas Denis d'Après de Mannevillette. (1707-1780). An hydrographer and explorer who drew the maps and wrote the sailing guides used by French explorers in the Indian Ocean. See his *Mémoire sur la navigation de la France aux Indes.* 1765. He also wrote a book on prevailing winds (1787).

12 A.N., Série marine B4, 317. The need for a base south, 12.9.1771; see also Kerguelen, *Reflections sur les avantages . . . procurer France Australe; Mémoire sur l'établissement d'une colone dans la France Australe* in Série marine B4, 317, p. 86-95.

13 Jacques Raimond de Grenier (1736-1803). Hydrographer and naval officer. See for his proposed new sea route. *Mémoire de la campagne de découvertes dans les mers des Indes.* Brest, 1770.

14 A.N. Série marine B4, 315.

15 Kerguelen-Tremarec, Y.D. de. *Relation de deux voyages.* Paris, 1782.

16 Kerguelen-Tremarec, *op cit;* A.N. Série marine B4, 317, *journal, Gros Ventre;* Série marine 4JJ 143, *journal Gros Ventre.*

17 Rosily stayed with Saint Allouarn until he arrived back at Mauritius. See B.N., N.A.F. 9438, p. 79. *Extract de mon journal depuis le 14 février que M. de Kerguelen me détachât dans sa chaloupe la Mouche, pour sonder en avant du Gros Ventre.*

18 B.N., N.A.F. 9438, p. 78, 9439, p. 91.

19 Dunmore, J. *French explorers in the Pacific.* Oxford, 1965, Vol. 1, p. 232 et seq.

20 A.N. Série marine B4, 317, *journal, Gros Ventre;* Série marine 4JJ, *journal, Gros Ventre.*

21 See chapter 7 above.

22 James Lind, a British naval doctor, recommended the use of citrus juice to prevent scurvy after the disaster of Anson's voyage around the world when 1051 of the 1955 men who left England to make the trip, died. France did not use anti-scorbutics of this sort.

23 François Etienne Rosily Mesros (1748-1832) ended up his career in the navy as director of the hydrographic office in the French navy. In this position he drew up the plans for France to survey and settle in western Australia. See Chapters 8-9 above.

CHAPTER 4

1 *A.M.* Série marine B4, 315 *La Touche's proposal, 1774.*

2 A.N. Série marine B4, 315 passim.

3 Jean François de Galaup, comte de la Pérouse (1741-88). See Milet-Mureau, N.L.A. *Voyage autour du monde de la Pérouse.* Paris, 1797. An original, well preserved copy of the plan for the expedition is in the manuscript collection in the municipal library at Rouen, Normandy.

4 Peter Nuyts, c1600- ? , made the first European survey of the coast of the Great Australian Bight in 1627 He was afterwards appointed governor of Formosa.

5 William Dampier (1651-1715). See his *New voyage around the world,* London, 1697 for his critical view of western Australia, and Marchant, L.R. "William Dampier, source materials for his Somerset years, 1651-1674" in *Early Days: Journal and Proceedings of the Royal Western Australian Historical Society,* Vol. 6, pt. 2, 1963, p. 41-47 for a critical comment on his perceptions.

6 *Encyclopédie ou dictionaire raisonné des sciences, des arts et des métiers.* Paris 1765, Vol. 8, p. 246, entry for Hollande, Nouvelle.

7 Fleuriot de Langle, P. *La tragique expedition de la Pérouse.* Paris, 1954.

8 Charles Pierre Claret de Fleurieu, (1783-1830) naval officer, hydrographer, then minister of the navy and colonies after the revolution, from 27 October 1790. Dismissed and detained during the revolutionary terror, he again rose to power after Napoleon took control of France.

9 A.N. Série marine BB4, 992; B. Mus. Hist. Nat. 46.I.

10 Hulot, E.G.T. *D'Entrecasteaux, 1737-1793.* Paris, 1894.

11 Dupetit-Thouars, F. *Mémoire addressé par la famille Dupetit-Thouars aux actionnaires et à l'équipage du Diligent axpédié à la recherche de M. de la Pérouse.* Paris, s.d.; A.N. Série marine BB4, 992.

12 A.N. Série marine BB4, 992; B. Mus. Hist. Nat. 46, I; 46, IV; 46, VI; 46X-XII.

13 A.N. Série marine BB4, 993, *notes a communiquer à messieurs les naturalists. 29.7.1791;* 5JJ, 4.

14 A.N. Série marine BB4, 992, instructions 16.9.1791.

15 See chapter 8 above.

16 Elisabeth Paul Edouard de Rossell (1765-1829), naval officer who accompanied D'Entrecasteaux and wrote the history of the voyage. He later served as director general of the office of charts and plans, where he helped shape the French restoration navy and empire.

17 Jean Baptiste Philibert Willaumez, 1761-1845, naval officer who rose to the rank of rear admiral in the restoration period, when he helped rebuild the French navy and empire.

18 Charles François Beautemps-Beaupré (1766-1854), hydrographic engineer who accompanied the D'Entrecasteaux expedition as cartographer. He was responsible for producing later charts and sailing directions.

19 Jacques Julien de la Billardière (1755-1834), naturalist who was inspired by Joseph Banks and later accompanied D'Entrecasteaux as one of the scientists sent to study the natural history of the places visited.

20 A topographical artist who accompanied the expedition as an artist.

21 A.N. Série marine 5JJ, 1-23 for journals kept on the ships.

22 La Billardière, J.J.H. de, *Relation du voyage.* Paris, anVIII.

23 Dillon, P. *Narrative and successful result of a voyage in the South Seas performed by order of the Government of British India, to ascertain the actual fate of La Pérouse's expedition.* London, 1829; A.N. Série marine BB4, 1003, Capt. P. Dillon.

24 For his journal see A.N. Série marine 5JJ, 13(11).

25 A.N. Série marine 5JJ, 5-13.

26 George Louis Leclerc Buffon (1707-1788), naturalist who achieved fame as a result of the publication of his three volume work *Natural History* (1749).

27 A.N. Série marine 5JJ1; 5JJ19 for an account of the Willaumez expedition to the islands and ashore in Esperance Bay.

28 A.N. Série marine BB4, 992, *Rapports au Comité du Salut Public.*

CHAPTER 5

1 On 20th December, 1792, the French revolutionary Convention decreed that the Committee of Public Instruction, whose members were hostile to Christianity, should prepare a report on a new revolutionary calendar. Gilbert Romme, a former professor of physics and then member of the revolutionary Convention worked with other scientists to plan a calendar. The new plan was for the year to be divided into twelve months with seasonal names. Each month was divided into three decades of ten days, with five extra days each year and six each leap year. The new year was commenced retrospectively on 22nd September, 1792, which was the beginning of the autumn equinox. Baudin used this revolutionary system of dating in the journals of the expedition. The system did not last after 1805. By then it fell into disuse.

2 A.N. Série marine BB4, 995; 5JJ, 24.

3 George Bass, 1771-c1803 recorded his observations of the flora and fauna in eastern Australia which was of interest in Britain.

4 Scott, E. *The Life of Captain Matthew Flinders.* Sydney, 1914.

5 A.N. Série marine BB4, 995 (1)-(3).

6 A.N. Série marine BB4, 995(1); A.N. Série marine 5JJ, 35, *Journal de mer de Baudin.*

7 A.N. Série marine BB4, 995 (2). *Rapport au premier consul, 3 ventôse an 10.* (22nd February, 1802).

8 A.N. Série marine 5JJ, 48 p. 27 when he was writing on Heirisson's expedition.

9 Scott, E. *Terre Napoleon: A history of French explorations and projects in Australia.* London, 1910.

10 A.N. Série marine BB4, 995, *Campagne de la Belle Angélique.*

11 Ibid, *Campagne de la Jardinière au Nouvelle Hollande par ordre de l'empereur Leopold II.*

12 Hyacinthe Yves Philippe Potetier de Bougainville, midshipman on the *Géographe.*

13 A.N. Série marine 5JJ, 35, 36.

14 Ibid.

15 Péron, F. *Voyage de découvertes aux terres australes.* Paris, 1807.

16 A.N. Série marine 5JJ 36-40 for Baudin's journals of the voyage on the *Géographe*; 5JJ 41-42 for Hamelin's journals of the voyage on the *Naturaliste.*

17 A.N. Série marine 5JJ 36-40, Journal entry 19th March, 1801. (28 Ventôse An 9).

18 A.N. Série marine 5JJ, 40.

19 Ibid.

20 Ante. Farcy Picquet of Lorient, sub-lieutenant appointed to the *Naturaliste,* but later transferred to the *Géographe.*

21 Pre Fs Bernier of La Rochelle, appointed astronomer on the *Naturaliste.*

22 Ch. Pierre Boulanger (or Boullanger) of Paris, appointed engineer-geographer on the *Géographe.*

23 Henri Desaules de Freycinet of Montélimar, appointed sub-lieutenant on the *Géographe.* His brother Louis served as sub-lieutenant on the *Naturaliste.*

24 Anselme Riedlé of Augsburg, appointed head gardener on the *Géographe.*

25 Louis Depuch (or Dupuche) of Jonzac, Gironde, appointed mineralogist on the *Géographe.*

26 Horace Bénédict de Saussure, 1740-99, regarded as one of the pioneer scholar mountaineers.

27 L. Ch Gaspard Bonnefoi or Bonnefoy of Laon, appointed midshipman on the *Géographe.* In the printed history of the voyage he is referred to as Montbazin.

28 Fois. Desiré Breton, a Creole from Guadaloupe, appointed midshipman on the *Géographe.*

29 Péron, F. *Voyage de découverts aux terres australes.* Paris, 1807.

30 F. Ante Boniface Heirisson or Herisson or Heyrisson of Madrid, appointed sub-lieutenant on the *Naturaliste.*

31 Le Bas de Saint Croix of Les Andelys, appointed Commander on the *Géographe.*

32 François Mel. Ronsard of St. Paul le Gauthier, Sarthe, appointed naval engineer on the *Géographe.*

33 René Maugé of Seine et Marne, appointed zoologist on the *Géographe.*

34 Js. Saint Cricq of Pau, Béarn, appointed sub-lieutenant on the *Naturaliste.*

35 Charles Alexandre Lesueur of Le Havre, first appointed as petty officer on the *Géographe,* and then after the desertions at Mauritius, appointed sketcher.

36 Péron, F. *Voyage de découvertes aux terres australes.* Paris, 1807. For a general history of native peoples written at the time see Babié, F. *Voyage chez les peuples sauvages, ou l'homme de la nature. Histoire morale des peuples sauvages dex deux continents, et des naturels des îles de la Mer du Sud.* Paris, 1801.

37 See esp. A.N. Série marine 5JJ, 56 Leschenault, *journal.*

38 There are two sketches of natives and canoes in Geographe Bay in the Lesueur collection at the Museum of Natural History at Le Havre. These have been incorrectly titled by someone else than Lesueur. There were no native canoes in Geographe Bay. See Leschenault journal, A.N. Série marine 5JJ, 56 p. 24 for a comment on the absence of canoes. The French did not expect to see these. They failed to find them near Esperance on the D'Entrecasteaux expedition. The Lesueur sketches at Le Havre appear to be of Tasmanian canoes.

39 Pre Bernard Milius (or Millius) of Bordeaux, appointed lieutenant commander on the *Naturaliste.*

40 Ths Timothée Vasse of Dieppe, appointed to the *Naturaliste.*

CHAPTER 6

1 A most unenthusiastic description of western Australia was given in French encyclopedias even after the return of the Baudin mission. See *Encyclopédie Méthodique: Géographie physique par M. Desmarest.* Paris 1811, Vol. 2, p. 742, entry for Nouvelle Hollande. In this the dangers of the coast were indicated, and the lack of fruits and nourishment pointed out, indicating it was no place on which to be cast away.

2 A.N. Série marine 5JJ, 36-40, Baudin journal, entry 29.5.1801; 11.6.1801; 12.6.1801; 13.6.1801; 9.3.1803. (9 prairial an 9; 22 prairial an 9; 23 prairial an 9; 24 prairial an 8:18 ventôse an 11).

3 Baudin did not keep to his plan. He quickly sailed past Cape Leeuwin in March 1803. See A.N. Série marine 5JJ, 36-40, Baudin journal, entry 9.3.1803. (18 ventôse an 11).

4 For the manuscript records of the voyage on the *Géographe* see A.N. Série marine 5JJ,25-26; 28-30; 33-40; 43-47; 53-54; 56.

5 Baudin noted that he had never seen a coast which presented such a disagreeable sight, providing no shelter even for a long boat. See A.N. Série marine 5JJ, 36-40, Baudin journal, entry 22.6.1801; (3 messidor an 9). Péron, F. *Voyage de découvertes aux terres australes sur le Géographe.* Paris, 1807.

6 Baudin used two chronometers, No. 27 and No. 35, to determine longitude. His assessment of longitude was made by averaging the readings of the two chronometers. Both of these had errors to the west. They were showing too small a reading. A reading on the north end of the island showed the longitude to be 109° 34′ 44″ east which made the error 1° 16′ 6″ west.

7 A.N. Série marine 5JJ, 36-40, Baudin journal, entry 8.7.1801. (19 messidor an 9).

8 Ibid, entry 19.7.1801. (30 messidor an 9).

9 The *Naturaliste* was a former store-ship.

10 For the manuscript records of the voyage of the *Naturaliste* see A.N. Série marine 5JJ, 27; 31; 41-42; 48-49; 53; 56.

11 A.N. Série marine 5JJ, 41-42, Hamelin journal.

12 The bearings taken from the ship were: to the north most point of Rottnest W 5° N; to the south most point of Rottnest W 27° S; to an island S 9° E; to another island S 20° E; to the south most part of the mainland S 33° E; to the north most part of the mainland N 12° E. See A.N. Série marine 5JJ, 41-42, Hamelin journal entry 14.6.1801. Distances and descriptions are not given. It is therefore not easy to identify the points except for Rottnest.

13 A.N. Série marine 5JJ, 41-42. Hamelin journal entry 15.6.1801. (25-26 prairial an 9.)

14 A.N. Série marine 5JJ, 41-42, Hamelin journal; 5JJ, 48, St. Cricq journal; 5JJ, 49, Freycinet journal; 5JJ, 53, Milius journal; 5JJ, 56, Heirisson journal.

15 A.N. Série marine 5JJ, 49 Freycinet journal.

16 Ibid.

17 A.N. Série marine 5JJ, 41-42 Hamelin journal entry 17.6.1801 (27-28 prairial an 9); 26.6.1801 (6-7 messidor an 9) for Bailly the mineralogist's and Heirisson's reports; 5JJ,56, Heirisson journal.

18 A.N. Série marine 5JJ, 41-42 Hamelin journal entry 24.6.1801 (4-5 messidor an 9) includes Milius and Le Villain report; 5JJ, 53 Milius journal.

19 A.N. Série marine 5JJ, 41-42 Hamelin journal entry 28.6.1801, (8-9 messidor an 9),

20 A.N. Série marine 5JJ, 56 Heirisson journal.

21 A.N. Série marine 5JJ, 36-40, Baudin journal entry 26.8.1801. (8 fructidor an 9).

22 A.N. Série marine 5JJ, 41-42 Hamelin journal, entry 26 July (6-7 thermidor an 9). Hamelin also nailed a Dutch flag to the post in honour of the first discoverers.

23 Flinders and Baudin met on 8 April 1802 at Encounter Bay, east of Kangaroo Island in South Australia.

24 A.N. Série marine 5JJ, 50, Freycinet journal, Casuarina, 23.9.1801 (1 vendémiaire an XI).

25 A.N. Série marine 5JJ, 36-40, Baudin journal; 18th February 1803 (29 pluviôse an II; 21 February 1803, 2 ventôse an II).

26 A.N. Série marine 5JJ, 36-40, Baudin journal, 20 February 1803 (1 ventôse an II); 50, Freycinet journal 18th February-1st March 1803 (29 pluviôse — 10 ventôse an XI).

27 A.N. Série marine 5JJ, 50, Freycinet journal 15.2.1803 (26 pluviôse an XI).

28 A.N. Série marine 5JJ, 36-40, Baudin journal 9th March 1803 (18 ventôse an XI).

29 A.N. Série marine 5JJ, 36-40, Baudin journal entry 11th March 1803 (20 ventôse an XI); 5JJ, 56, Bonnefoi journal.

30 A.N. Série marine BB4, 996 esp. report dated 29.5.1803 (9 prairial an XI) written at Timor about western Australia.

CHAPTER 7

1 Charles Matthieu Isidore Ducaen (1769-1832), governor of the French establishments in India, 1803-1811. See Prentouth, *L'Ile de France sous Decaen, 1803-1810.* Paris, 1901.

2 This included live specimens which went on show in France after the expedition returned.

3 Nemours, A. *Histoire de la captivité de la mort de Toussant Louverture.* Nancy, 1929.

4 Charles Alexandre Léon Durand de Linois (1761-1848) went to the Indian Ocean with Decaen in 1803 to take the initiative against Britain in the Indian Ocean. Britain responded by despatching a force to take Mauritius and its dependencies in 1810.

5 Robert Surcouf, 1773-1827. A rich ship owner and trader who went to the Indian Ocean in 1807, on board the *Revenant*, a ship of 18 cannon with a hundred and ninety men, where he maintained a French presence as a privateer based on Mauritius. See Cunat, C. *Histoire de Robert Surcouf capitaine de corsaire.* Paris, 1842.

6 Sugar was first produced from beet in Germany in the first half of the eighteenth century. Production of this commodity was stepped up in Napoleonic Europe after the loss of other sources. See Deerr, N. *The History of Sugar.* London, 1949-50.

7 A.N. Série marine BB4, 998.

8 B.N. N.A.F. 9443.

9 Kergariou, A. de. *La mission de la Cybèle en Extrême-Orient, 1817-1818.* Paris, 1914.

10 Armand Emmanuel du Plessis, duc de Richelieu, 1766-1822, Prime Minister of France from 1815 to 1818 and from 1820-21. He exercised control over foreign affairs in his early

ministry and was responsible for achieving a respected position for France in Europe after Napoleon.

11 Pierre Victor, baron Malouet, 1740-1814, served in the administration of the colonies during the *ancien régime*. After going abroad in the early part of the revolution, he returned to serve Napoleon and later the restored Bourbons. He served as Minister of the Navy from May to September, 1814.

12 Pierre Barthélemy, baron Portal (1765-1845). See Gervain, baronne de, *Le Baron Portal: un ministre de la marine et son ministère sous la restauration*. Paris, 1898.

13 Péron, F. et Freycinet, L. de, *Voyage de découvertes aux terres australes*. Paris, 1807-1816.

14 A.N. Série marine BB4, 998, note 8.4. 1815.

15 A.N. Série marine BB4, 998.

16 A.N. Série marine BB4, 998; 5JJ, 68.

17 A.N. Série marine BB4, 998.

18 A.N. Série marine 5JJ, 62A.

19 A.N. Série marine BB4, 999 instructrions sanitaire.

20 For the journals of the journey see A.N. Série marine 5JJ, 58-79; B. Mus. His. Nat. 474.

21 A.N. Série marine 5JJ, 68, notes Nouvelle Hollande; 72 Fabré journal.

22 A.N. Série marine 5JJ, 68, 70.

23 Arago, J. *Narrative of a Voyage around the World*. London, 1823.

24 *Ibid*, p. 181.

25 *Ibid*, p. 182.

26 A.N. Série marine BB4, 998, letter from Timor 17.10.1818 indicating that western Australia had no resources to refit ships and refresh crews.

CHAPTER 8

1 From the mid-Napoleonic period, after the publication of Jean Baptiste Say's *Traité d'économie politique* (1803), the French school of classical economists who had laid the foundations of modern economic science, was overshadowed by a group of socialist writers who laid the foundations of scientific socialism. These socialist theorists such as Saint Simon and Fourier had no interest in colonies. They were pre-occupied with their domestic drive to discredit the revolutionary belief in the "rights of man", which they wanted replaced with the contrary belief that man had functions to fulfil in society; and with an analysis of wealth they believed was the product of the labour of others than the wealthy. In these circumstances the discussion of colonial systems and theories of colonization was left primarily to British writers such as Wakefield. Britain thus became the theoretical leader in this field, and the model to follow while the new emergent French school turned attention to co-operative production and nationalization.

2 A.N. Archives des Colonies Série H1, dossier G (VII), Conseil des ministres, décision, 25.1.1819.

3 *Moniteur*, No. 365, 31.12.1790; p. 1; No. 60, 1.3.1791, p. 1. Governor Philip was reported as saying that the situation was "truly alarming".

4 See e.g. "Loi qui ordonne que la peine de déportation sera désormais pour la vie entière, du 5 Primaire, an II" in *Lois et actes du gouvernement*. Paris, 1807, Vol. 8, p. 79-80.

5 *Moniteur*, No. 157, 6.6.1791, p. 5.

6 *Moniteur*, passim.

7 *Moniteur*, No. 76, 17.3.1791, p. 2. [book news].

8 *Code pénal 1791*, art. 1, which listed deportation as one of the eight punishments in France; art. 29, 30, which stated the place of deportation was to be decreed.

9 *Moniteur*, No. 115, 24.4.1792; No. 162, 11.6.1793; No. 206, 25.7.1793; No. 263, 20.9.1793; No. 143, 3.11.1793; No. 62, 22.11.1793; No. 122, 21.1.1795; No. 20, 9.4.1795; No. 73, 3.12.1797; No. 210, 19.4.1798. Guiana, Madagascar, Sierra Leone and St. Dominique were all suggested.

10 *Moniteur*, No. 206, 25.7.1793 for the debate between Danton and Robespierre on the matter.

11 A.N. Archives des Colonies, Série H1, dossier 9, Citoyen Rollet, *Danger et inconvenience des bagnes, projet de replacement*, 14.10.1792.

12 *Moniteur,* No. 74, 4.12.1797.

13 *Moniteur,* No. 43, 3.11.1793.

14 Gervain, baronne de, *Un ministre de la marine et son ministère sous la restauration, le baron Portal.* Paris, 1898; Portal, baron, *Mémoires du Baron Portal.* Paris, 1846.

15 A.N. Archives des Colonies, Série H1, dossier 9.

16 A.N. Archives des Colonies, Série H1, dossier 1.

17 A.N. Archives des Colonies, Série H1, dossier 9, Le conseil des ministre. *Décision,* 25.1.1819; *Rapports du comité de déportation.*

18 A.N. Archives des Colonies, Série H1, dossier 9;H22, Procès verbaux.

19 See chapter 9 below.

20 For his life see Blosseville, E. de, *Jules de Blosseville.* Evereux, 1854; Larronde, N., *A la Mémoire de Jules de Blosseville.* Paris, 1935; Mancy, Jarry de, *Jules de Blosseville,* np, 1835; Passy, L., *Le Marquis de Blosseville, souvenirs.* Evereux, 1898.

21 Blosseville, E.P. marquis de, *Histoire de la colonisation et des établissements de l'Angleterre en Australie.* Evereux, 1859; Blosseville, E.P. marquis de, *Histoire des colonies pénales de l'Angleterre dans l'Australie.* Paris, 1831; La Société Libre d'Agriculture, Sciences, Arts et Belles-lettres de l'Eure. *Recueil des travaux de E. de Blosseville,* p. 246.

22 A.N. Série marine 5JJ, 82. Blosseville, journal.

23 B.N., N.A.F. 6785, p. 30. Blosseville, J. *Projet d'une colonie pénale sur la côte S.O. de la Nouvelle Hollande.*

24 B.N., N.A.F. 6785, p. 48. Blosseville, J. *Côtes S.O. de la Nouvelle Hollande, Plan d'exploration.*

CHAPTER 9

1 Dumore, J. *French explorers in the Pacific.* Oxford, 1969, vol. 2, p. 109-177.

2 A.N. Série marine BB4, 1000, plaquette 4, 19.6.1822.

3 A.N. Série marine BB4, 1000; 1023 (notes sur Chili); 5JJ 82.

4 Gaffarel, P. *La politique coloniale en France de 1789* à *1830.* Paris, 1908; Hardy, G. *La mise en valeur du Sénégal de 1817* à *1854.* Paris, 1921.

5 See e.g. Dumore, J., *French explorers in the Pacific.* Oxford, 1969, vol. 2. Faivre, J.P., *L'expansion Française dans la Pacifique de 1800* à *1842.* Paris, 1953.

6 A.N. Série marine BB4, 1000, plaquette 4, *Mémoire pour Duperrey.*

7 *Ibid.*

8 A.N. Série marine BB4, 1000, plaquette 4, itinerary 20.2.1822.

9 *Ibid.*

10 *Ibid.*

11 B. Serv. Hist. Duperrey, dossier personnel.

12 A.N. Série marine BB4, 1000, Duperrey, Teneriffe, 30.8.1822.

13 For journals of the voyage see A.N. Série marine BB4, 1000; 5JJ 80-87; B. Mus. Nat. Hist. 1793.

14 B. Mus. Nat. Hist. 1793, Lesson journal 14.11.1823.

15 A.N. Série marine BB4, 1000, Duperrey, 15.2.1824.

16 A.N. Série marine BB4, 1001, mémoire 22.8.1827.

17 For journals of the voyage see A.N. Série marine BB4, 1001; 5JJ 88-98.

18 A.N. Série marine BB4, 1001, Rosily, itineraire, 1.2.1824.

19 A.N. Série marine 5JJ 88, report on voyage to Sydney.

20 *Ibid.*

21 A.N. Série marine BB4, 1001, mémoire 22.8.1827.

22 B.N., N.A.F. 6785. Blosseville, J. *Projet d'une colonie pénale à la Nouvelle Zélande.* 31.12.1828: 1.1.1829.

23 A.N. Série marine BB4, 1010, 1012; B.N., N.A.F. 9446, 9502. See also Marchant, L.R. "The French discovery and settlement of New Zealand 1769-1846: a bibliographical essay on naval records in Paris" in *Historical Studies, Australia and New Zealand*, vol.10, no.40, May 1963, p.511-518.

24 The British colonial authorities in Australia were not warned about a specific French threat until March 1826. See U.K.C.O. Bathurst to Darling, 1.3.1826, private.

25 Although the British Foreign Office in 1819 was seeking information about the state of crime and punishment in Europe (see F.O. 27 vol 200, circular to European ambassadors, 29.3.1819), the British authorities had no idea that the French were planning for the establishment of a convict colony. The Foreign Office despatches were concerned mostly about France in South America and the Mediterranean.

26 A.N. Série marine BB4, 1002, 2.3.1826.

27 A.N. Série marine BB4, 1002, 8.4.1826.

28 For journals of the voyage see A.N. Série marine BB4, 1002; 5JJ, 99-102; B. Mus. Hist. Nat. 104.

29 Marchant, L.R. "The French Discovery and Settlement of New Zealand 1769-1846" in *Historical Studies Australia and New Zealand* vol. 10 no. 40, May 1963, p.511-518.

BIBLIOGRAPHY

Manuscript Records

General

Bibliothèque du Service Historique de la Marine.

Dossier personnel, Duperrey, D'Urville etc.

Gonneville, Bino Paulmier de

Bibliothèque Nationale. Département des Manuscrits.

Nouvelles acquisitions Françaises —

9439: Mers australes . . . relation du voyage de Paulmier de Gonneville etc.

Bouvet de Lozier, Jean Baptiste Charles

Archives Nationales

Série Marine B4 —

45 Campagnes 1738-1739 mers australes (Lozier Bouvet)

77 Campagnes 1757 mers australes (Lozier Bouvet)

Bibliothèque du Service Historique de la Marine.

Mss vol 105^3, 1-18.

Bibliothèque Nationale. Département des Manuscrits.

Nouvelles acquisitions Françaises —

9341.340 Bouvet de Lozier mémoire.

9407.1 Copie de la lettre de Lozier Bouvet à M. Duvalaër, directeur de la Compagne des Indes 8.2.1755. etc.

9438.12 Mémoire sur les Terres Australes.

9439.27 Extrait du voyage.

Dufresne, Marc Joseph Marion

Archives Nationales

Série Marine B4 —

317 Série supplémentaire. 1768-1795. Voyages de découvertes et d'observations scientifiques: Marion Dufresne; Kerguelen.

Série Marine 4JJ —

142 (18) Journal du vaisseau le Mascarin et de la flûte le Marquis de Castries Ct. Marion Dufresne de l'île de France, Nouvelle Zélande . . . par Jean Roux, officier du Mascarin, 1776. 1771-1773.

Série Marine 5JJ —

142 (19) Journal du voyage de Marion et du Clesmeur aux Terres Australes par Jean Roux. 1771-1773.

Bibliothèque du Service Hydrographique de la Marine

A5708 Pottier de l'Horme. Voyage de M. de Surville . . . à la Mer du Sud. (et) Duclesmeur. Relation d'un voyage dans les Mers Australes et Pacifique commencé en 1771 sous le commandement de M. Marion du Fresne . . . et achevé sous celui de M. Duclesmeur.

Bibliothèque Nationale, Département des Manuscrits.

Nouvelles acquisitions Françaises —

9437 Fol. 111-180: Relation d'un voyage dans les mers australe et pacifique commencé en 1771 par M. Marion Dufresne, Capitaine de Brulot, et achevé en 1773 par Duclesmeur, Garde de la Marine. Fol. 181-290: Journal du voyage fait sur le vaisseau du Roy le Mascarin commandé par M. Marion . . . par le sieur Roux.

Saint Allouarn, François Alesno Comte De

Archives Nationales.

Série Marine B4 —

317 Extrait du journal du vaisseau Gros Ventre; Extrait de mon journal depuis le 14 fevrier que M. de Kerguélen me détachat dans la chaloupe La Mouche pour sonder avant du Gros Ventre, par Rosily.

Série Marine 4JJ —

143 (26) Extrait du journal de la flûte Gros Ventre, St. Allouarn, aux Terres Australes, 1772.

Bibliothèque Nationale, Département des Manuscrits.

Nouvelles acquisitions Françaises —

9438 Rosily, extrait de mon journal; Extrait du journal du vaisseau Gros Ventre.

D'Entrecasteaux, Joseph Antoine Bruni

Archives Nationales.

Série Marine B4 —

315 Série supplémentaire. 1776-1787. Voyages de découvertes et d'observations scientifiques.

Série Marine BB4 —

992-994 Expédition de d'Entrecasteaux:

992 Expédition de d'Entrecasteaux à la recherche de La Pérouse. Correspondance des Ministres et instructions divers pour l'expédition. 1791. Notes du ministère 1792-1793 . . . États majors . . . Documents sur la proposition de l'expédition.

Inventaire . . . Affaire Hesmivy d'Auribeau (Rapports au Comité du Salut Public).

993 Lettres et mémoires de Hesmivy d'Auribeau Ct. la Recher-

che, puis l'Espérance, après la mort de d'Entrecasteaux. Lettres de Rossel . . . Pièces annexes; analyses de la carte de la Nouvelle Hollande . . . Voyage de la flûte le Gros Ventre . . .

994 Comptabilité de l'expédition d'Entrecasteaux. États d'appointments . . .

Série Marine 5JJ —

1-23 Voyage de d'Entrecasteaux sur la Recherche et l'Espérance à la recherche de la Pérouse.

1 Observations de longitude depuis la départ de Brest; observations et calculs de Bertrand; observations de Willaumez, Gicquel, Rault, H. Lambert, Bonvouloir, sur la route; observations météorologiques (1793); code de signaux; cahiers de variations observées (1791-1793); positions géographiques de divers points de l'Océan Pacifique; vocabulaires polynesiens†, journaux de route; relèvements et vues (Nouvelle-Zélande, Bouero, îles des Trois-Rois, île de la Vandola, etc). (1791-1793).

2. Relèvements et vues, suite (Timor, Nouvelle-Calédonie, Nouvelle-Hollande, îles Arsacides); liste de plantes et graines du jardinier (1791-1793).

3 Relèvements, vues et observations (Célèbes, Nouvelle-Hollande, Bouton); marche des montres; observations de Tongatabou à Balade; 2e cahier du récit de Keyt à Onin (s.d.); note sur l'établissement des Hollandais dans les Moluques (1792); notes informes diverses (1791-1793).

4 Observations, calculs, calques divers; copies de lettres de d'Entrecasteaux, communiquées par Van Swinden, député batave, en l'an VII; inventaire des papiers de d'Entrecasteaux; correspondances diverses au sujet de l'impression de son voyage; préparation de la publication; esquisses au crayon de M. Piron; notes de M. de la Billardière, botaniste; papiers provenant de divers membres de l'expédition (1792, an VII).

5:1 Journal des relèvements faits à bord de l'Espérance (1792-1793).

5:2 Journaux de route de l'Espérance (2 séries, incomplètes) (1791-1793); journaux de de la Seinie, sur l'Espérance (1793-1794); de Lamotte, ibid; extrait du journal de Williaumez (1793); cahiers de visites medicales (octobre 1791-juin 1794) (1791-1794).

6:1-2 Cahiers et tables de route de la Recherche (1791-1794); journal anonyme (1793-1794); journaux de relèvements (1793-1794); observations faites de l'île Bouro à l'île Bouton; carnet de rade de Sourabaya (1794); registres

† Includes a vocabulary of Australian Aboriginal dialects.

astronomiques de Brest au Cap (1791-1792); observations de montres (1792-1793); cahier d'angles horaires et de hauteurs méridiennes (1793) (1791-1794).

7:1 Journal de l'Espérance, commandant Huon de Kermadec, par M. de Lusunçais, lieutenant de vaisseau (1791-1792).

7:2-8 Journal de Huon de Kermadec, commandant de l'Espérance:

7: 2 (1791-1792).

7: 3 Idem (1792).

7: 4 Copie de 7:2 (1791-1792).

7: 5 journal original? (1791).

7: 6 Idem (1791-1792).

7: 7 Table de loch (1791).

7: 8 (1792).

8 Journal de mer de Fitz, sur l'Espérance, du Cap aux îles d'Océanie et aux Indes hollandaises (1792-1793).

9:1-2 Journal de mer de Rossel, sur la Recherche (1793).

10 Journal de mer de Longueruë, élève de la marine, sur la Recherche, de Brest à Sourabaya (1791-1793).

11 Journal de mer de Merité, à bord de la Recherche (1791-1792).

12 Journal de mer de Crestin, sur la Recherche (avec un cahier pour août) (1792).

13:1 Journal de mer de Delamotte du Portail, sur l'Espérance (1791-1794).

13:2 Journal de mer de Laignel, enseigne surnuméraire sur l'Espérance (1791-1792).

13:3-8 Journal de mer de d'Auribeau, sur l'Espérance (1791-1794)

13:9 Journal météorologique de d'Auribeau, sur l'Espérance (1791-1794).

13:10 Journal de Le Danseur, sur l'Espérance (1791-1793).

13:11 Journal de Le Grand, sur l'Espérance (1791-1792).

13:12 Journal de Jurien, sur l'Espérance (1791-1793).

13:13-16 Journal de Trobriand, sur l'Espérance (1791-1793).

13:17 Journal de Rault, sur l'Espérance (1791-1792).

13:18 Journal de Boynes, sur l'Espérance (1792-1793).

13:19 Journal des malades, Espérance (1791-1794).

14 Observations d'angles horaires à Balade (Nouvelle Calédonie); traversée de Balade à l'île Waigiou; observations à Boni, île Waigiou, sur divers points de la Terre de Van Diémen, traversée du port du Sud de l'île de Van Diémen à Tongatabou; traversée de Tongatabou à Balade; du port du nord à Amboine; observations à Amboine; traversée d'Amboine au port de l'Espérance, Terre-Nuitz;

observations au port de l'Espérance; au port du Sud; observations de Sainte-Croix-de-Ténériffe au Cap; au Cap; observations de l'enseigne Gigquel sur la Recherche (1791- an VI); observations médicales sur une blessure par flèche à Sainte-Croix-de Ténériffe (1793); analyses de relèvements et descriptions diverses; découvertes de groupes d'îles par les capitaines Marshall et Gilbert (1788); journaux émanant de la Recherche (1er pilote Rault), du Naturaliste (Duval d'Ailly, enseigne); observations sous forme inutilisable (1788-1794).

14 bis. Cahiers et papiers divers; calculs, observations de toute espèce, provenant principalement de Bonvouloir (1791-1793).

15 Registre d'observations par Laignel (1791-1793).

15 bis. Registre des observations faites à Sourabaya par Bonvouloir, sur la Recherche (1794).

16 Registre des observations astronomiques du Cap à l'île Waigiou, faites par Pierson, sur l'Espérance (1792).

16 bis. Registre d'observations diverses (distances et hauteurs) (1792)

17 Registre de la marche des montres par Pierson, sur l'Espérance (1792).

18 Observations astronomiques à terre (Terre de Van Diémen et Sourabaya) (1792-1794).

19-21 Observations de Beautemps-Beaupré (Relèvements et croquis, à bord de la Recherche (avec un journal de route pour 1791-1795) (1792-1795).

22 Cahiers et agenda d'observations à bord de la Recherche; observations astronomiques de Rossel (1792-1793); 34 cahiers d'observations astronomiques de 1792; notes de Pierson: observations de longitudes; cahiers de comparaisons (1792-1793).

23 Observations astronomiques de Rossel; de Faure; calculs de Fitz; observations de montres, de baromètre, par le 1er pilote Rault (1792-1794).

Série Marine 6JJ —

203 Voyage de D'Entrecasteaux:

2 Voyage de D'Entrecasteaux sur l'Espérance et la Recherche. Routes établis par Beautemps-Beaupré, ingenieur hydrographe; vues par Piron; cartes de la Terre de Nuyts; de la terre Van Diémen; . . . Nouvelle Zélande . . . etc.

3 Voyage de D'Entrecasteaux. Routes et Cartes établis par Beautemps-Beaupré . . . vues de côtes . . . cartes Nouvelle Hollande . . . Nouvelle Zélande (etc.) 1792-1793.

Bibliothèque Centrale du Muséum National d'Histoire Naturelle.

46:I-XIII	Voyage de d'Entrecasteaux à la Recherche de La Pérouse, 1791-1797.
46:I	Observations générales de la Société d'Histoire Naturelle sur le voyage à entreprendre; lettres concernant les personnes à choisir pour le voyage; expose des préparatifs nécessaires pour les naturalistes voyageurs autour du monde.
46:II	Système artificiel pour la classifications des larves; des insectes par Ant. Riche.
46:III	Observations de M. Le Blond.
46:IV	Observations sur le choix des minéralogistes et leur recherches par M. Besson.
46:V	Instructions sur les parages de Kamtchatka par M. Patrin, 21 janvier 1790.
46:VI	Instructions partielles pour les voyageurs naturalistes; instructions sur la minéralogie . . . par M. Richard.
46:VII	Tableau systématique divers par Brugère; espèces de la classe des chenilles du système des larves.
46:VIII	Journal d'observations sur les vers, 26 fevrier 1792; sur les oiseaux en anatomie et en histoire naturelle.
46:IX	Notice de mes collections d'objects d'historie naturelle qui m'ont été enlevées par Rossel . . . La Billardière.
46:X	Exposé de préparatifs nécessaires pour le naturalistes voyageurs autour du monde par Ant. Riche: notes et souvenirs relatifs à la météorologie et à la physique.
46:XI	Notes à l'usage du jardinier destiné à faire le voyage projeté pour l'avancement des sciences et la découverte de M. de La Pérouse; mémoire pour diriger le jardinier dans les travaux de son voyage autour du monde.
46:XII	Correspondance relative au jardiner Lahaye; lettre de recommendation de Thouin, jardinier en chef du Jardin des Plantes; lettres de Lahave à Thouin.
46:XIII	Note des graines recoltées dans le voyage autour du monde par le citoyen Lahaye de 1791 à 1797.
265	Observations faites par Riche pendant le voyage de d'Entrecasteaux.
1041	Journal abrégé du voyage d'Entrecasteaux par le citoyen Beaupré, suivi d'un journal du même voyage par un matelot, 1791-1794.

Bibliothèque Nationale, Département des Manuscrits.

Nouvelles acquisitions Françaises —

9347: Fol.65-106 D'Entrecasteaux, 1787.

Baudin, Nicolas Thomas

Archives Nationales

Série Marine BB4 —

995-997	Expédition de Baudin:
995:2	Expédition du Géographe en Nouvelle Hollande. Rapports et arrêts des consuls et du Ministre de la Marine; instructions du Muséum; décrets impériaux. An VI - 1810.
995:3	Lettres diverse écrites du ministère concernant l'expédition, au Cre. Baudin, au Cre. Hamelin, aux savants de l'expédition; au v.a. Rosily, à l'ingénieur Ronsard; à la commission de l'Institut chargé de préparer l'expédition, aux ministres; au Consul Lebrun; au Secrétaire d'Etat Marer; aux administrateurs de l'Ile de France; aux administrateurs et préfets maritimes; à des officiers et personnages divers. An VIII- an XII.
995:4	Rapports et lettres du Cr de vau Baudin, mort en Brumaire an XII, remplacé par le Cre de Fte Milius. An VII - an XII.
995:5	Rapports, lettres et journal du Cre. de vau Hamelin sur le Naturaliste. An IX - 1803.
996	Expédition du Cre. Baudin aux Terres Australes.
996:1	Correspondances diverses concernant l'expédition et la publication de les resultats. An VIII - 1811.
996:2	Correspondence du Ministre de la Marine, le Cre. Baudin et les administrateurs de l'Ile de France. An IX - XII.
996:3	Correspondance et inventaires concernant le Géographe et le Naturaliste 1804.
997	Expédition du Cre. Baudin aux Terres Australes.
997:1	Personnel et frais de l'expédition; lettres de demandes, de congé, de démission, de réclamation, états divers. An VIII -1815
997:2	Passeports et lettres de recommendation fournis au Cre. Baudin par les puissances étrangères (Angleterre, Hollande, Danemark, Espagne, Suède, Portugal, Prusse).

Série Marine 5JJ —

24-57	Voyage de Baudin aux Terres Australes sur le Géographe et le Naturaliste:
24	Inventaire des journaux et documents provenant de l'expédition Baudin; inventaire des cartes et objets remis par Boullanger, ingénieur-géographe sur le Géographe (IX-XII); relèves des points du Géographe; dépenses vivres et comptabilité de l'expédition; brouillon d'un récit de la traversée du Havre à Téneriffe (VIII); correspondance de Decrès, ministre de la marine, avec Fleurieu, sur l'origine et les resultats de l'expédition (VIII-XII); rapport au ministre du c. de f. Milius, commandant du Géographe (XII); lettres de Baudin et Milius à Decrès et à Fleurieu

(IX-XII); lettres addressées aux gouverneurs des différentes colonies espagnoles d'Amérique pour accréditer l'expédition; notes relatives à la Nouvelle-Hollande; aux Indes hollandaises; projets d'itinéraires, dont un de Fleurieu; état des cartes et plans transmis par Lescallias, ordonnateur de la marine à Corfou, à Baudin (VII); état des fournitures faites par Fonchevrueil à l'expédition; factures; lettres d'Hamelin, commandant le Naturaliste, Milius, Faure, géographe du Naturaliste, Freycinet, Saint-Cricq (X); brouillons de Bernier, astronome du Géographe (ans VII-XII).

25 Journaux de bord du Géographe (vend. IX. therm. XI).

26A Journaux nautiques du Géographe (ans IX-XI).

26B Tables de loch et journaux de l'ingénieur-géographe du Géographe (ans IX-XI).

27A-B Journaux nautiques et de loch du Naturaliste (vend. IX-prair. XI).

28-30 Journal de mer et tables de loch de Ronsard, Officier du génie maritime et lieutenant de vaisseaux sur le Géographe (ans IX-XI).

31 Relèvements des côtes par Louis de Freycinet, sur le Géographe (IX) et la Naturaliste (XI) (ans IX-XI).

32 Journal géographique de Louis de Freycinet, sur la goélette la Casuarina (an IX).

33 Observations astronomiques de Henri de Freycinet, sur le Géographe (ans X-XI).

34 Journal de navigation de Henri de Freycinet, sur le Géographe (an XI).

35 Journal historique et de navigation du capitaine de vaisseau N. Baudin, commandant l'expédition (ans VIII-IX).

36 Journal de mer de Baudin (an IX).

37 Journal de mer de Baudin (an IX).

38 Journal de mer de Baudin (ans IX-X).

39 Journal de mer de Baudin (ans X-XI).

40 Journal de mer de Baudin (an XI).

40: 2-4 Rapport d'ensemble, copie (incomplet, à partir du chapitre VII, nos 2, 3 et 4; 1 feuille écrite seulement pour 4) (ans IX-X).

41 Journal de mer d'Hamelin, commandant le Naturaliste (an IX).

42 Journal de mer d'Hamelin, commandant le Naturaliste (ans IX-XI).

43 Journal de la marche des montres, par Boullanger, ingénieur-géographe, au retour (an XII).

44 Journal des observations et sondes par les ingénieurs-géographes Boullanger et Faure (ans IX-XI).

45 Comparaison des montres par Bernier, astronome sur le Géographe (ans IX-X).

46 Mise au net du Journal de Bernier (ans IX-X).

47 Journal d'observations astronomiques de Bernier (ans IX-XI).

48 Journal historique et de navigation de l'enseigne de vaisseau Saint-Cricq, sur le Naturaliste (ans IX-XI).

49 Journaux nautiques de Louis de Freycinet, sur le Naturaliste (ans IX-XI).

50 Journaux nautiques de Louis de Freycinet, sur la Casuarina (an XI).

51 Relèvements au lavis des côtes des terres australes (1 a 224) (s.d.).

52 Notes et brouillons de Bernier, astronome, et Boullanger, ingénieur-géographe (s.d.).

53 Observations de Gicquel; routes établies par J.J. Ransonnet; journal géographique de Henri de Freycinet; rapports de Boullanger, ingénieur, Maurouard, aspirant, Ronsard, Picquet, Louis de Freycinet, Milius; catalogues de livres de Baudin; brouillon de journal; lettres anglaises addressées à Baudin; relèvements par Boullanger et Faure, ingénieurs-géographes; feuillets détaches d'un casernet de timonerie. Journal du Naturaliste de la Baie des Chiens Marins à Timor (ans IX-XI).

54 Tableaux de loch, relèvements et observations astronomiques, par Henri de Freycinet (ans X-XI).

55 Journal nautique de Gicquel, sur le Géographe (ans IX-XI).

56 Observations astronomiques de Louis de Freycinet (sur la Casuarina, an IX); journal de Bernier, astronome, sur le Naturaliste; relèvements par Henri de Freycinet, Ronsard, Bonnefoy, Ransonnet; journal du Hérisson; extraits de la rélation de Leschenault, botaniste; journal nautique de Brevedent, aspirant de 2e classe; journal de sondes sur la Casuarina; table de loch (therm. XI); journal de Maurouard, aspirant de 1re classe; journal de Faure (ans IX-XI).

Bibliothèque Nationale, Département des Manuscrits.

Nouvelles acquisitions Françaises —

9440-9441 Baudin Journal:

9440 Journal de Bord du Commandant Baudin, 1779-1802.

9441 Journal de Bord du Commandant Baudin, 1802-1804 (ans IX-XI).

Muséum d'Histoire Naturelle, Le Havre.

Nouvelle Hollande — Histoire Naturelle. Topographic générale de le Baie du Géographe sur la côte ouest — du 10 au 27 Prairial an IXe — (Péron).

Histoire Naturelle — Topographie géné de l'île Maria sur la côte orientale de la Terre de Diémen — Ventose an X^{e} — (Péron).

Île Maria — Observations anthropologiques — (Péron).

Île Maria — Observations de physique et d'histoire naturelle. Ventose an X^{e} — (Péron).

Île Maria — Suite des observations de physique et d'histoire naturelle. Ventose an X^{e} — (Péron).

Zoologie animaux observés pendant la traversée de l'Île de Timor à la Terre de Diémen du 22 Brumaire au 23 Nivose an X^{e} — (Péron).

Péron et Lesueur. Voyage aux Terres Australes — Reptiles. Sauriens — 10 folders. Cheloniens — 2 folders. Batraciens — 3 folders.

Notes tenues par Le C^{en} Stanislav Levillain zoologiste dans l'exped^{n} de découvertes — sur la Corvette le Naturaliste, Depuis le 25 florial an 9e époque du commencement de la Compagne jusqu'au 22 Brumaire an 10e.

Freycinet, Louis Claude Desaules de

Archives Nationales

Série Marine BB4 —

998-999	L'expédition de Freycinet sur l'Uranie:
998	Expédition du Ct de Vau de Freycinet sur l'Uranie autour du monde.
998:1	Historique sommaire de l'expédition du 17.9.1817 au 19.5.1820.
998:2	Rapports au roi sur l'expédition et la publication. 1816-1841.
998:3	Décision du ministre. 1821-1842.
998:4	Lettres: circulairement à Louis de Freycinet. 1816-1844.
998:5	Lettres du Cre de Vau de Freycinet. 1815-1841.
998:6	Lettres adressées au ministre sur l'expédition de l'Uranie (administrateurs de la marine). 1816-1841.
998:7	Rapports de l'Institut (instructions et résultats). 1817-1833.
999	Registre contenant copie des instructions au Cre de Vau de Freycinet; des passeports . . . état et plan de la corvette l'Uranie; rôles états et inventairs divers.

Série Marine 5JJ —

58-79	Voyage de L. de Freycinet avec l'Uranie et la Physicienne dans l'Ocean Pacifique.
58: A-B	Journal des routes de l'expédition 1817-1820: A (1817-1820), B (1820).

59: A-B	Journal des marées (1817-1820).
60: A-F	Journal météorologique 1817-1820: A (1817-1818), B (1818), C (1818-1819), D (1819), E (1819-1820), F (1820).
61: A-B	Tables de loch tenues par les officiers (1817-1820).
62: A-D	A. Plan des travaux exécutés par l'expédition; B. minutes diverses de L. de Freycinet; C. journaux de montres (1817-1820); D. journal d'observations barométriques à terre (1817-1820).
63	Journaux de bord de Lamarche, lieutenant de vaisseau de l'Uranie (1817-1820).
64: A-D	Journaux d'observations magnétiques par le commandant (1817-1820).
65: A-B	Journaux d'observations magnétiques par les officiers (1819-1820).
66:A-E	Journal géographique de Duperrey, sur le Géographe (1818-1820): A (1818), B (janvier 1819), C (juin 1819), D (août 1819), E (1820).
67	Journal géographique de Labiche et Bérard, sur le Géographe (1818).
68	Instructions reçues par L. de Freycinet; journaux divers: élève de marine Alph. Pellion (1817-1820); Ferrand (1817-1820); Dubaut (1820; élève de marine Guérin (1817-1820); élève de 1re classe Raillard (1819); A. Gabert, secrétaire du commandant (1817-1818) (1817-1820).
69:A-D	Journal de Bérard, géographe sur l'Uranie (1817-1820).
70	Observations de marées à Rawak, île de Guam; observations du lieutenant de vaisseau Duperrey; notes de Paquet et Fleury; notes sur les Sandwich, Timor, les Carolines, Port Jackson, Waigiou, Baie des Chiens Marins, etc.; par Dubois et Jeanneret; journal nautique de Labiche (1817-1820).
71: A-D	Table de loch de Fabré, sur la Physicienne (1817-1820): A (1817-1818), B (1818-1819), C (1819-1820), D (1820).
72	Journal de navigation de Fabré (1817-1820).
73	Journal de bord de Fabré (1817-1820).
74	Journal d'observations de Fabré (1818-1818).
75: A-D	Journaux d'observations astronomiques des officiers (1817-1820): A (1817-1818), B (1818-1819), C (1819-1820), D (1820).
76: A-B	Journaux d'observations astronomiques des élèves (1817-1820): A (1817-1820), B (1820).
77: A-B	Journaux de timonerie (1817-1820).
78: A-B	Journaux de loch tenus par les élèves (1817-1820).
79	Journal de Laborde.

Série Marine 6JJ —

6 Voyage de Freycinet sur l'Uranie et la Physicienne. Routes; plans et croquis par Fabré; Baillard; Bérard; Labiche; Duperrey.

Bibliothèque Centrale du Muséum National d'Histoire Naturelle.

474 Manuscrits du docteur Quoy, chirugien-major; notes prises à bord de la corvette l'Uranie pendant le voyage autour du monde, 1818.

Proposal for a Penal Colony

Archives Nationales

Archives des colonies. Série H.

1. Transportation et déportation, études preliminaires; traduction de documents sur la déportation à Botany Bay, régime penitentiare dans les colonies australiennes et en général dans les colonies étrangères, mémoirs sur la choise d'un lieu de déportation.

3. Transportation et déportation, études preliminaires, projets de transportation dans diverses colonies au Sénégal, à Madagascar, aux Indes, projets des commissions préparent la loi sur la déportation 1823-1852.

Série Marine 5JJ

82 Journal nautique de J. de Blosseville, etc. (9 cahiers).

Bibliothèque Nationale. Département des Manuscrits.

Nouvelles acquisitions Françaises —

6785.30 Recueil de documents sur les colonies Françaises et sur la traité des nègres 1797-1830. Projet d'une colonie penale sur la côte sud ouest de la Nouvelle Hollande 1826-1848. Côtes sud ouest de la Nouvelle Hollande, plan d'exploration 1826.

9445.168 J. de Blosseville.

9446 Océanie et mers australes. Établissement du Baron Thierry à la Nouvelle Zélande 1825-1864.

9502.243 Les Français à la Nouvelle Zélande, 1825-1847.

Evreux, Normandy. Les Archives d'Evreaux.

Serie III F 269 Jules de Blosseville mss.

Duperrey, Louis Isidore

Archives Nationales.

Série Marine BB4 —

1000 Expéditions autour du monde du Ct de Frégate Duperrey sur la Coquille:

(1): Rapports et décisions concernant l'expédition.
(2): Lettres et instructions du Ministre de la Marine.
(3): Lettres du Duperrey
(4): Lettres de divers.
(5): Voyage autour du monde.

Série Marine 5JJ —

80 - 87	Voyage du lieutenant de vaisseau Duperrey, sur la Coquille:
80: A-B	Journaux nautiques de la Coquille (1822-1825).
81	Vues des côtes et relèvements; observations des marées (13 cahiers); observations d'inclinaison (1822-1824).
82	Voyage autour du monde de la Coquille, entre autres Brésil, Tahiti, îles de la Société, les Moloques, Nouvelle-Zélande, îles Carolines, Chili, Perou (Lima et l'Independance, le général de l'armée du sud), le Cap Horn, la Patagonie, les terres australes, îles Malouines, Brésil (décret de l'Empereur Pierre 1er) (vie economique, politique, missions, etc.) Journaux divers: notes de l'enseigne de vaisseau Bérard, de l'enseigne de vaisseau Lottin; notes de Jacquinot (4 cahiers); journal de la Coquille, par de Blois (2 cahiers), journal nautique de J. de Blosseville et materiaux divers (9 cahiers, dont un vocabulaire).
83	Vues et relèvements topographiques (12).
84: A-D	Observations d'angles horaires et la latitude (1822-1825): A (1822-1823), B (1823), C (1823-1824), D (1825).
85	Journal de comparaison et marches diverses des montres (1822-1825).
86	Registre d'observations de distances lunaires (1822-1825).
87: A-D	Registre d'observations d'azimuths (1822-1825): A (1822-1823), B (1824-1825).

Bibliothèque Centrale du Muséum d'Histoire Naturelle.

354	Voyage autour du monde sur la Coquille, années 1822, 1824 et suivantes. Collection R.P. Lesson. Animaux conservées dan l'alcool, mollusques et crustaces, ornithologie . . .Collections rapportées du Brésil, de Tahiti, du Chili, de la Nouvelle Hollande, de la Nouvelle-Zélande . . .
654-658	Papiers et manuscrits d'Adolphe Theodore Brongniart, 1801-1876. 658: Botanique de la Coquille. Voyage autour du monde.

1602 Journal d'un voyage autour du monde entrepris sur la corvette de S.M. la Coquille sous les ordres de M. Duperry . . . par J. Dumond D'Urville, 1822-1825. Observations de botanique et d'entomologie.

1793 Manuscrit d'une partie du voyage de la Coquille rédigé par P.R. Lesson, 22 mars, 1823 à avril, 1824.

Bibliothèque du Service Hydrographique de la Marine.

A4211 Atlas. Voyage autour du monde sur la corvette la Coquille commandé par M. Duperrey, Jules Le Jeune dessinateur.

A4211 Le Jeune, Jules. Journal sur la Coquille.

Bougainville, Hyacinthe

Archives Nationales.

Série Marine BB4 —

1001 Expon. autour du monde du Bon. de Bougainville (Thétis et Espérance) mers du l'Inde et de la Chine. 1: Instructions au Bon de Bougainville. 1824. 2: Lettres du Bon de Bougainville, sur la Thétis, 1824-1852. 3: Rapports, lettres et décisions concernant la publication du voyage de Bougainville. 1827-1828. 4: Documents anciens concernant Bougainville . . . journal de navigation . . . 1763-1773.

Série Marine 5JJ —

88 à 98 Voyage de Bougainville sur la Thétis et l'Espérance:

88: Correspondence touchant la publication du voyage; rapports divers de Bougainville; mémoire des fournitures faites a l'expéditions; lettres de l'Ens. de V. Lapierre; . . . côtes de Brésil; . . . Santiago; Valparaiso.

89: Tableaux généraux d'observations de latitudes et longitudes; résultats des observations astronomiques; marche des montres; sondes à Port Jackson; aus îles Arramba; observations astronomiques à bord de la Thétis dans les mers d'Asie; le Pacifique et l'Atlantique. 1824-1826.

90: Relèvement (provenant en partie de la publication en gravure).

91: Journaux de bord de la Thétis. 1823-1826.

92: Journaux de bord de l'Espérance (manque 32). 1823-1826.

93: Observations diverses faites à bord de la Thétis. 1824-1825.

94: Données et resultats des observations astronomiques faites à bord de l'Espérance par . . . De Nourquer des Campes. 1823-1826.

95: Routes de l'Espérance. 1824-1826.

96: Observations faites à bord de l'Espérance pour déterminer la déclinaison par le chef de timonerie Lugan. 1824-1826.

97: Observations diverses (longitudes; latitudes; montres; declinaison; etc.) par Fabré. 1826.

98: Observations diverses. 1825.

Dumont D'Urville, Jules Sebastien Cesar

Archives Nationales.

Série Marine BB4 —

507:2 Recherche de La Pérouse (Inde et Océanie). Expédition de la Bayonnaise, commandée par Legoarant de Tromelin: instructions à Dumont d'Urville; et l'Astrolabe; documents annexes . . . 1826-1828.

1002 Expédition autour du monde de Dumont d'Urville (sur l'Astrolabe).

- 1: Rapports, notes et décisions concernant l'expédition et la publication des résultats. 1825-1837.
- 2: Lettres du ministre à divers. 1825-1836.
- 3: Pièces comptables divers. 1825-1833.
- 4: Lettres addressées au ministre par divers, en particulier par le Cre de Vau Dumont d'Urville (rapports compris). 1825-1841.

Série Marine 5JJ —

99-102 Voyage de Dumont d'Urville sur l'Astrolabe:

99 Tables de loch (1826-1827).

100: A-R Tables de loch des officiers (1825): 100 A—I. En défecit, A,B,C,D,E,F,G,H,I,J,K,L,M,N,O,P,Q,R.

101: A-I Régistres des montres (1826-1829): A (1826), B (1826-1827), C (1827), D (1827), E (1827), F (1827)), G (1827-1828), H (1828-1829), I (1829).

101: J-L Régistres d'azimuths 1826-1828, J (1826-1827), K (1827), L (1827-1828).

101: M-N Régistres de latitudes (1826-1827), A (1826-1827), B (1827-1828).

101: O Régistres des montres (1826-1829).

101: P Régistre de comparaison (1826-1829).

101: Q Journal de physique (1826-1829).

101: R Journal de navigation des îles Curtis aux Moloques, suivi d'une description des planches (1-58).

102: A-B Vues des côtes et dessins à la mine de plomb.

102: C Cahiers de relèvements et de vues (7 par Gressier, 6 par Guilbert, 5 par Paris, 12 par Lottin).

Série Marine 6JJ —

7 Voyage de Dumont d'Urville sur l'Astrolabe. Cartes calques; routes; plans et vues de E. Paris; Gressier; Guilbert; Cottin; Lauvergne . . . Nouvelle Zélande etc. 1826-1829.

Bibliothèque Centrale du Muséum Nationale d'Histoire Naturelle.

62 Voyage de l'Astrolabe. Botanique, 1826-1829. Manuscrit de R.P. Lesson.

104-109: Voyage de l'Astrolabe, 1826-1829.

104 Tome I: texte. Voyage de découverte de l'Astrolabe sous le commandement de M. le capitaine Dumont d'Urville, de 1826 à 1829.

105 Tome II: Zoologie.

106 Tome III: Poissons.

107 Tome IV: Zoologie.

108 Tome V: Mollusques.

109 Tome VI: Mammifères; reptiles.

124-174 Manuscrits provenant d'Antonine-Étienne-Renaud-Augustin Serres, 1786-1868.

153 Leçons sur les races humaines.

165 Notes sur les races humaines. Rapport sur la collection d'Anthropologie recueillie pendant l'expédition de l'Astrolabe et de la Zelée par M. Dumoutier.

839-840 Voyage de l'Astrolabe, 1826.

Bibliothèque Nationale, Département des Manuscrits.

Nouvelles acquisitions Françaises —

9445 Fol 189-220: Dumont d'Urville.

Books

General

Bériot, A., *Grands voiliers autour du monde: les voyages scientifiques 1760-1850.* Paris, 1962.

Brosses, C. de, *Histoire des navigations aux Terres Australes.* 2 vols. Paris, 1756.

Callender, J., *Terra Australia Cognita, or voyages to the Terra Australis or southern hemisphere during the 16th, 17th and 18th centuries.* London, 1766-68.

Collingridge, G., *The Discovery of Australia.* Sydney, 1895.

Dahlgren, E.W., *Les relations commerciales et maritimes entre la France et les côtes de l'océan Pacifique.* Paris, 1909.

Dowling, J., *Soixante ans de navigations Françaises autour de l'Australie, 1768-1828.* Thèse, Aix en Province, 1955, unpublished.

Dunmore, J., *French Explorers in the Pacific.* 2 vols. Oxford, 1965-69.

Faivre, J.P., *L'expansion Française dans la Pacifique de 1800 à 1842.* Paris, 1953.

Hanotaux, G.A.A. et Martineau, A., *Histoire des colonies Françaises et de l'expansion de la France dans le monde.* 6 vols. Paris, 1851.

Howe, S.E., *Les grands navigateurs à la recherche des épices*. Paris, 1939.

La Roncière, C. de, *Histoire de la marine Française*. 5 vols. Paris, 1899-1920.

La Roncière, C. de, *Origines du Service Hydrographique de la Marine*. Paris, 1916.

Major, R.H., *Early voyages to Terra Australis*. London, 1859.

Marguet, F.P., *Histoire générale de la navigation du XVe au XXe siècle*. Paris, 1931.

Maupertius, P.L.M. de, *Lettre sur le progrès des sciences*. Paris, 1752.

Nicolas, L., *Histoire de la marine Française*. Paris, 1949.

Prévost, A.F., *Histoire générale des voyages*. 19 vols. Paris, 1746-70.

Rainaud, A., *Le continent austral: hypothèses et découvertes*. Paris, 1893.

Schefer, C., *Recueil de voyages et de documents pour servir à l'histoire de la géographie, depuis le XIIIe jusqu'à la fin du XVIe siècle*. 16 vols, Paris, 1882-97.

Scott, E., *Terre Napoleon: a history of French explorations and projects in Australia*. London, 1910.

Thévenot, M., *Relations de divers voyages curieux*. 4 vols. Paris, 1663-72.

Verne, J., *Historie générale des grands voyages et des grands voyageurs*. Paris, 1879.

Early Searches for Terre Australes

Avazec, A. d', *Note sur le première expedition de Béthencourt aux Canaries*. Paris, 1846.

Avazec, A. d', *Notice de découvertes faites au moyen âge dans l'océan Atlantique antérieure aux grandes explorations portugaises du XVe siècle*. Paris, 1845.

Bontier, P., *Histoire de la première descouverte et conqueste des Canaries faite dès l'an 1402 par messire Jean de Béthencourt, escrite du temps mesme par Pierre Bontier*. Paris, 1630.

Bréard, C., *Documents relatifs à la marine normande et à ses armaments au XVIe et XVIIe siècles*. Rouen, 1889.

Bréard, C., *Histoire de Pierre Berthelot, pilote et cosmographe du roi de Portugal aux Indes Orientales . . . né en Normande en MDC, mort à Achen en MDCXXXVIII*. Paris, 1889.

Guillaume de Rubrouk, *Relation des voyages en Tartary de Fr Guillaume de Rubruquis, Fr Jean du Plan Carpin, Fr Ascelin et autre religeux de S. François et S. Dominique*. Paris, 1634.

La Roncière, C. de, *La Découverte de l'Afrique au moyen âge: cartographes et explorateurs*. 2 vols. Cairo, 1924-25.

La Roncière, C. de, *Découverte d'une relation de voyage datée du Touat et décrivant en 1447 le bassin du Niger*. Paris, 1919.

La Roncière, C. de, *Les navigations Françaises au XVe siècle*. Paris, 1896.

La Roncière, C. de, *Les precurseurs de la Compagnie des Indes Orientales, la politique coloniale des Malouines*. Paris, 1913.

Rémusat, A., *Mémoires sur les relations politiques des princes chrétiens, et particulièrement des rois de France, avec les empereurs Mongols.* Paris, 1824.

Yule, H., *Cathay and the way Thither: being a collection of medieval notices on China.* London, 1866.

Wolff, P., *Une famille de XIIIe au XIVe siècle: les Ysalguier de Toulouse.* Paris, 1942.

Wolff, P., *Commerces et marchands de Toulouse vers 1350 vers 1450.* Paris, 1954.

Gonneville, Binot Paulmier De

Avazec, A. d', *Compagne de navire l'Espoir de Honfleur 1503-1505. Re lation authentique du voyage du capitaine de Gonneville ès nouvelles terre des Indes publiée intégralement pour le première fois avec une introduction et des éclaircissements.* Paris, 1869.

Boissas, E., *Binot Paulmier dit le capitaine de Gonneville, son voyage, sa descendance.* Caen, 1912.

Bréard, C., *Notes sur la famille du capitaine Gonneville, capitaine normand du XVIe siecle.* Rouen, 1885.

Julien, C.A., et al. *Les Français en Amérique pendant la première moitié du XVIe siècle: textes des voyages de Gonneville etc.* Paris, 1946.

Julien, C.A. *Les Français en Amérique pendent la deuxième moitié du XVIe siècle.* Paris 1953.

Paulmier de Courtonne, Jean Chanoine de Lisieux, *Mémoires touchant l'établissement d'une mission chrestienne dans le troisième monde, autrement appellé la terre australe, meridionales, antartique et inconnues . . . par un ecclesiastique originaires de cette mesme terre.* Paris, 1663.

Paulmier de Gonneville, B. *Voyage du capitaine Paulmier de Gonneville au Brésil 1503-1505, annoté par C.A. Julien.*

Explorations After Gonneville to Bouvet de Lozier, and Exotic Literature

Alfonce, J., *Les voyages avanteureux du capitaine Jan Alfonce, sainctogeois.* Poitiers, 1559.

Anthiaume, A., *Cartes marines, constructions navales voyages de découvertes chez les Normands, 1500-1560.* Paris, 1916.

Anthiaume, A., *Evolution et enseignement de la science nautique en France, et principalement chez les Normands.* Paris, 1920.

Asseline, D.R., *Les antiquités et chroniques de la ville de Dieppe.* Dieppe, 1874.

Desmarquets, C., *Mémoires chronologiques pour servir à l'histoire de Dieppe et à celle de la navigation Française.* Paris, 1785.

Estancelin, L., *Dissertation sur les découvertes faites par les navigateurs Dieppois.* Abbeville, s.d.

Estancelin, L., *Recherches sur les voyages et découvertes des voyageurs nor mands en Afrique, dans les Indes orientales et en Amérique, suivi d'observations sur la marine, le commerce et les établissements coloniaux des Français . . . journal du voyage de Jean Parmentier de Dieppe à l'île de Sumatre en l'année 1529.* Paris, 1832.

Gaffarel, P., *Les Français au delà des mers. Les découvreurs Français du XIV*e au XVe siècle. Paris, 1888.

Gaffarel, P., *Jean Ango.* Rouen, 1889.

Guérin, E., *Ango et ses pilotes d'après des documents inedits tires des archives de France, de Portugal et d'Espagne.* Paris, 1901.

Hellot, A., *Jean Ango et sa famille d'après de nouveaux documents.* Dieppe, s.d.

Hellot, A., *Un Grand marchand de Dieppe an XIVe siècle: les inscriptions de la chapelle Saint Sauveur le Longueil.* Rouen, 1878.

Hervé, M.R. "Australia in French geographical documents of the renaissance", in *Royal Australian Historical Society Journal and Proceedings*, vol.41, 1955, p.23-38.

Howe, S.E. *Premiers essais de pénétration Anglais en Afrique occidentale.* Paris, 1939.

Julien, C.A. *Les voyages de découverte et les premiers établissements, XIVe - XVe siècle.* Paris, 1948.

Mollat, M., *Le commerce maritime normand à la fin du moyen âge, étude economique et sociale.* Paris, 1952.

Parmentier, J., *Description nouvelle des merveilles de ce monde et de la dignité de l'homme, composée en rithme francoyse en manière de Jan Parmentier faisant sa dernière navigation avec Raoul son frère, en l'îsle Taprobane, autrement dicte Sumatra.* Paris, 1531.

Exotic, Utopian Novels

Foigny, G. de, *La terre australe connue, c'est à dire la description de pays inconnu jusqu' ici, de ses moeurs et de ses coutumes, par Mr Sadeur, avec les avanteurs qui la conduiserent en ce continent . . . réduites et mises en lumière par les soins et la conduite de G. de F.* Vannes, 1676.

Foigny, G. de, *Les avantures de Jacques Sadeur dans le découverte et le voyage de la terre australe.* Paris, 1692.

Foigny, G. de, *Nouveau voyage de la terre australe, contenant les coutumes et les mouers des Australiens, leur religion, leurs exercises, leurs études, leurs guerres, les animaux particuliers à ce pays.* Paris, 1693.

Vairasse, D. d'Alais, *L'histoire des Sevarambes, peuples qui habitent une partie du troisième continent communement, appellé la terre australe (australie) contenant un comte (conte) exact du gouvernement, des moeurs, de la religion et du language de cette nation jusqués aujourd huy inconnue aux peuples de l'Europe.* 5 vols, Paris, 1677-79.

Dufresne, Marc Joseph Marion

Ly Tio Fane, M., *Mauritius and the Spice Trade: the Odyssey of Pierre Poivre*. Port Louis, Mauritius Archives, 1958.

Montemont, Albert, *Relation du Voyage Autour du Monde de 1785 à 1788 par La Pérouse. Relation du Voyage de Marion en 1771-1772*. Paris, 1855.

Rochon, Abbé Alexis Marie de, *Crozet's Voyage to Tasmania, the Ladrone Island and the Philippines in the Years 1771-1772*, translated by H. Ling Roth . . . with a preface and a brief reference to the literature of New Zealand by Jas Bowe. London, 1891.

Rochon, Abbé Alexis Marie de, *Nouveau Voyage à la Mer du Sud, Commencé sous les Ordres de M. Marion . . . et Achevé après la Mort de cet Officier sous ceux de M. le Chevalier Duclesmeur . . . Cette Relation a été Rédigée d'après les Plans et Journaux de M. Crozet. On a Joint à ce Voyage un Extrait de celui de M. de Surville dans les mêmes Parages*. Paris, 1783.

Kerguelen-Tremarec, Yves Joseph De

Grenier, Vte Jacques Raymond de, *Mémoire de la campagne de découvertes dans les mers des Indes, par le Cher Grenier . . . ou il propose une route qui abrège de 800 lieues la traversée de l'Isle de France à la côte Coromandel et en Chine*. Brest, R. Malassis, 1770.

Grenier, Vte Jacques Raymond de, *Relation de deux voyages dans les mers australes et des Indes faits en 1771, 1772, 1773 et 1774*. Paris, Knapen et fils, 1782.

D'Entrecasteaux, Joseph Antoine Bruny

Burney, J., *A Memoir of the Voyage of d'Entrecasteaux in Search of La Pérouse*. London, 1820.

Cordier, H., *La Mission de M. le chevalier d'Entrecasteaux à Canton en 1787, d'après les archives du Ministre des Affaires Etrangères*. 1911.

Hulot, E.G.T., *D'Entrecasteaux, 1737-1793*. Paris, 1894.

Jurien de la Gravière, Jean Baptiste Edmond. *Souvenirs d'un amiral, mémoire du vice-amiral Jurien de la Gravière*. Paris, Hachette, 1860.

La Billardière, Jacques Julien Houtou de, *An Account of a Voyage in Search of La Pérouse in the Years 1791, 1792 and 1793*. Translated from the French. London, 1800. 2 vols, plates.

La Billardière, Jacques Julien Houtou de, *Relation du Voyage à la Re cherche de la Pérouse, fait par Ordre de l'Assemblée Constituante pendant les Années 1791, 1792 et pendant la 1e la 2e Années de la République Française par le Cen La Billardière*. Paris, an VIII, 2 vols, atlas.

La Billardière, Jacques Julien Houtou de, *Voyage in Search of La Pérouse performed by order of the Constituent Assembly*. Translated from the French. London, 1800, plates.

Rossel, Elizabeth Paul Edouard de, *Voyage de D'Entrecasteaux envoyé à la Recherche de la Pérouse . . .* redigé par M. de Rossel. Paris, 1808. 2 vols, atlas.

Baudin, Nicolas

Baudin, Nicolas Thomas, *The Journal of Post Captain Nicolas Baudin*, translated by C. Cornell. Adelaide, 1974.

Bouvier, R. et Maynial, E., *Une Aventure dans les mers australes: l'expedition du commandant Baudin 1800-1803*. Paris, 1947.

Péron, François et Freycinet, Louis de, *Voyage de Découvertes aux Terres Australes exécuté . . . sur les Corvettes le Géographe, le Naturaliste et la Goelëtte le Casuarina, pendant les Années 1800, 1801, 1802, 1803 et 1804, Rédigé par M.F. Péron et Continué par M. Louis Freycinet*. Paris, 1807-1816, 2 vols., atlas.

Péron, François et Freycinet, Louis de, *Voyage de Décourvertes aux Terres Australes fait par Ordre du Gouvernment sur les Corvettes le Géographe, le Naturaliste et la Goëlette Casuarina pendant les Années 1800, 1801, 1802, 1803 et 1804, Historique Rédigé par M.F. Péron et Continué et Augmenté par M. Louis de Freycinet*. Paris, 1824, 4 vols., atlas.

Freycinet, Louis Claude Desaules De

Arago, Jacques, *Promenade Autour du Monde pendant les Années 1817, 1818, 1819 et 1820, sur les Corvettes du Roi l'Uranie et La Physicienne, commandées par M. Freycinet*. Paris, 1822, 2 vols.

Bassett, F.M., *Realms and Islands: the World Voyage of Rose de Freycinet in the Corvette Uranie, 1817-20 from her Journal and Letters*. London, 1962.

Freycinet, Louis Claude Desaulces de, *Voyage Autour du Monde . . . execute sur les Corvettes de S.M. L'Uranie et la Physicienne pendant les Annees 1817, 1818, 1819 et 1820 . . . publie par M. Louis de Freycinet*. Paris, 1824-1844. 9 vols, 4 vols. atlas.

Freycinet, Rose Desaulces de, *Journal d'après le manuscrit original accom pagné de notes par Charles Duplomb*. Paris, 1927.

Penal Colony

Blosseville, E. de, *Histoire de la colonization pénale et les établissements de l'Angleterre en Australia*. Évreaux, 1859.

Blosseville, E. de, *Jules de Blosseville*. Évreaux, 1854.

Mancy, Jarry de, *Jules Blosseville*. Paris, 1835.

Passy, L., *Le Marquis de Blosseville: souvenirs*. Évreaux, 1890.

Portal, Baron P.B., *Mémoires*. Paris, 1846.

Duperrey, Louis Isidore

Duperrey, Louis Isidore, *Voyage Autour du Monde Exécuté par Ordre du Roi sur la Corvette de Sa Majesté la Coquille pendant les Années 1822, 1823, 1824 et 1825*. Paris, 1825-1830. 7 vols., 4 vols. atlas.

Lesson, René Primevère, *Journal d'un Voyage Pittoresque Autour du Monde exécuté sur la Corvette la Coquille commandée par M.L.J. Duperrey, pendant les Années 1822, 1823, 1824, 1825*. Paris, 1830.

Lesson, René Primevère, *Voyage Medical Autour du Monde exécuté sur la Corvette du Roi, la Coquille, commandé par M.L.J. Duperrey, pendant les Années 1822, 1823, 1824 et 1825 . . . suivi d'un Mémoire sur les Races Humaines répandues dans l'Oceanie, la Malaisie et l'Australie par R.P. Lesson*. Paris, 1829.

Bougainville, Hyacinthe Yves Philippe Potentien de

Bougainville, H.Y.P.P. de, *Journal de la navigation autour du globe de la Thétis et l'Espérance pendant les années 1824, 1825 et 1826*. 2 vols. Paris, 1837.

Dumont D'Urville, Jules Sebastien César

Dumont D'Urville, J. *Enumeratio plantarum quas in insulis Archipelagi aut littoribus Ponti-Euxini, annis 1819 et 1820, collegit atque detexit J. Dumont D'Urville*. Paris, 1822.

Dumont D'Urville, Jules Sebastien César, *Voyage Autour du Monde. L'Astrolabe par Dumont d'Urville*. Limoges, 1881.

Dumont d'Urville, Jules Sebastien César, *Voyage de la Corvette l'Astrolabe exécuté par Ordre du Roi pendant les Années 1826, 1827, 1829 sous le commandement de M.J. Dumont d'Urville*. Paris, 1830-1834, 15 vols, 7 vols. atlas.

Index